Diatom Cultivation for Biofuel, Food and High-Value Products

Scrivener Publishing
100 Cummings Center, Suite 541J
Beverly, MA 01915-6106

Diatoms: Biology and Applications

Founding Editors: Richard Gordon and Joseph Seckbach
Series Editor: Janice L. Pappas (jlpappas@umich.edu)

There are about 100,000 species and as many papers written about them since their discovery three hundred years ago. The literature on diatoms shows about 60,000 papers during the period 2006-2023. In this context, it is timely to review the progress to date, highlight cutting-edge discoveries, and discuss exciting future perspectives. To fulfill this objective, this Diatom Series provides a comprehensive and reliable source of information on diatom biology and applications as well as to enhance interdisciplinary collaborations required to advance knowledge and applications of diatoms.

Publishers at Scrivener
Martin Scrivener (martin@scrivenerpublishing.com)
Phillip Carmical (pcarmical@scrivenerpublishing.com)

Diatom Cultivation for Biofuel, Food and High-Value Products

Edited by

Vandana Vinayak

School of Applied Sciences, Dr. Hari Singh Gour Vishwavidhyalaya (A Central University), Sagar, Madhya Pradesh, India

and

Richard Gordon

Gulf Specimen Marine Lab & Aquarium, Florida, USA

Scrivener Publishing

WILEY

This edition first published 2025 by John Wiley & Sons, Inc., 111 River Street, Hoboken, NJ 07030, USA and Scrivener Publishing LLC, 100 Cummings Center, Suite 541J, Beverly, MA 01915, USA

For more information about Scrivener publications please visit www.scrivenerpublishing.com.

Wiley Global Headquarters
111 River Street, Hoboken, NJ 07030, USA

For details of our global editorial offices, customer services, and more information about Wiley products visit us at www.wiley.com.

Library of Congress Cataloging-in-Publication Data

ISBN 978-1-394-17448-5

Cover image provided by Dr Vandana Vinayak and Urvashi Soni
Cover design by Russell Richardson

Set in size of 11pt and Minion Pro by Manila Typesetting Company, Makati, Philippines

Printed in the USA

10 9 8 7 6 5 4 3 2 1

Contents

Part II: High-Value Products 79

Preface

Diatoms are emerging as pivotal players in addressing some of the most pressing environmental and energy challenges of our time. This book delves into the multifaceted world of diatoms, exploring their cultivation, harvesting, and diverse applications. By harnessing the potential of diatoms, we can develop sustainable solutions that benefit both industry and the environment, contributing to a healthier planet for future generations.

The cultivation and harvesting of diatoms present unique challenges and opportunities. This book provides a comprehensive overview of the technological limitations and needs in this field, highlighting innovative methods such as the use of chitosan as a natural flocculant. These advancements are crucial for optimizing the efficiency and effectiveness of microalgae harvesting processes, its advantages over synthetic flocculants, and its potential for large-scale applications.

Diatoms, with their intricate cell wall structures and biochemical compositions, offer immense potential for various applications, including biofuel production, bioremediation, and even space exploration. This book examines the technoeconomic prospects of diatom cultivation, the design and implementation of algal reactors, and the potential of diatoms as a source of biofuel and other value-added products. Diatoms are unique among microalgae due to their silica-based cell walls, known as frustules, which exhibit intricate nanostructures. These frustules not only provide mechanical protection but also have potential applications in nanotechnology and materials science.

One of the key features of diatoms is their ability to produce valuable biochemical compounds. This book provides an in-depth analysis of the pigments and proteins found in *Phaeodactylum tricornutum*, exploring their potential applications in various industries. Diatoms are known for their diverse and unique biochemical compositions, which include pigments such as chlorophylls, carotenoids, and fucoxanthin. These pigments have potential applications in the food, pharmaceutical, and cosmetic industries due to their antioxidant and anti-inflammatory properties.

Biofuel production from diatoms is a promising avenue for sustainable energy. This book explores the methods for lipid extraction from diatoms, the economic feasibility of biofuel production, and the industrial aspects of this process. Diatoms are considered one of the most promising sources of biofuels due to their high lipid content and rapid growth rates.

The potential of diatoms extends beyond Earth, with applications in long-term space missions and life support systems. This book discusses the testing of diatom growth under microgravity conditions and the potential use of diatoms in space applications. Diatoms have been proposed as a potential component of life support systems for long-term space missions due to their ability to produce oxygen and remove carbon dioxide.

Genetic engineering plays a crucial role in enhancing the lipid production of diatoms. This book delves into the genetic modifications and stress conditions that can increase oil production in diatoms, providing insights into the future prospects of this field. Genetic engineering techniques, such as CRISPR-Cas9 and RNA interference, have been used to modify the metabolic pathways of diatoms to enhance lipid production.

The book discusses the potential of these techniques for improving the efficiency and yield of diatom-based biofuels, as well as the challenges and ethical considerations associated with genetic engineering. Further, the work highlighted in this wide-ranging book is divided into three parts:

Part I: Culture Methods. This section introduces the foundational aspects of microalgae and diatom cultivation. Chapter 1 discusses the adaptation of chitosan-based harvesting methods for microalgae flocculation, addressing the challenges and technological limitations in microalgae cultivation and harvesting. The chapter delves into the various methods of microalgae harvesting, including centrifugation, filtration, and flocculation, with a particular focus on the use of chitosan as a natural flocculant. It explores the mechanisms of chitosan flocculation, its advantages over synthetic flocculants, and its potential for large-scale applications. Chapter 2 explores the trends, scope, and technoeconomic prospects of diatom cultivation, including the design and implementation of algal reactors and the potential of diatoms as a source of biofuel and other value-added products. The chapter discusses the cultivation of diatoms in photobioreactors and open ponds, highlighting the advantages and limitations of each system. It also examines the potential of diatoms in the biofuels industry, focusing on their high lipid content and rapid growth rates. The economic feasibility of diatom-based biofuel production is analysed, considering factors such as feedstock availability, production costs, and market demand. Chapter 3 provides an in-depth analysis of the pigments and proteins found in

Phaeodactylum tricornutum, exploring their potential applications in various industries. It discusses the diverse and unique biochemical compositions of diatoms, including pigments such as chlorophylls, carotenoids, and fucoxanthin, and their potential applications in the food, pharmaceutical, and cosmetic industries. The chapter also examines the protein content of diatoms, including enzymes and bioactive peptides, and their potential uses in nutraceuticals and functional foods.

Part II: High-Value Products. This section covers advanced applications and innovative techniques in the field of diatoms and microalgae. Chapters 4 and 5 examine the nature and applications of diatom cell walls, including their purification processes and industrial uses. The chapters discuss the silica frustules of diatoms, which exhibit intricate nanostructures and have potential applications in nanotechnology and materials science. They explore the methods for extracting and purifying diatom frustules, as well as their potential for creating high-value products. Chapter 6 discusses the biochemical engineering of diatoms for health and biorefinery concepts, highlighting the potential of diatoms in producing biofuels and other high-value products. The chapter examines the genetic modifications and stress conditions that can enhance lipid production in diatoms, providing insights into the future prospects of this field. Chapter 7 focuses on the metabolic and transcriptomic stress and engineering of diatoms to enhance lipid production, exploring the stress conditions that can increase oil yield. The chapter delves into the genetic engineering techniques, such as CRISPRCas9 and RNA interference, that have been used to modify the metabolic pathways of diatoms to enhance lipid production. It discusses the potential of these techniques for improving the efficiency and yield of diatom-based biofuels, as well as the challenges and ethical considerations associated with genetic engineering. Chapter 8 examines the potential of diatoms in space applications, including long-term space missions and life support systems. The chapter discusses the challenges of cultivating diatoms in space, including the effects of microgravity on growth and metabolism, and explores the potential benefits of using diatoms for cultivation within bio-regenerative life support system (BLSS). Chapter 9 discusses diatom as prospective green anode material.

Part III: Low-Value Products. This section highlights the environmental and industrial applications of diatoms for low value products. Chapter 10 discusses diatom cell disruption and milking via a nanobiorefinery for biofuel production, utilizing techniques like pulsed electric fields, high pressure homogenisation, ultrasonication, etc. Chapter 11 discusses genetic engineering and metabolic engineering in diatoms for oil production. Chapter 12 explores the use of diatoms for heavy metal bioremediation, discussing the mechanisms and challenges involved. The chapter examines the high capacity of diatoms for accumulating heavy metals, making them suitable for bioremediation applications. It explores the mechanisms of heavy metal uptake by diatoms, including biosorption and bioaccumulation, and discusses the challenges and limitations of using diatoms for bioremediation. Chapter 13 discusses diatoms for biofuel in biofilm reactors, it discusses transesterification of diatom oil and parameters for optimisation and Chapter 14 discusses diatom harvesting for lipid production like bubble wrap (Bubble Farming), oozing, mechanical pressure, pulse electric field etc.

We are pleased to present this book, which documents the latest advancements in the cultivation, harvesting value added products from diatoms, and applications of microalgae and diatoms. By fostering an interdisciplinary exchange of ideas and actions, we hope to contribute to the innovative design of high-performing systems that provide practical solutions to global challenges. We trust that this book will be a valuable resource for researchers, industry professionals, and anyone interested in the potential of microalgae and diatoms.

There are four general methods for agricultural scale growth of diatoms and other microalgae: Runways; Photobioreactors; Bubble Farming; Biofilms. These vary widely in ease to contaminate; requirement for flat land; gas barrier; water retention; nutrients requirement; scalability; running energy, and cost.

The editors and Clifford Merz have been developing Bubble Farming at a laboratory scale. It needs outdoor, agricultural testing to see if it needs adjustments for variable weather conditions, such as bubble size, UV screening, wind and bird resistance, humans and animals popping bubbles by stepping on it, and when to harvest and reusability. Crops that could grow under air pillows need exploring. Here we have discussed Bubble Farming in Chapters 1 and 14. A combination of Bubble Farming and Biofilm cultivation can be imagined. For example, a large surface area could be provided by 0.3mm diameter steel ball bearings, contained in flat air pillows with a thin layer of water with nutrients. Using directed light to guide motile diatoms and concentrate them is also worth consideration, as this could require no energy.

Vandana Vinayak
Richard Gordon
January, 2025

Acknowledgements

Vandana Vinayak is extremely thankful to enthusiastic team members and students of Diatom Nanoengineering and Metabolism Laboratory and Indo French Project CEFIPRA grant no PPMB 7133/2020 and Dr Hari Singh Gour Vishwavidyalaya (A Central University), Sagar, Madhya Pradesh, India. She is thankful to Professor Richard Gordon and Scrivener Publishers for this opportunity.

Last but not the least her parents, family, and brother Shaheed Major Kapil Vinayak for kept her joyful in this journey.

Richard Gordon is grateful to the authors for contributing this basic material for all who endeavor to grow diatoms.

Part I
CULTURE METHODS

1

Adaptation of Chitosan-Based Harvesting Methods for Flocculation of Microalgae

Mainavi Patel[1], Hirak Parikh[2] and Gayatri Dave[1*]

[1]P D Patel Institute of Applied Sciences, Charotar University of Science and Technology, Anand, Gujarat, India
[2]Department of Earth and Environmental Systems, Indiana State University, Terre Haute, Indiana, USA

Abstract

This chapter addresses the potential of microalgae, focusing on diatoms because of their distinctive features and high nutritional value. Single-celled photosynthetic microalgae have been around for three billion years. They have also been used as a food source for humans over 2000 years, with *Spirulina* gaining global popularity as a dietary supplement. Diatoms' high lipid content makes them an excellent source for biofuel production, offering a renewable and environmentally friendly alternative to conventional fossil fuels.

The cultivation of microalgae poses challenges due to their small size, low settling rate, and susceptibility to damage during harvesting. Various harvesting methods, such as physical (mechanical), chemical, and biological techniques, have been explored to address these challenges. Flocculation, a widely used method, involves the formation of aggregates (flocs) to separate microalgae from the growth medium or water. The natural polysaccharide chitosan has gained attention as an eco-friendly and effective flocculant for microalgae harvesting. Studies have shown the potential benefits of combining chitosan with other techniques, such as electroflotation, electrolysis, and magnetic nanoparticles, to enhance harvesting efficiency and reduce processing costs. The chapter emphasizes the significance of interdisciplinary research to enhance microalgae cultivation and harvesting methods, unlocking their full potential for sustainable applications in various industries.

***Keywords*:** Chitosan, harvesting methods, flocculation, microalgae

**Corresponding author*: gayatridave.bt@charusat.ac.in

Vandana Vinayak and Richard Gordon (eds.) Diatom Cultivation for Biofuel, Food and High-Value Products, (3–20) © 2025 Scrivener Publishing LLC

1.1 Microalgae

Microalgae are single-celled photosynthetic organisms residing in aquatic habitats [1.1]. Microalgae have been around for three billion years, and their exceptional potential in a variety of industries, including food, feed, fuel, and pharmaceuticals, has drawn the attention of both researchers and businesses [1.2]. Since microalgae cells include carbohydrates, proteins, lipids, vitamins, antioxidants, and necessary trace elements, they have great nutritional significance for people and animals [1.3] [1.4].

The Chinese first used microalgae (*Nostoc*) as a food source over 2,000 years ago. Eventually, *Spirulina* gained global popularity as a food supplement due to its exceptional nutritional properties. *Spirulina* is composed of 62% amino acids and is a source of vitamins A, B1, B2, B12, and phytopigments like carotenoids and xanthophylls [1.5]. Additionally, it contains essential fatty acids, including linolenic acid, which the human body cannot synthesize.

Microalgae biomass is exceptionally rich in lipids; under certain growth conditions, some strains can accumulate 20–50% triacylglycerols, while others can produce up to 80% lipids [1.6]. Microalgae are an excellent source for producing biofuel due to their high lipid content, higher growth rates, adaptability to environmental conditions and capacity to grow on non-arable ground [1.7] [1.8]. Microalgae-based biofuels also benefit from being renewable and emitting fewer greenhouse gases than traditional fossil fuels.

Diatoms belonging to the Bacillariophyta group have unique siliceous cell walls with intricate patterns and shapes known as frustules [1.9]. They are found in aquatic and semi-aquatic habitats, including freshwater and marine water. Diatoms' remarkable diversity, ecological significance and biotechnological potential are fascinating [1.10].

1.2 Microalgae Cultivation and Challenges

Microalgae exhibit rapid growth rates; their high photosynthesis rate and ability to store significant byproducts make them suitable for industrial feedstock [1.11]. Furthermore, microalgae cultivation does not require fertile soil, extensive freshwater supplies or herbicides, distinguishing them from other crops [1.12]. Open and closed culture systems have been exploited for maximum cultivation of microalgae. An open cultivation system involves growing microalgae in outdoor environments exposed to natural sunlight and air. These include raceway ponds, shallow ponds and natural open ponds. However, these systems are more susceptible to contamination, temperature fluctuations and other environmental variables.

On the other hand, closed systems such as enclosed bioreactors or photobioreactors entail growing microalgae in a controlled environment. These systems precisely control light intensity, temperature, nutrient concentrations and carbon dioxide levels. As a result, closed systems tend to produce higher biomass and have a lesser risk of contamination (Figure 1.1) [1.13] [1.14].

In specialized microalgae groups, the diatoms encompass diverse species with varying growth requirements. Each species may have specific temperature, light, nutrient, and pH preferences [1.15] [1.16]. Identifying the optimal conditions for diatom growth and understanding the particular needs of target species is crucial for successful cultivation. Also, silica supplementation in cultivating vessels is essential for optimum diatom growth [1.17]. Usually, diatoms are cultivated for two purposes:

1. Laboratory-level research studies
2. Aquaculture and bio-fuel production.

Xenic culture cultivates diatoms with other microorganisms [1.18]. This method allows the study of diatoms in a more natural environment, as they often coexist with other natural microbes. Xenic cultures can be established by collecting water samples from natural habitats, such as freshwater or

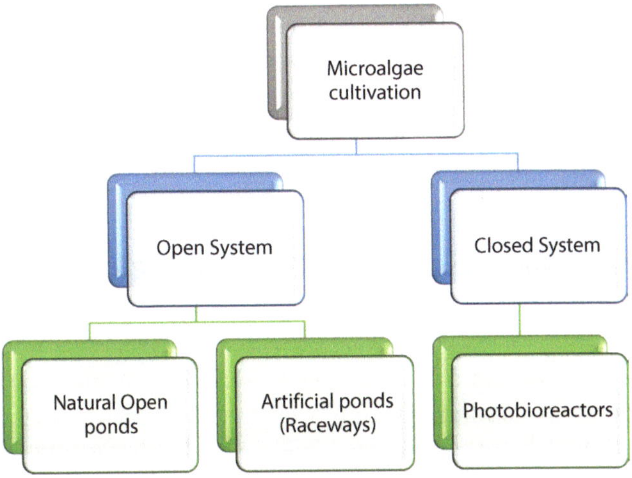

Figure 1.1 Methods for microalgae cultivation.

marine environments, and inoculating diatoms and the associated microbial community in culture media.

On the other hand, Axenic cultures are grown without any other metabolizing cells and can be used to study the pure physiology and metabolism of the diatom itself. This can be particularly useful for studying the biochemical pathways and genetic regulation of diatom growth and metabolism [1.19].

Obtaining biomass is easy if you have the proper seed culture, a suitable cultivation medium, and a light source. However, there are two significant difficulties at two different stages of growth, such as seeding and harvesting:

1) The environmental sampling often contains diatom cells closely surrounded by grazers. Also, grazers are of similar size to diatoms and hence cannot be separated by filtration, limiting the representation of all of taxas during the enrichment and thereby further cultivation. For this reason, a method for pooling diatom fractions from heterogeneous natural water samples is required.
2) After cultivation, the biomass must be collected from diluted water for further application.

However, despite the immense potential of microalgae, there are still significant challenges associated with their large-scale biomass production. These challenges include optimizing cultivation techniques, such as selecting suitable strains, optimizing growth conditions, and enhancing cultivation efficiency to achieve higher yields [1.20]. Scaling up production processes and establishing cost-effective harvesting and extraction methods are crucial focus areas. Furthermore, ensuring the sustainability of microalgae cultivation, such as minimizing water and nutrient requirements, preventing contamination, and managing waste byproducts, remains an ongoing challenge.

1.3 Microalgae Harvesting: Technological Limitations and Needs

Harvesting microalgae from reaction vessels such as closed bioreactors to open ponds is another mammoth task. Several factors make harvesting difficult, which are summarized below. Small cell size and low settling rate: Microalgae cells are typically tiny, ranging from 1–10 micrometer in diameter. Their small size and low settling rate make separating them from the growth medium or water difficult. Traditional gravitational force-mediated sedimentation methods could be more effective, requiring developing innovative harvesting techniques [1.21].

The second key reason is that biomass is often found in dilute concentrations scattered in large volumes of water, making the harvesting process even more challenging. Efficient methods are needed to concentrate the biomass before harvesting to reduce energy and cost requirements [1.22].

Cell fragility and sensitivity: Microalgae cells are delicate and can easily rupture or damage during harvesting. Mechanical methods like centrifugation or filtration can cause cell breakage, biomass loss and reduced product quality. Gentle and non-invasive harvesting techniques must be developed to maintain cell integrity and preserve the valuable components of microalgae. Furthermore, the grazers, particularly protozoa and dinoflagellates, negatively impact the biomass yield. Effective separation and purification techniques are necessary to ensure a clean and uncontaminated final product [1.23].

Many microalgae harvesting methods are effective at small scales but struggle to scale up. Commercial production requires scaling up the harvesting process while maintaining efficiency and cost-effectiveness. Additionally, a harvesting method choice can affect the environment. Energy and chemicals may be consumed in some methods, posing environmental concerns. Developing sustainable and eco-friendly harvesting techniques that minimize the ecological footprint is essential.

Addressing these challenges requires interdisciplinary research and development efforts, encompassing engineering, biology, chemistry, and environmental science. By overcoming these obstacles, the efficient and sustainable harvesting of microalgae biomass can be achieved, unlocking the full potential of microalgae for various applications, including biofuels, food, feed, and pharmaceuticals [1.24].

Despite these challenges, ongoing research and technological advancements continue to address these limitations and unlock the full potential of microalgae for biomass production. With further exploration and innovation, microalgae show promise as a sustainable and valuable resource for multiple industries, contributing to a more sustainable and greener future.

1.4 Harvesting Methods

Dewatering of microalgal biomass is a crucial step for producing feed, fuel and other high-value products. It has been reported that 20–30% of microalgal biomass cost is due to harvesting costs [1.25]. Although several methods are available, appropriate harvesting methods should be selected based on the end products and properties of microalgal strains. Present harvesting methods include physical (mechanical), chemical, and biological methods. Sedimentation, centrifugation and filtration are the standard physical techniques, along with a few electrical practices, such as flotation and electrolysis. Due to the small size and colloidal stability of microalgal cells in suspension, sedimentation is not viable [1.23]. Centrifugation, commonly applied for high-value products, is costly and cannot be employed for large volumes. It has been reported that separating microalgae by settling and centrifugation consumes harvesting energy of 1MJ/kg of dry biomass, which is more than that required for wood (0.7–0.9 MH/kg) [1.26]. Filtration can harvest larger species, such as *Spirulina*; however, smaller species, such as *Chlorella,* may fail to get purified.

Flocculation is another prominent technique for dewatering microalgal biomass [1.22] [1.27]. This method involves the formation of aggregates (flocs) by separating the solute particles from the solution occurring through one or combined mechanisms: (i) charge neutralization, where the charged microalgal cell surface is neutralized by incorporating oppositely charged ions, polymers or colloids, (ii) electrostatic patch, where binding of opposite charges results in patches, (iii) bridging, where a bridge is formed between two microalgal cell surfaces by the use of polymers and (iv) sweeping, where the smaller particles get entrapped into large flocs as they settle, aiding their quick removal from the liquid [1.28].

Chemical flocculating or coagulating agents have been studied extensively for biomass dewatering. The majority of the two chemical flocculants observed were inorganic and organic. Frequently utilized inorganic flocculants are ferric chloride, aluminum sulfate, ammonia and ferric sulfate [1.29] [1.30]. Uduman *et al.* reported three inorganic flocculants' optimum dosage and pH: lime, alum and ferric sulfate. The report suggested that alum and ferric sulfate gave better flocculating ability than lime. Unlike alum and ferric sulfate, the optimum dose of lime was almost twice the quantity. However, lime flocculation would be favorable if the suspension included magnesium or calcium ions [1.31]. Zhu *et al.* showed that $Al_2(SO_4)_3$ efficiently increased flocculating *Chlorella vulgaris* cells with the achieved concentration of 2.5g/L. The flocculating efficiency, as well as the optimum dosage required, varies from cell to cell. Although inorganic flocculants provide good harvesting efficiency, the high cost of chemicals, high dosage and metal ion contamination in biomass residues limits their application for human consumption [1.32].

Organic flocculating agents, such as polyacrylamide, polyacrylic acid, and polyamine, are commonly utilized. Although these water-soluble polymers can be highly effective in low dosages, their toxic monomeric residues lead to questionable consumption due to health and environmental consequences [1.33].

Scientists have recently tried replacing conventional chemical flocculants with nontoxic, eco-friendly, biodegradable, or natural polymers. Biopolymers such as chitosan, tannins, gums, cellulose, and alginates have taken hold.

1.5 Chitosan as Natural Flocculant

Chitosan is a naturally appearing linear polysaccharide made of ß-linked D-glucosamine and N-acetyl D-glucosamine [1.34]. Commercially, chitosan is derived from the waste product of shrimp, chitin (Figure 1.2). Further deacetylation of the acetamido group converts chitin into chitosan [1.35]. The product obtained is a positively charged primary amine group at low pH. Industrially, chitosan is highly resourceful. It is used in biomedical engineering, pharmaceuticals, dentistry, biotechnology, and the food industry [1.36]. Apart from these applications, it is also widely applied in water and wastewater treatment. The biodegradability, nontoxicity and reusability of chitosan make it an ideal flocculant [1.37] [1.38]. In a low pH solution, the positively charged amine group interacts with the negatively

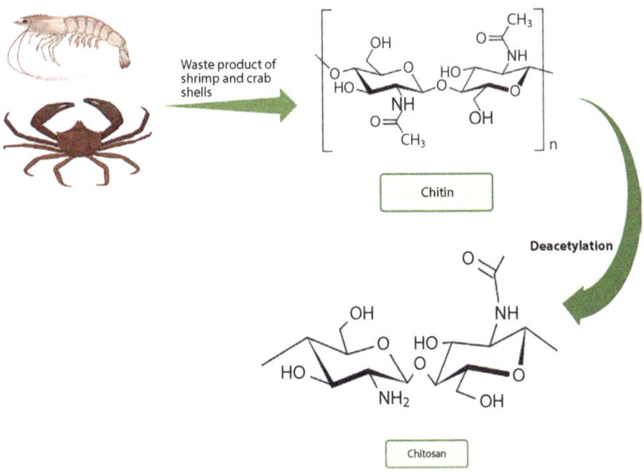

Figure 1.2 Commercial production of chitosan from shrimp and crab waste.

charged microalgae surface, which induces flocculation utilizing charge neutralization or bridging mechanisms [1.39].

Usually, chitosan adsorbing on a diatom may attach to another diatom particle's bare surface, forming bridges through patching and charge neutralization mechanisms. When flocculation and adsorption processes are correlated, tuning the process to obtain the desired floc characteristics is possible. Adsorption and flocculation must coincide with functioning rather than as an independent mechanism.

Furthermore, the molecular dynamic simulation study for chitosan and diatom cell surface protein silaffin by Patel *et al.* elaborated that the chitosan-silaffin interaction mainly involves the silaffin's PRO276, VAL278, THR280, SER281, ALA283, SER285, GLN287, GLN290, GLU296, SER310, SER311, ARG312, PRO313, ASP315, and ARG346, the majority of which are the polar residues with an average of 40% contact across the simulation time (Figure 1.3) [1.40]. Here, the $-NH_2$ and -OH functional groups of the chitosan play a crucial role in forming the specific H-bond with the protein residues. The interaction is non-stabilizing, suggesting that chitosan is an ideal diatom flocculant. Also, unlike the other microalgae, Patel *et al.* applied the chitosan flakes instead of a chitosan solution and observed a pH-independent flocculation of diatoms. They are projecting the lone chitosan flakes for diatom flocculation.

Yang *et al.* reported the harvesting efficiency of four flocculants (Tanfloc, Cationic starch, Chitosan, *Moringa oleifera*) for *Chlorella vulgaris* and *Scenedesmus obliquus* [1.41]. In their observation, they stated that the pH of culture media affected the harvesting efficiency of flocculants. As chitosan is insoluble in water under alkaline conditions, maximum harvesting efficiency was noted only up to pH 7.0 (90%). Biomass concentration also affected the optimum dosage of flocculants. When biomass was 0.3 g/L, optimum dosage of chitosan was 3 mg/L for *C. vulgaris* and 1 mg/L for *S. obliquus*. Even if the biomass concentration was increased to 1.5 g/L, of all the four flocculants, chitosan had the highest harvesting efficiency in less dosage. Furthermore, they have noted that, although chitosan has low dosage requirements, due to the narrow activity pH range, the applicability of chitosan is limited for the selected microalgae.

Hadiyanto *et al.* performed flocculation of *Chlorella* sp. using chitosan as a bio-flocculant. They utilized response surface methodology (RSM) with a central composite design (CCD) to improve the flocculation efficiency and biomass recovery parameters. In their study, 400 ml of culture with a 12 mg/ml concentration was tested with varying pH, chitosan concentration, cationic inducer and stirring speed. According to their study, chitosan

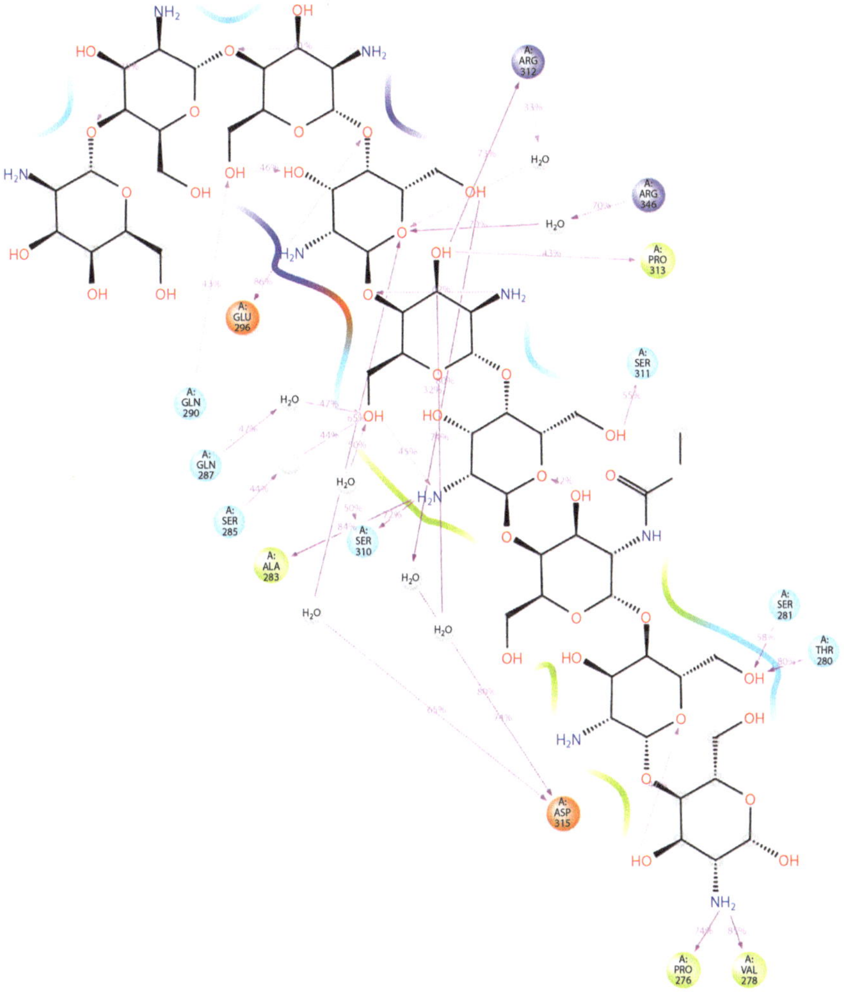

Figure 1.3 Chemical interaction of chitosan and diatom's cell surface protein silaffin (purple dashed line indicates H-bond).

at a concentration of 100 mg/ml was required to promote flocculation for 15 minutes, followed by adding 50 mg/ml $ZnCl_2$ (cationic inducer) at a pH of 8 and stirring at 240 rpm. A 96.12% flocculation efficiency and 312.5 mg biomass recovery were achieved, which served as a better alternative to synthetic flocculants [1.42].

A recent study by Islam *et al.* explored chitosan-mediated microalgae harvesting, followed by producing high-quality bio-crude from harvested biomass. A co-culture of *Scenedesmus* sp. and *Chlorella* sp. cultivated in domestic wastewater was utilized for the experiment. The optimization of biomass harvesting was achieved using response surface methodology (RSM) to find the interrelation between independent and response variables. The study concluded that chitosan-harvested microalgae co-culture showed >95% harvesting efficiency at a 17–26 mg/L dosing rate, more than 11 minutes of mixing and >75 rpm mixing speed [1.43].

Due to the limitations of chitosan's reactivity and processability, researchers have made several modifications worldwide. The efficiency of chitosan can be increased by physical or chemical alteration. Methods such as electroflotation, electrolysis and using magnetic nanoparticles in conjunction with chitosan have been derived to enhance recovery efficiency. Apart from these, the synergistic effects of chitosan and inorganic flocculants have been studied by Vu *et al.* [1.44].

1.6 Chitosan in Conjunction with Other Physicochemical Methods

1.6.1 Electroflotation and Mechanical Stirring

Electroflotation relies on generating gas bubbles at an electrode surface, typically through water electrolysis. The gas bubbles attach to the targeted algae cells, causing them to rise to the surface and form a floating layer, which can then be easily separated from the liquid medium. Once the algal cells have accumulated at the liquid surface, the foam layer is collected using various methods. The foam can be mechanically skimmed off or siphoned, and the harvested algal biomass can then undergo further processing, such as dewatering or drying, for subsequent applications.

Flocculation combined with electroflotation for varying pH and cell concentrations for *Chlorella* sp. was studied by Zhou *et al.*, who noted a rise of 5% in the flocculation efficiency of chitosan. According to the report, chitosan had >90% flocculation efficiency at pH slightly acidic to neutral at a concentration range of 8–12 mg/L. On the other hand, with an increase in microalgal concentration, the dosage of chitosan needed to be increased. For 0.63 g/L biomass, >85% flocculating efficiency was observed at 12 mg/L chitosan concentration. Although chitosan alone gave appropriate flocculating efficiency, stirring enhances the formation of flocs. Thus, optimum stirring parameters were determined for 0.5 g/L biomass density with 10 mg/L chitosan concentration. According to that study, a high stirring rate reduces the mixing time and damages the cells; hence, flocculation efficiency can be improved at 90–120 rpm with a mixing time of 30 s [1.45].

1.6.2 Electroflocculation

Electrocoagulation is an electrochemical process that uses metal electrodes to destabilize and aggregate algal cells in a liquid medium. The metal electrodes typically release metal ions (e.g., aluminum or iron). These ions neutralize the negative charge on the algal cell surfaces, causing them to coagulate and form larger flocs. These flocs are then separated by adding chitosan or other flocculants.

Zhu *et al.* applied an electric current to the system to enhance the charge neutralization capacity of chitosan. According to their report, lone Chitosan (4 mg/L) had only 16.9% efficiency, whereas when an electric current of 0.6A was supplied, flocculation efficiency was raised to approximately 65%. Due to the imposed voltage, microalgae cell surface charges escalated, which was observed by the shift in zeta potential [1.46].

In a study by Fayad *et al.*, electro-coagulation-flocculation (ECF) in batch mode was investigated for microalgae *Chlorella vulgaris*. The recovery was studied for various stirring speeds, electrode material, current density, pH and inter-electrode distance. In their observations, they noted that aluminum electrodes proved to be more efficient than iron electrodes. After studying the zeta potential, the mechanism of harvesting was elucidated, which showed that at pH 4 and 6, the mechanism was charge neutralization, whereas it was sweeping at pH 8 [1.47].

In yet another report, the potential for ECF was investigated for the recovery of *Scenedesmus almeriensis*. Here, the authors studied both batch and continuous modes of harvesting. According to the reports, the continuous mode was found to be more realistic to minimize the required residence time. At a conductivity of 5 $mS.cm^{-1}$, 12 V voltage and 23 $mA.cm^{-2}$ current density, 98% biomass recovery was achieved. Residence times of 3 to 6 s were enough to recover 85–95% of the biomass, resulting in a reduction of specific energy consumption up to 0.2 $kWh.kg^{-1}$ [1.48].

Hou *et al.* reported a novel technique integrating alkali-induced flocculation and electrolysis, named salt-bridge electroflocculation (SBEF). The method was developed to harvest *Nannochloropsis oculata*. The method provided a high harvesting efficiency of 90.4% under 300mA in 45 minutes. The energy consumption was also reduced to 1.50 Wh/g biomass. A 3.85-fold increase (2.75 to 10.59 μm) in mean floc size was also observed, which led to ease in filtration of flocs from SBEF. Also, SBEF did not bring any external flocculants to biomass, which may affect the algal metabolites [1.49].

1.6.3 Synergistic Effects of Chitosan and Inorganic Flocculants

The maximum harvesting efficiency was observed at a high dosage when inorganic flocculants (ferric chloride and aluminum sulfate) were used as flocculants per liter of *Chlorella vulgaris* suspension by Vu *et al.* [1.44]. At 450 mg/L concentration, ferric chloride provided 85% efficiency, while aluminum sulfate provided 80% efficiency. Chitosan and synthetic cationic polymer Flopam™ were also tested. At 450 mg/L, chitosan had only 60% efficiency, whereas Flopam had more than 95% efficiency. Synergistic effects were investigated to enhance chitosan's flocculation efficiency. An inorganic flocculant was paired with chitosan to double flocculate. A 57% increase was recorded in ferric chloride-chitosan dual flocculation, while a 24% rise was observed when chitosan was paired with the aluminum sulfate [1.44].

1.6.4 Integrated Flocculation

The use of nanoparticles has been widely accepted lately for various applications. In a report by Augustine *et al.*, chemo-magnetic and integrated chitosan-magnetic nanoparticles were utilized to flocculate *Parachlorella kessleri*-I. Magnetic nanoparticles and chitosan-magnetic nanoparticles were synthesized by the chemical co-precipitation method. Initially, the effect of pH was studied for the flocculation of *P. kessleri*-I. Later, the impact of pH on different flocculants at a uniform concentration was tested, in which chitosan (75 mg/L) and chitosan-MNPs at pH 8.5 gave 85% and 80% harvesting efficiency, respectively. After identifying the optimum pH of chitosan and CS-MNP, the effect of concentration at a stagnant pH was studied. At pH 8.5, chitosan yielded 85% flocculating efficiency at a concentration of 50 mg/L and >90% at 100 mg/L. Chitosan-coated magnetic nanoparticles showed similar results, where at a concentration of 100–150 mg/L, 85% efficiency was noted. On the other hand, flocculants, such as $FeCl_3$, $ZnCl_2$, $Al_2(SO_4)_3$ and $ZnSO_4$, offered 70–75% efficiency at lower concentrations. Apart from efficiency, speed and cost, another significant advantage of using chitosan-magnetic nanoparticles was the reusability of the flocculant and the culture medium [1.50].

1.7 Comparison of Different Harvesting Methods

Different harvesting agents and methods have been summarized above. It can be stated that the flocculation efficiency of each compound and each technique depends on the culture organism, its biomass and the culture medium pH. As studied, the mechanism behind flocculation/coagulation is charge neutralization and bridging. Methods such as electroflotation, electrolysis, dual flocculation and chemo-magnetic nanoparticles were reported to shift harvesting efficiency. It was observed that electroflotation increased the efficiency to a certain extent, whereas the efficiency observed for electrolysis was only 65%. When the dual flocculation studies were carried out, it was observed that the synergistic effects of inorganic flocculants with chitosan improved the efficiency. Similarly, instead of using chitosan alone, magnetic nanoparticles coated with chitosan would be an effective alternative with high flocculation efficiency. Although the solitary effect of chitosan was sufficient for flocculation, synergistic effects and utilization of magnetic nanoparticles would be helpful as the dosage requirement would reduce and, simultaneously, the toxicity and energy for the processing would decrease.

In addition, diatoms' flocculation can be challenged by their intricate silica cell walls, which may require special considerations to prevent damage to the frustules during flocculation.

1.8 Conclusion

As the need for harvesting microalgae increases, biopolymer-chitosan utilization becomes essential. Although the flocculating ability of chitosan is sufficient, new methods developed should be utilized in combination such that the dosage requirement of chitosan can be eased. The flocculant should be selected based on the properties of the microalgal culture to be harvested. Chitosan residues after biomass flocculation are biodegradable and nontoxic; hence, the medium can be reused, reducing nutrient requirements. Alternatives for chitosan can be the consumption of Tanfloc or Flopam™, which have been reported to provide high harvesting efficiency at low dosages.

Most studies on title flocculants for microalgae demand specific chitosan pretreatments; however, recent reports have suggested that owing to direct interaction between silaffin and chitosan, the chitosan flakes can directly be applied for flocculation of diatoms from a raw water sample. The developed process is highly accurate and precise for diatoms, leaving microalgae and protozoa in the supernatant and giving flocs within 60 minutes.

Furthermore, most other methods are selective towards particular microalgal cells. The authors of this chapter see a need for two sets of flocculants. One group should be selective for specific algal cells and modified according to cell chemistry. At the same time, the other flocculants should have non-selective interaction and can flocculate the different types of algal cells. The chitosan solution at a specific pH could serve the purpose of selective flocculant. In contrast, chitosan flakes can be employed as a less particular flocculating agent for pooling more diatoms from environmental samples.

References

[1.1] Vuppaladadiyam, A. K., Prinsen, P., Raheem, A., Luque, R., & Zhao, M. (2018). Microalgae cultivation and metabolites production: A comprehensive review. *Biofuels, Bioproducts and Biorefining*, 12(2), 304–324. https://doi.org/10.1002/bbb.1864

[1.2] Ananthi, V., Balaji, P., Sindhu, R., Ki\m, S.-H., Pugazhendhi, A., & Arun, A. (2021). A critical review on different harvesting techniques for algal based biodiesel production. *Science of The Total Environment*, 780, 146467. https://doi.org/10.1016/j.scitotenv.2021.146467

[1.3] Sathasivam, R., Radhakrishnan, R., Hashem, A., & Abd_Allah, E. F. (2019). Microalgae metabolites: A rich source of food and medicine. *Saudi Journal of Biological Sciences*, 26(4), 709–722. https://doi.org/10.1016/j.sjbs.2017.11.003

[1.4] Ansari, F. A., Guldhe, A., Gupta, S. K., Rawat, I., & Bux, F. (2021). Improving the feasibility of aquaculture feed by using microalgae. *Environmental Science and Pollution Research*, 28(32), 43234–43257. https://doi.org/10.1007/s11356-021-14989-x

[1.5] Branyikova, I., & Lucakova, S. (2021). Technical and physiological aspects of microalgae cultivation and productivity—*Spirulina* as a promising and feasible choice. *Organic Agriculture*, 11(2), 269–276. https://doi.org/10.1007/s13165-020-00323-1

[1.6] Hu, Q., Sommerfeld, M., Jarvis, E., Ghirardi, M., Posewitz, M., Seibert, M., & Darzins, A. (2008). Microalgal triacylglycerols as feedstocks for biofuel production: perspectives and advances. *The plant journal*, 54(4), 621-639.

[1.7] Slade, R., & Bauen, A. (2013). Micro-algae cultivation for biofuels: Cost, energy balance, environmental impacts and future prospects. *Biomass and Bioenergy*, 53, 29–38. https://doi.org/10.1016/j.biombioe.2012.12.019

[1.8] Faried, M., Samer, M., Abdelsalam, E., Yousef, R. S., Attia, Y. A., & Ali, A. S. (2017). Biodiesel production from microalgae: Processes, technologies and recent advancements. *Renewable and Sustainable Energy Reviews*, 79, 893–913. https://doi.org/10.1016/j.rser.2017.05.199
[1.9] Kale, A., & Karthick, B. (2015). The diatoms. *Resonance*, 20(10), 919-930.
[1.10] Leynaert, A., Fardel, C., Beker, B., Soler, C., Delebecq, G., Lemercier, A., ... & Heggarty, K. (2018). Diatom frustules nanostructure in pelagic and benthic environments. *Silicon*, 10(6), 2701-2709.
[1.11] Khan, M. I., Shin, J. H., & Kim, J. D. (2018). The promising future of microalgae: current status, challenges, and optimization of a sustainable and renewable industry for biofuels, feed, and other products. *Microbial cell factories*, 17(1), 1-21.
[1.12] Enamala, M. K., Enamala, S., Chavali, M., Donepudi, J., Yadavalli, R., Kolapalli, B., Aradhyula, T. V., Velpuri, J., & Kuppam, C. (2018). Production of biofuels from microalgae—A review on cultivation, harvesting, lipid extraction, and numerous applications of microalgae. *Renewable and Sustainable Energy Reviews*, 94, 49–68. https://doi.org/10.1016/j.rser.2018.05.012
[1.13] Suparmaniam, U., Lam, M. K., Uemura, Y., Lim, J. W., Lee, K. T., & Shuit, S. H. (2019). Insights into the microalgae cultivation technology and harvesting process for biofuel production: A review. *Renewable and Sustainable Energy Reviews*, 115, 109361. https://doi.org/10.1016/j.rser.2019.109361
[1.14] Veerabadhran, M., Natesan, S., MubarakAli, D., Xu, S., & Yang, F. (2021). Using different cultivation strategies and methods for the production of microalgal biomass as a raw material for the generation of bioproducts. *Chemosphere*, 285, 131436. https://doi.org/10.1016/j.chemosphere.2021.131436
[1.15] Parkinson, J., & Gordon, R. (1999). Beyond micromachining: the potential of diatoms. *Trends in biotechnology*, 17(5), 190-196.
[1.16] Seckbach, J., & Gordon, R. (Eds.). (2019). *Diatoms: fundamentals and applications*. John Wiley & Sons.
[1.17] Pahl, S. L., Lewis, D. M., Chen, F., & King, K. D. (2010). Growth dynamics and the proximate biochemical composition and fatty acid profile of the heterotrophically grown diatom *Cyclotella cryptica*. *Journal of applied phycology*, 22, 165-171.
[1.18] Barreto Filho, M. M., Walker, M., Ashworth, M. P., & Morris, J. J. (2021). Structure and long-term stability of the microbiome in diverse diatom cultures. *Microbiology Spectrum*, 9(1), e00269-21.
[1.19] Chauton, M. S., Winge, P., Brembu, T., Vadstein, O., & Bones, A. M. (2013). Gene regulation of carbon fixation, storage, and utilization in the diatom *Phaeodactylum tricornutum* acclimated to light/dark cycles. *Plant physiology*, 161(2), 1034-1048.

[1.20] Rakesh, S., Tharunkumar, J., Sri, B., Jothibasu, K., & Karthikeyan, S. (2020). Sustainable cost-effective microalgae harvesting strategies for the production of biofuel and oleochemicals. *Highlights in bioscience*, *3*(July), 1-8.

[1.21] Ummalyma, S. B., Gnansounou, E., Sukumaran, R. K., Sindhu, R., Pandey, A., & Sahoo, D. (2017). Bioflocculation: an alternative strategy for harvesting of microalgae–an overview. *Bioresource technology*, 242, 227-235.

[1.22] Labeeuw, L., Commault, A. S., Kuzhiumparambil, U., Emmerton, B., Nguyen, L. N., Nghiem, L. D., & Ralph, P. J. (2021). A comprehensive analysis of an effective flocculation method for high quality microalgal biomass harvesting. *Science of The Total Environment*, 752, 141708. https://doi.org/10.1016/j.scitotenv.2020.141708

[1.23] Singh, G., & Patidar, S. K. (2018). Microalgae harvesting techniques: A review. *Journal of Environmental Management*, 217, 499–508. https://doi.org/10.1016/j.jenvman.2018.04.010

[1.24] Branyikova, I., Prochazkova, G., Potocar, T., Jezkova, Z., & Branyik, T. (2018). Harvesting of microalgae by flocculation. *Fermentation*, 4(4), 93.

[1.25] Christenson, L., & Sims, R. (2011). Production and harvesting of microalgae for wastewater treatment, biofuels, and bioproducts. *Biotechnology advances*, 29(6), 686-702.

[1.26] Morrissey, K. L., Keirn, M. I., Inaba, Y., Denham, A. J., Henry, G. J., Vogler, B. W., Posewitz, M. C., & Stoykovich, M. P. (2015). Recyclable polyampholyte flocculants for the cost-effective stering of microalgae and cyanobacteria. *Algal Research*, 11, 304–312. https://doi.org/10.1016/j.algal.2015.07.009

[1.27] Milledge, J. J., & Heaven, S. (2013). A review of the harvesting of micro-algae for biofuel production. *Reviews in Environmental Science and Bio/Technology*, 12(2), 165–178. https://doi.org/10.1007/s11157-012-9301-z

[1.28] Blockx, J., Verfaillie, A., Thielemans, W., & Muylaert, K. (2018). Unravelling the Mechanism of Chitosan-Driven Flocculation of Microalgae in Seawater as a Function of pH. *ACS Sustainable Chemistry & Engineering*, 6(9), 11273–11279. https://doi.org/10.1021/acssuschemeng.7b04802

[1.29] Matter, Bui, Jung, Seo, Kim, Lee, & Oh. (2019). Flocculation Harvesting Techniques for Microalgae: A Review. *Applied Sciences*, 9(15), 3069. https://doi.org/10.3390/app9153069

[1.30] Wang, Q., Oshita, K., & Takaoka, M. (2021). Flocculation properties of eight microalgae induced by aluminum chloride, Chitosan, amphoteric polyacrylamide, and alkaline: Life-cycle assessment for screening species and harvesting methods. *Algal Research*, 54, 102226. https://doi.org/10.1016/j.algal.2021.102226

[1.31] Uduman, N., Qi, Y., Danquah, M. K., Forde, G. M., & Hoadley, A. (2010). Dewatering of microalgal cultures: A major bottleneck to algae-based fuels. *Journal of Renewable and Sustainable Energy*, 2(1), 012701. https://doi.org/10.1063/1.3294480

[1.32] Zhu, L., Li, Z., & Hiltunen, E. (2018). Microalgae *Chlorella vulgaris* biomass harvesting by natural flocculant: Effects on biomass sedimentation, spent medium recycling and lipid extraction. *Biotechnology for Biofuels*, 11(1), 183. https://doi.org/10.1186/s13068-018-1183-z
[1.33] Shaikh, S. M. R., Hassan, M. K., Nasser, Mustafa. S., Sayadi, S., Ayesh, A. I., & Vasagar, V. (2021). A comprehensive review on harvesting of microalgae using Polyacrylamide-Based Flocculants: Potentials and challenges. *Separation and Purification Technology*, 277, 119508. https://doi.org/10.1016/j.seppur.2021.119508
[1.34] Kumaran, J., Singh, I. S. B., & Joseph, V. (2021). Effective biomass harvesting of marine diatom *Chaetoceros muelleri* by chitosan-induced flocculation, preservation of biomass, and recycling of culture medium for aquaculture feed application. *Journal of Applied Phycology*, 33(3), 1605–1619. https://doi.org/10.1007/s10811-021-02369-4
[1.35] Renault, F., Sancey, B., Badot, P.-M., & Crini, G. (2009). Chitosan for coagulation/flocculation processes – An eco-friendly approach. *European Polymer Journal*, 45(5), 1337–1348. https://doi.org/10.1016/j.eurpolymj.2008.12.027
[1.36] Chen, G., Zhao, L., Qi, Y., & Cui, Y.-L. (2014). Chitosan and Its Derivatives Applied in Harvesting Microalgae for Biodiesel Production: An Outlook. *Journal of Nanomaterials*, 2014, 1–9. https://doi.org/10.1155/2014/217537
[1.37] Ahmad, A. L., Mat Yasin, N. H., Derek, C. J. C., & Lim, J. K. (2011). Optimization of microalgae coagulation process using Chitosan. *Chemical Engineering Journal*, 173(3), 879–882. https://doi.org/10.1016/j.cej.2011.07.070
[1.38] Yin, Z., Chu, R., Zhu, L., Li, S., Mo, F., Hu, D., & Liu, C. (2021). Application of chitosan-based flocculants to harvest microalgal biomass for biofuel production: A review. *Renewable and Sustainable Energy Reviews*, 145, 111159. https://doi.org/10.1016/j.rser.2021.111159
[1.39] Demir, I., Blockx, J., Dague, E., Guiraud, P., Thielemans, W., Muylaert, K., & Formosa-Dague, C. (2020). Nanoscale Evidence Unravels Microalgae Flocculation Mechanism Induced by Chitosan. *ACS Applied Bio Materials*, 3(12), 8446–8459. https://doi.org/10.1021/acsabm.0c00772
[1.40] Patel, M., Parikh, H., & Dave, G. (2023). Chitosan flakes-mediated diatom harvesting from natural water sources. *Water Science and Technology*, 87(7), 1732-1746.
[1.41] Yang, Z., Hou, J., & Miao, L. (2021). Harvesting freshwater microalgae with natural polymer flocculants. *Algal Research*, 57, 102358. https://doi.org/10.1016/j.algal.2021.102358
[1.42] Hadiyanto, H., Widayat, W., Christwardana, M., & Pratiwi, M. E. (2022). The flocculation process of *Chlorella* sp. using Chitosan as a bio-flocculant: optimization of operating conditions by response surface methodology. *Current Research in Green and Sustainable Chemistry*, 5, 100291.

[1.43] Islam, M. B., Khalekuzzaman, M., Kabir, S. B., & Hossain, M. R. (2022). Shrimp waste-derived Chitosan harvested microalgae for the production of high-quality biocrude through hydrothermal liquefaction. *Fuel*, 320, 123906.

[1.44] Vu, H. P., Nguyen, L. N., Lesage, G., & Nghiem, L. D. (2020). Synergistic effect of dual flocculation between inorganic salts and Chitosan on harvesting microalgae *Chlorella vulgaris*. *Environmental Technology & Innovation*, 17, 100622. https://doi.org/10.1016/j.eti.2020.100622

[1.45] Zhou, W., Gao, L., Cheng, W., Chen, L., Wang, J., Wang, H., Zhang, W., & Liu, T. (2016). Electroflotation of *Chlorella* sp. Assisted with flocculation by Chitosan. *Algal Research*, 18, 7–14. https://doi.org/10.1016/j.algal.2016.05.029

[1.46] Zhu, L., Pan, G., Xu, H., Kong, L., Guo, W., Yu, J., Mortimer, R. J. G., & Shi, W. (2021). Enhanced chitosan flocculation for microalgae harvesting using electrolysis. *Algal Research*, 55, 102268. https://doi.org/10.1016/j.algal.2021.102268

[1.47] Fayad, N., Yehya, T., Audonnet, F., & Vial, C. (2017). Harvesting of microalgae *Chlorella vulgaris* using electro-coagulation-flocculation in the batch mode. *Algal research*, 25, 1-11.

[1.48] Inostroza, C., El Bahraoui, N., Rivera-Tinoco, R., & Acién, F. G. (2022). Uses of electro-coagulation-flocculation (ECF) for the pre-concentration of microalgae biomass. *Process Biochemistry*, 122, 1-7.

[1.49] Hou, Y., Liu, C., Liu, Z., Han, T., Hao, N., Guo, Z., ... & Chen, F. (2022). A Novel Salt-Bridge Electroflocculation Technology for Harvesting Microalgae. *Frontiers in Bioengineering and Biotechnology*, 10, 902524.

[1.50] Augustine, A., Tanwar, A., Tremblay, R., & Kumar, S. (2019). Flocculation processes optimization for reuse of culture medium without pH neutralization. *Algal Research*, 39, 101437. https://doi.org/10.1016/j.algal.2019.101437

2

Diatoms Cultivation: Trends, Scope and Technoeconomic Prospects

Anshuman Rai[1,2], Nirmala Sehrawat[1], Mukesh Yadav[1], Varruchi Sharma[3], Vikas Kumar[4] and Anil K. Sharma[5*]

[1]Department of Bio-Sciences and Technology, M.M.E.C., Maharishi Markandeshwar (Deemed to be University), Mullana-Ambala, Haryana, India
[2]Forensic Science Laboratory, Haryana, Madhuban, Karnal, India
[3]Department of Biotechnology & Bioinformatics, Sri Guru Gobind Singh College, Chandigarh, India
[4]International Medical School, University of International Business, Almaty, Kazakhstan
[5]Department of Biotechnology, Amity School of Biological Sciences, Amity University Punjab, Mohali, India

Abstract
Diatoms are one of the most promising microalgae for about 25% of CO_2 fixation. They are also known as primary producers of our Mother Earth. The main value-added products for which they have gained importance in the commercial market are due to their pigment fucoxanthin, which has high antioxidant, medicinal, therapeutic and pharmaceutical properties. Besides this, they alone provide 30% of the crude oil at the global level; therefore, their techno-economic cultivation holds promise and importance. Because of the oil abundance in them, they are a rich source of biofuel, also known as Diafuel™ (biofuel from diatoms). The cultivation practices used to grow diatoms, however, are not cost-effective, which raises the cost of the value-added products derived from them. The open raceway ponds and closed photobioreactors have their own pros and cons. The current chapter discusses the cultivation practices generally used to cultivate microalgae and diatoms economically. This is needed to make the biofuel to be used commercially to compensate for the requirements of renewable energy.

***Keywords*:** Diatoms, cultivation, photobioreactor, biofuel, fucoxanthin

**Corresponding author*: anibiotech18@gmail.com

Vandana Vinayak and Richard Gordon (eds.) Diatom Cultivation for Biofuel, Food and High-Value Products, (21–50) © 2025 Scrivener Publishing LLC

2.1 Introduction

The benefits of diatoms, an important group of microalgae, have lately surprised the scientific community with enormous industrial potential in biotechnology, biomedical, nanotechnology and environmental technologies. They are the most suitable alternative to fossil fuels because of their high lipid production capacity, higher growth rate, short generation time and universal aquatic habitats [2.1]. Diatoms fix up to 20% of the global carbon dioxide, which is partitioned into carbohydrates or lipids; particularly triacylglycerols (TAGs), making up to 25% of dry biomass [2.2].

Their fatty acid profile is the most exclusive feature of diatoms, containing various lengths of fatty acid chains such as polyunsaturated fatty acids (PUFA), a feature deficient in chlorophytes or plants. The notable fatty acids, such as 14:0, 16:0, 16:1 and 20:5, originate from diatoms, whereas C18 is found in traces while C24 is a norchloestane marker found only in diatoms [2.3]. Medium-chain fatty acids produce less-viscous biodiesel [2.4]. Carotenoids, such as fucoxanthin and didinoxanthin, participate as natural food as well as feed in the pharmaceutical and cosmeceutical industry. A list of very-long-chain PUFAs, such as arachidonic acid (ARA) and eicosapentaenoic acid (EPA), serves as an indispensable source of fundamental omega-3 fatty acids, which is partitioned into TAGs and found in most diatoms for pharmaceutical and nutraceutical applications [2.5]. Constitutionally, lipids, i.e., saturated and unsaturated fatty acids, can predict the quantity of fuel present in diatoms followed by its extraction [2.6]. When nutrient medium and physicochemical factors are manipulated along with cultivation methods, a significant variation in lipid production was found, and thus represents a solution to increase the lipid yield. Such variation increases the growth period of diatoms, which can contrive overall production per unit area. For instance, biofuels have provided better properties than traditionally used fuels under nitrogen starvation conditions [2.7]. Under nitrogen starvation conditions, *Chaetoceros gracilis* can account for 70% of the total volume on a per weight basis [2.8]. With decreasing chlorophyll content, TAG and other lipid levels start increasing in microalgae, which can be employed for higher production of biofuel; this study was proven upon testing diatom *Thalassiosira pseudonana* [2.9]. Diatoms are freely accessible to water, carbon dioxide, and other nutrients, and thus allow the conversion of lipid up to 85% of their total weight, thereby producing 30 times more oil per unit area than oilseed crops [2.10]. Diatom components can be extracted through two-phase extraction, filtration, chromatography and transesterification [2.11].

The best suitable indicators for the screening in biodiesel production are biomass and lipid productivity, which was confirmed in the studies by Joseph *et al.* in which the five diatom species studied were as follows: *Chaetoceros calcitrans, Thalassiosira* sp., *Skeletonema* sp., *Navicula* sp. and *Nitzchia* sp. [2.12]. A liquid fuel energy equal to 1015 BTUs (~1018 kJ) could be produced from 200 kHa of open pond cultivation, providing efficient results proving the potential of power generated using diatoms [2.13].

An exquisite research study, which involved manipulating the nutrition, physicochemical environmental conditions, cultivation methods and extraction protocol, provided a significant amount of lipid production. In reference to studies conducted by Milano *et al.* which concluded that biofuels have better properties than traditional fuels in nitrogen starvation [2.7], a study supporting this view showed that under nitrogen starvation conditions, *Chaetoceros gracilis* can account for 70% of total volume on a per weight basis [2.8]. A study conducted by Guiry showed how 60% of algal strains were diatoms which survived harsh natural conditions and provided outstanding performance in lipid productivity, hence proving that diatom-based microalgae biofuel can serve as an efficient power generation source and can be commercialized on a larger scale by optimizing conditions for economic feasibility [2.14]. *Chaetoceros* sp., *Skeletonema* sp. and *Thalassiosira* sp. are the major classes of diatom species exploited due to their greater dry cell weight yielding capacity, implying better lipid productivity and enhancing the quantity of biofuel harvested [2.4]. Studies show that diatom species, namely *Pharodactylum tricornutum* and *Synedra ulna,* are the classes with least dry cell weight, implying that the harvesting of biofuel from these species can be a tedious process that leads to a lower quantity of diafuel [2.15] [2.16]. Several studies which talk about species of diatoms with their relative lipid content derived via their dry cell weight are mentioned in Table 2.1. The FAME profile establishes the suitability of the feedstock for biodiesel production ascertained by vital factors depicting the quality of biodiesel and its characteristics like cloud point, cetane number, plugging point, cold filter, iodine value, and kinematic viscosity [2.17]. Fatty acid methyl esters (FAME) assessment through Gas chromatography and mass spectroscopy (GC–MS) showed higher proportions of palmitoleic acid (C16:1; ~18–29%) and oleic acid (C18:1; ~26–35%) with saturates and monounsaturates of C16–C18 up to 90%, confirming the superior biodiesel quality property, which is desirable [2.18].

Table 2.1 Biomass and lipid productivity under different modes of cultivation for different species of diatoms.

Diatoms	Mode of cultivation	Organic carbon source	Biomass productivity (g/Ld)	Lipid productivity (g/Ld)	Reference
Phaeodactylum tricornutum	Mixotrophy	Glucose	0.053	0.038	[2.32]
		Acetate	0.020	0.017	
		Starch	0.023	0.024	
P. tricornutum	Mixotrophy	Glycerol	*0.713	–	[2.33]
		Acetate	*0.587	–	
		Glucose	*0.555	–	
P. tricornutum UTEX 640	Mixotrophy	Glycerol	0.511	–	[2.34]
		Fructose	0.379	–	
		Glucose	0.245	–	
P. tricornutum	Mixotrophy	Soluble fractions of potato	^ 4.1×10^6	–	[2.35]
Chaetoceros sp.	Mixotrophy	Glucose	*0.217	*0.22	[2.4]
Thalassiosira sp.	Mixotrophy	Glucose	-	*0.52	[2.4]
Skeletonema sp.	Mixotrophy	Glucose	-	*0.44	[2.4]

*biomass/lipid concentration g L^{-1} ; ^number of cells L^{-1} d^{-1}

2.2 Cultivation Strategy and Production

Depending on the cultivation type, the biomass retained in the microalgae also varies accordingly. The three major types of cultivation are autotrophic, heterotrophic and mixotrophic-based cultivations which vary on the basis of their trophodynamic properties, i.e., source and duration of growth factors provided to the feedstock [2.19]. The autotrophic microalgae transform the light energy along with organic source to chemical energy during photosynthesis [2.20]. The autotrophic mode has been the most predominant mode of microalgal cultivation [2.21]. The heterotrophic microalgae growth depends upon exogenous organic matter consumption during absence of light through fermentation or aerobic respiration, generally resulting in increased cell density and cell growth [2.22]. Mixotrophic microalgae can be considered as a synergistic combination of the latter two, wherein it achieves greater biomass via potential strategy of resource utilization mostly useful during commercialization or for mass production needs [2.23]. Mixotrophic growth modes tend to maximize resource utilization by eliminating problems associated with limited light availability, thus leading to enhanced growth rate, biomass, and lipid productivity [2.24]. The biomass yield was highest in mixotrophic diatoms (139.16 ± 3 mg/L), followed by autotrophic (127.04 ± 4 mg/L) and heterotrophically grown diatoms (95.22 ± 3 mg/L) [2.18]. A summary of their growth pattern for different cultivation modes is presented in Table 2.2. Diatoms are metabolically flexible organisms with proven characteristics of surviving

Table 2.2 Different diatom species and their relative lipid content in dry cell weight (DCW).

Sl. no.	Diatom species	Lipid (%) DCW	Reference
1.	*Achnanthes delicatulum hauckiana*	29.8	[2.36]
2.	*Achnanthes* sp.	27.7	[2.37]
3.	*Aulacoseria ambigua*	19.7	[2.38]
4.	*Bacillaria paradoxa*	33.61	[2.39]
5.	*Cocconeis peltoides*	20.9	[2.36]
6.	*Chaetoceros curvisetus* *Chaetoceros muelleri* *Chaetoceros calcitrans*	14.86 33.6 39.8	[2.40] [2.41]
7.	*Diatoma* sp.	15.76	[2.36] [2.42]
8.	*Cocconeis* sp.	31.8	[2.43]
9.	*Cyclotella cryptica*	27.0	[2.13]
10.	*Melosira* sp.	14.75	[2.44]
11.	*Nitzschia dissipata* car. *media*	37.5	[2.44]
12.	*Pharodactylum tricornutum*	10.7 18.7	[2.15] [2.40]
13.	*Synedra ulna*	7.58	[2.16]
14.	*Tryblionella navicularis*	24.2	[2.36]

(Continued)

Table 2.2 Different diatom species and their relative lipid content in dry cell weight (DCW). (*Continued*)

Sl. no.	Diatom species	Lipid (%) DCW	Reference
15.	*Nitzschia punctata*	16.0	[2.5]
16.	*Skeletonema costatum* *Skeletonema sp.*	21.0 31.8	[2.40]
17.	*Thalassiosira pseudonana*	20.6	[2.40]
18.	*Thalassiosira* sp.	52	[2.4]
19.	*Skeletonema* sp.	44	[2.4]
20.	*Chaetoceros* sp.	22	[2.4]
21.	*Cylindrotheca* sp.	16	[2.45]
22.	*Nitzschia* sp.	16	[2.45]
23.	*Phaeodactylum tricornutum*	18-57	[2.45]
24.	*Thalassiosira weissflogii*	19.5	[2.46]
25.	*Phaeodactylum tricornutum*	14.3	[2.47]
26.	*Chaetoceros* sp.	44-47	[2.48]
27.	*Thalassiosira* sp.	29-32	[2.48]

different growth modes of autotrophy, heterotrophy and mixotrophy. Diatom species *Amphora* sp., *Navicula saprophila, Phaeodactylum tricornutum, Cyclotella cryptica,* and *Nitzschia* sp. are reported to grow under both heterotrophic and mixotrophic nutrition modes [2.25]. Higher biomass achieved (139.16 ± 3 mg L^{-1}) in mixotrophy shows potential industrial utilization. From the perspective of biodiesel production using mixotrophically grown diatoms, the results elucidate higher proportions of saturates and MUFAs, especially C16:0, C16:1, and C18:1 up to ~91%, which renders superior quality biodiesel [2.18] (Table 2.2).

The aspects in fatty acid profile (FAP) and lipid yield (LY) of microalgae during the synthesis as well as harvesting process have immense changeability. Research studies highlight the effects of algae-algae co-cultivation having a comparatively higher FAP quality and improving the flocculation efficiency. Enhancing FAP&LY via the harvesting process could measurably be an effective breakthrough technology for cost-efficient microalgae-based biodiesel production [2.26].

Designing a culture system with the addition of flue gas algae cultivation for microalgae harvesting is another crucial issue to take into account [2.27]. The choice of culture system is largely influenced by the cultivation's ultimate product. To obtain maximum output and lower total costs, an appropriate system should be efficiently constructed. There are several culture conditions accessible, including photobioreactors, closed systems, and open pond systems [2.28] [2.29]. Photobioreactors are significantly less financially feasible than open ponds. It is advantageous to grow species like *Chlorella* in an open pond because they are well suited to wastewater and a particular limit of CO_2. Additional steps, such as the addition of antibacterial or antifungal medicines, should be followed to decrease bacterial and fungal infection. The first step in making this technology feasible is choosing the right strain of microalgae, which should be chosen based on its suitability to grow in the presence of industrial wastewater and flue gases as well as in open pond culture conditions [2.30]. Only then can a successful technology be developed for combined remediation and biodiesel production. There is now no method that can handle all the demands of 280 algal cultivation. If large-scale production is the main concern, an open pond is unquestionably the greatest choice, but it needs to be modified to allow for consistent flue gas addition and proper mixing [2.31].

Thorough research investigations are needed to design an economical large-scale harvesting technique. Physical approaches to dewater massive microalgal suspension have high energy requirements and time-consuming qualities, paving the way for the development of an effective harvesting methodology.

2.3 Design and Implementation of a Prototype Algal Reactor

Algae are now well known for potential applications in various industries. Algae cultivation has been reported to obtain various benefits, including alternate healthy foods, pigments, secondary metabolites, wastewater treatments [2.49–2.51]. Also, algae have been investigated for production of bioactive compounds such as polysaccharides, hydrocarbons, biologically active compounds, molecules having therapeutic potential, fuel production, etc. [2.50–2.54]. Cultivation of microalgae has several advantages, including industrially important products, cost-effective growth of algal biomass, utilization of sunlight and others. Algal cultivation has already been focused on obtaining various products that have commercial value. The targeted algae can be grown in a suitable and efficient bioreactor. In a broad sense, a bioreactor is the manufactured or engineered device or system capable of supporting a biologically active environment [2.55]. Broadly speaking, bioreactors are used for cultivation of desired microbes under controlled conditions to obtain biomass and product as well. The design of bioreactors range from a small, lab-scale shake flask culture to large-scale bioreactors, including large open pond and engineered bioreactors [2.55]. Carvalho *et al.* discussed the microalgal reactors along with their performances. The authors discussed tubular reactors along with their configurations, including airlift, bubble column, horizontal tubular reactor, and helical tubular reactor. Airlift and bubble column reactors are examples of vertical tubular reactors (VTR). These are generally composed of significantly transparent polyethylene or glass-based tubes for better light penetration and also manufactured using cost-effective materials [2.51] [2.55]. These reactors are considered a comparatively simple form of bioreactors, while the corresponding technology is considered primitive [2.51] [2.56]. An airlift reactor is a form of bubble reactor having internal draft-tube that favors the gas–liquid mass transfer and mixing [2.55]. In an airlift reactor, air leads to circulation and mixing of media. Horizontal tubular reactors (HTR) have also been reported for algal culture. Flat-plate reactors and fermenter-type reactors have also been discussed [2.51]. Flat-plate reactors (FPR) are conceptually designed to make efficient use of sunlight and, therefore, narrow panels are usually built so as to attain high area-to-volume ratios. Pruvost *et al.* have also discussed photobioreactors for large-scale production of algal biomass [2.57]. The authors concluded that open systems are comparatively simple while the closed photobioreactors (PBRs) are considered complex systems. Various designs of photobioreactors have been reported with improved efficiency. Posten has reviewed the design principles of photobioreactors for cultivation of microalgae [2.58]. The author discussed the designs along with the basic limiting factors like light distribution, use of light/dark cycles, mass transfer, carbon dioxide supply, oxygen removal and energy demand. Similarly, Xu *et al.* have also reviewed the microalgal bioreactors [2.59]. The authors presented open-air cultivation systems and closed cultivation systems. The major focus in designing the photobioreactors includes the maximal use of sunlight, cost-effective operation, easy control and other relevant parameters. Photobioreactors (PBRs) have been developed to overcome the problems associated with open-pond systems [2.55]. The PBRs can be classified depending on the design and mode of operation. Scientists are still working on the development of improved and efficient designs of bioreactors for algal cultivation. Major considerations for photobioreactor design and operation for mass cultivation of microalgae have been discussed by Cañedo and Lizárraga [2.60]. Efficient lighting processes, efficient supply of carbon dioxide and removal of oxygen, improved gaseous exchange, and energy efficient systems are the major focal points [2.58] [2.61]. Fiber optics-based photobioreactors have already been proposed for delivering light to microalgal cultures inside the reactor as an alternative to using glass tubes with outside radiation. Hincapie and Stuart have reported the development of fiber optics-based photobioreactors for growing algae using internal illumination. The proposed design was on the basis of the air-lift principle. The authors used optical fibers for distribution of photons inside the culture media and controlled the light cycle as well as its intensity. The developed reactor was studied by culture of an algae strain of *Chlorella* sp. and the proposed design was suggested to be used in an effective as well as economic way for carbon dioxide mitigation technologies and also in the production of algal biomass for biofuel and other bioproducts [2.61]. In recent years, significant progress has been observed in bioengineering and biotechnology for efficient mass production of microalgae [2.59]. Different types of closed culture systems (photobioreactors) have been revealed as promising for large-scale production of microalgal culture for various purposes. Still, further improvement is required in the design, operation, and efficiency of bioreactors for large-scale culture of microalgae in a cost-effective manner.

2.4 Potential of Diatoms as a Source of Biofuel with Value-Added Products

Diatoms are a photosynthesizing algae with a siliceous skeleton. They are found in fresh and marine waters, soils, and almost anywhere moist, which is most suited for these organisms. These are among the most promising microorganism types, which are capable enough in producing biofuel. The production of biofuels is due to its dense presence and rapid doubling of biomass. Diatoms are mainly divided into two main groups which are demarcated on the basis of shape of the frustule. The first one is the centric diatoms and the second type is pennate diatoms; which are bilaterally symmetric in nature [2.62].

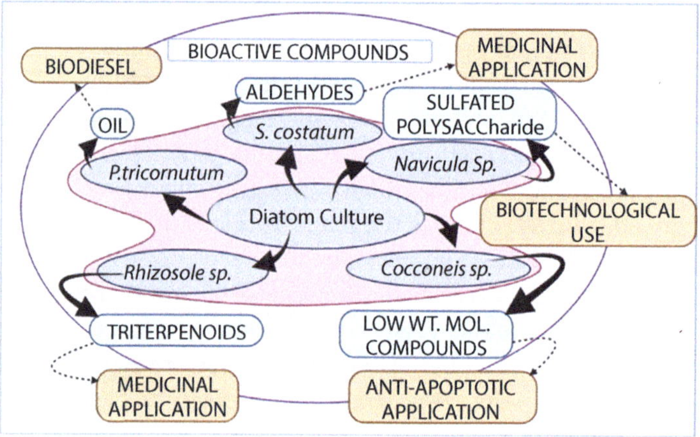

Figure 2.1 Probable bioactive compounds extracted from diatoms along with applications.

The diatom biomass can be put to profitable use by using silicate. Sterols, acyl lipids, waxes, etc., are identified from diatoms. There are approximately 200 genera of existing diatoms and about 100,000 living species. Also, these contain an extensive lipid variety, membrane-bound polar lipids, free fatty acids and triglycerides, etc. [2.62] [2.63]. Virtually any biological material, including wood, grass, any crop types, trees, agricultural waste (sugarcane, rice, beet, wood chippings, canola oil, corn oil, palm oil, etc.) or even animals (animal fat), can be used to constitute biofuels, which are considered as an alternative to fossil fuels in the current scenario. Biofuels, which are derivatives of animal and plant materials processing, by their very nature are renewable energy sources as hydrocarbon fuel is a renewable source of diesel [2.64] [2.65]. Figure 2.1 highlights the probable bioactive compounds extracted from diatoms along with the underlying applications.

2.4.1 Diatoms in the Biofuels Industry

Some of the common diatom species which have biofuel potential are: (i) *Thalassiosira pseudonana*, (ii) *Cocconeis* sp., (iii) *Lampriscus* sp., (iv) *Gyrosigma balticum*, (v) *Cyclotella cryptica*, (vi) *Nitzschia* sp., (vii) *Thalassiosira weissflogii*, (viii) *Achnanthes* sp.

These diatoms have distinctive cell walls that is made up of nanostructured silica which are between 10 and 50 μm in size, that have also been faithfully reproduced through generation by genetically controlled assembly processes [2.8] [2.41] [2.66].

2.4.2 Medical Applications

Diatoms are a major class of phytoplankton that account for about 40% of the ocean carbon fixation. Due to their valuable properties in reference to human health, these microbes are used to synthesize biofuels, biosilica, enzymes, mineral materials, etc. [2.67] [2.68]. The marine benthic diatoms, which are abundant and ubiquitous in terms of ecosystems and are sensitive to nutrient variation, have the capability to induce leukaemia cell death and modulate blood platelet activity [2.66]. The antibacterial and antifungal compounds are the major components present in FA extracts. *Skeletonema costatum* (a dominant species which is responsible for red tide outbreaks) has been shown to have antibacterial growth inhibitory activity against *Vibrio* sp. (a pathogen of fish and shellfish) [2.69]. *Skeletonema costatum* (a marine and occasionally brackish diatom found in coastal areas which forms large-scale algal blooms in eutrophic coastal waters and estuaries) has been shown to have promising anti-tumor activity against lung cancer in humans along with anti-HIV effects as well, which are mostly attributed to the presence of polyunsaturated sesterpenes in them [2.66] [2.70].

2.5 Industrial Aspects of Diatoms as a Source of Biofuel

Carbon dioxide and related pollutant mitigation is an environment-friendly aspect of diatoms. Apart from CO_2 mitigation, some micro-algal cultures, including diatoms, have the capability to remove biological NOx from combustible gases [2.8]. With the advancement of processes, management of diatom nutrients has been used effectively for wastewater treatment [2.68]. The use of diatoms in liquid fuel production mitigates pollutants on a large scale [2.71]. In thermal processing, the organic carbon in the diatoms is converted into biocrude. Moreover, nutrients remaining in the post-processed water were also reused. On the other hand, in non-thermal processing the processed proteins are used as an animal feed [2.8] [2.72].

In the present scenario, the development of environment-friendly and cost-effective fuel is a major challenge which can be met by using diatoms as a source for fuel production. The biofuels are alternative sources for chemical energy. Recently, many countries in the world are using biodiesel as an alternative energy source for which the marine algae are a highly suitable source for the production of biodiesel [2.65] [2.73].

2.5.1 Biomedical Industrial Aspects

The marine derivatives have the potential to protect humans against diverse chronic, viral and bacterial diseases. The presence of diatoms have great therapeutic potential against diseases [2.74]. Diatoms are involved in the defense mechanism due to their ability to produce chemical substances, therapeutic proteins and polysaccharides against pathogenic organisms, which are quite effective in chronic disease management [2.75] [2.76]. Algal extracts have polysaturated fatty acids, eicosapentaenoic acid, etc., which have a key role in the antimicrobial defense mechanism [2.73] [2.77] [2.78]. The use of diatom frustules, along with other bio-inspired alternatives for biomedical applications, have become quite helpful in the medical fraternity. In reference to efficient use of diatoms as drug delivery systems, there are already available delivery tools such as nanogels, liposomes, carbon nanotubes, etc. The complex silicified cell wall of diatoms, including proper defined surface area, thermal stability, and biocompatibility, have made diatoms more promising for use in drug and gene delivery systems. This includes diatom species, such as *Thalassiosira weissflogii* sp. and *Coscinodiscus concinnus* sp., which have the potential to act as probable drug carriers [2.79–2.81].

2.6 Economic Feasibility Assessment

Diatoms are the most valuable unicellular algae belonging to Bacillariophyta, which are mainly characterized by their siliceous cell wall, along with a peculiar yellow-brown pigmentation. These are the significant producers of lipids and biomass, due to which they attain more attention for varying commercial applications. Biodiesel and aviation fuels are a few of the commercial applications which are leading to diatoms becoming more economically beneficial in the near future [2.82]. However, biofuel procurement is a hazardous process, which requires further improvements in the whole downstream process. In the present scenario, the technologies used for harvest and recovery of microalgae are the same that are used for water purification. Therefore, some of technical glitches need to be sorted out in relation to microalgal harvesting for making this a better economically feasible process [2.83] [2.84].

2.7 Biochemical Composition

Diatoms are a phytoplankton species particularly belonging to the class Bacillariophyceae. These are generally unicellular organisms characterized by their cell wall (frustule) made up of silica with fast growth rate and possessing high lipid content. The main factors controlling microalgae growth and chemical composition are light, nutrients, temperature and pH [2.66]. Diatoms contain a wide variety of lipids, including membrane-bound polar lipids, triglycerides, and free fatty acids. Compounds, such as sterols, waxes, and acyl lipids, have also been identified. Increased lipid concentrations within different species of diatoms have been observed by the modification of nutrient availability and other requisite growth conditions. These organisms have great potential to be an alternative renewable feedstock for the sustainable production of carbon-based fuels [2.8]. Diatoms contain the following important constituents:

a) Pigments

Diatoms are commonly photosynthetic organisms due to the presence of essential pigments in their cells. Their pigments typically consist of two groups: (i) chlorophylls and (ii) carotenoids. Chlorophylls are capable of absorbing light energy and its conversion into chemical energy in the form of chemical energy-rich compounds, i.e., carbohydrates or lipids. Diatoms consist of two kinds of chlorophylls, i.e., Chl *a* and Chl *c*. Among these, Chlorophyll *a* plays a key role in the conversion of photochemical energy during the photosynthetic process in other photosynthetic organisms. And Chl *c* has a similar function to Chl *b* in higher plants and green algae. The most abundant among of Chl *c* types in diatoms are Chl *c*1 and Chl *c*2. Chlorophylls show major absorption bands in visible light ("red" [Q-] band and "blue" [Soret or B-] band) due to extended π-delocalization at the edge of their cyclic tetrapyrrole (porphyrin) skeleton. However, the absorption values are mainly dependent on their organic extraction solvent. These characteristics are commonly used for the practical identification of chlorophylls by the UV-visible spectroscopy measurement.

Carotenoids are lipophilic compounds made up of isoprene units (terpenoid) that usually have a yellow, orange or red color. These compounds show intense UV absorption in the range of 400–500 nm, in which their maximum absorbance is generally observed in the region of 470 nm [2.85]. Carotenoids are divided into groups on the basis of the presence of oxygen in their structure. The carotenoids having presence of oxygen (hydroxyl or keto-) are classified as xanthophyll groups, including astaxanthin, fucoxanthin, antheraxanthin, diadinoxanthin, zeaxanthin, violaxanthin, and diatoxanthin. The xanthophylls are relatively hydrophilic in nature due to the presence of oxygenated functional groups at the end of their rings. And the carotenoids with only hydrocarbon structure are known as carotenes, i.e., α-carotene, β-carotene, and lycopene [2.85]. Diatoms contain seven kinds of carotenoids, including β-carotene, antheraxanthin, fucoxanthin, diadinoxanthin, diatoxanthin, zeaxanthin and violaxanthin. Among these, β-carotene, an orange-yellowish color compound, is used for food or nutritional supplements because it is a precursor of vitamin A (retinol). And fucoxanthin is a component of fucoxanthin-chlorophyll protein (FCP) complexes in diatom's cell, which is responsible for light-harvesting during the photosynthesis process. The two main species of diatoms for the production of fucoxanthin are *Phaeodactylum tricornutum* and *Isochyris galbana*. However, the pigment compositions, as well as the content, greatly depend on the environmental conditions and the species of diatoms [2.86] [2.87].

The carotenoids have the ability to absorb light and quench excess energy in photosynthetic metabolism. The carotenoid fucoxanthin participates in the light-harvesting process together with chlorophyll *a* (Chl *a*) and chlorophyll *c* (Chl *c*). Some other carotenoids help in protection of photosynthetic components of the cell from photo-oxidation and stabilize the integrity of the lipid membrane from peroxidation. Carotenoids, such as astaxanthin, canthaxanthin, and echinenone, play an important role in photoprotection of the cell's photosystem from the oxidative-induced damages. Other carotenoids, such as α-/β-carotene, lutein, violaxanthin, zeaxanthin, and neoxanthin, also act as photosynthetic components [2.117] [2.88] [2.89]. The carotenoids are attractive pigments in the pharmaceutical and cosmetics industries because of their potential bioactivities. The market interest in carotenoid pigments is mainly for β-carotene, astaxanthin, zeaxanthin, lutein, lycopene, and canthaxanthin. Currently, fucoxanthin also has a growing market because it has potential benefits for anti-obesity functional foods, and in anticancer and anti-inflammatory agents. Chlorophylls also possess a potential market and have been used as a control in geriatric patients, dietary supplements, antimutagenic and anticarcinogenic agents. Although some synthetic pigments are already available, the market demand for natural pigments is still increasing due to their unique properties which cannot be found in the artificial ones, as well as the consumer's demand for healthy natural sources [2.87] [2.88].

b) Lipids and their derivatives

Microalgae utilize atmospheric carbon dioxide, water and sunlight to accumulate lipids through the photosynthetic process, especially in the form of an ester of glycerol and fatty acids (triglyceride). The most commonly synthesized fatty acids have chain lengths that range from C16–C18. Fatty acids are either saturated or unsaturated, and unsaturated fatty acids may vary in the number and position of double bonds on the carbon chain backbone. In general, saturated and mono-unsaturated fatty acids are predominant [2.6]. The major component of diatom's lipids is triglycerides of fatty acids. Generally, diatoms contain 15–25% DW of fatty acids that could be increased up to 25–45% DW of biomass in a specific environment having a low amount of nitrogen. The nitrogen deficiency suppresses the productivity of microalgae biomass since it is required in the biosynthesis of amino acids as well as protein for the growth of the cells. In response to nitrogen deprivation, diatoms reprogram several metabolic pathways and store carbon in the form of 1,3-β-d-glucan (chrysolaminarin), changing their storage patterns in favor of neutral lipid accumulation [2.90]. It has been reported earlier that molecular species of triglycerides, e.g., diacylglycerols, palmitic acid, stearic acid, oleic acid, increase significantly, leading the accumulation of lipids of triglycerides in their biomass for their energy storage under nitrogen-deficient conditions. Lipid accumulation can also be increased from 28% to 68% DW in diatoms due to silicon starvation without suffering the growth productivity of diatoms. This is because silica has no significant effect on the physiological growth rates of diatoms compared with that in the nitrogen starvation [2.64].

Lipids of diatoms have similar potency to be converted into multiple valuable chemicals (plant oils, glycerol) due to the transesterification reaction. The occurrence and extent of lipid content or triacylglycerol (TAG) production in diatoms is highly specific to the species or strain involved and its genetic makeup. The oleaginous algae including diatoms produce only small quantities of TAG under optimal growth or favorable environmental conditions [2.118]. However, the synthesis and accumulation of large amounts of TAG accompanied by considerable alterations in lipid and fatty acid composition occur in the cell when oleaginous algae are placed under stress conditions such as nutrient starvation, salinity, growth-medium pH, temperature, light intensity, growth phase and/or aging of the culture. The increase in total lipids in aging algal cells or cells maintained under various stress conditions primarily contain neutral lipids, i.e., TAGs. The reason for this increase is the shift in lipid metabolism from membrane lipid synthesis to the storage of neutral lipids. *De novo* biosynthesis and conversion of certain existing membrane polar lipids into triacylglycerols may also contribute to the overall increase in TAG content [2.6]. Diatoms have great potential for the production of green diesel (biohydrocarbons) via catalytic deoxygenation reaction of their lipids [2.6] [2.91]. However, the viability of microalgae species as biodiesel feedstock depends on the optimization of biomass and neutral lipid productivity [2.92]. Presence of high-density lipid content renders algae (e.g., *Chlamydomonas*, *Nannochloropsis*, *Schizochytrium*, *Chlorella*, *Nitzschia*) attractive for biofuels production. Lipid content in algae can be as high as 50% (dry weight) with higher percentages of total dry weight (~60%) found as protein and starch. These fractions of microalgae can be used to produce biodiesel, bioethanol, biohydrogen, bioplastics, and other products, while simultaneously contributing to CO_2 mitigation [2.93]. Lipids isolated from microalgae (diatoms) were reported to have an energy content of 35,800 kJ kg-1 or nearly 80% of the average energy contained in crude petroleum. Hence, microalgae are expected to have promising productivity for alternative biofuel (biodiesel) feedstock, even compared with the oleaginous crops [2.88].

c) Carbohydrates

Carbohydrates (primarily starch) are another valuable component of the algal cell. Dry weight content of carbohydrates in algae generally ranges from 20–40% of total cell mass. Certain components, such as starch, can be easily converted to ethanol by hydrolysis and fermentation. Currently, the starch used for ethanol production is obtained from food crops such as corn, wheat, sorghum, and others. For algae, high-starch strains, including *Chlorella vulgaris* (with 37% dry weight starch), are being studied for potential use has high-yield feedstocks. Algal starch is known to be readily fermentable by yeast and, therefore, is being intensely studied for use in ethanol production [2.94].

d) Silica

Diatoms have a unique capability to utilize the soluble silicic acid in sea waters and metabolize it to build their silicate cell walls, called frustules, having porous structure. The frustules are the major component of diatoms besides lipids that may be raised up to 50–60% DW of diatom's biomass [2.95]. However, the shapes and the physical characteristics of the frustules are varied depending on the species. For example, *Coscinodiscus* sp. is a dish-like diatom with diameter of 60–200 μm with the inner pore size of 900-950 nm. *Pseudostaurosira trainorii* has a similar shape with diameter and pore size of 4–5 μm and 150–200 nm, respectively. *Talassiosira* sp. and *Skeletenome* sp. are two centric diatoms with 10 μm size and 64–156 m^2 g^{-1} of surface area. The biosynthesis process of these silicified materials is complex and involves multiple silicon transport mechanisms. Herein, the dissolved silicic acid is absorbed into the cell and intracellularly transported by silicic acid transporters [2.96]. The formed internal soluble silicon pool is then transported into the silica deposition vesicle, which subsequently condensates the soluble silicic acid into the insoluble amorphous silica of the cell walls catalyzed by specific proteins (silaffins) and polyanionic peptides (silacidins) [2.80]. The hard silica cell walls are beneficial in providing mechanical protection of their cells from their zooplankton predators, high-pressure conditions and collision by solid particles in seawater. The silica cell wall also protects the membrane cells from the penetration of viruses and bacteria. Meanwhile, their porous architectures have multiple advantages for diatoms, such as providing the fluid dynamic structure for easy movement in water environment, enhancing the optical scattering for optimized light as well as energy harvesting, and optimizing molecular sieving for nutrient uptake [2.97]. Owing to these characteristics, the hierarchical porous structure of the diatom's frustules has potential applications for nano-biotechnology industries, drug delivery systems and biosensors. Diatoms are a promising resource for the sustainable production of high-purity silica as a potential alternative natural source for the manufacture of hierarchical porous silica materials. The use of diatoms offers minimum impact on environment as well as health issues [2.88] [2.98].

2.8 Feedstock Availability Assessment

Microalgae possess great biotechnological potential for the production of a wide variety of compounds such as polysaccharides, lipids, proteins, carotenoids, and other pigments, vitamins, steroids, among others. Diatoms have been used for feeding animals and humans. They are highly efficient in producing a wide variety of chemical components such as fatty acids, pigments and other biologically active compounds of interest. Also, they are already used for their high production of fatty acids for aquaculture and human consumption. The exploitation of microalgae as a protein source has led to increased interest in their use (e.g., *Spirulina, Chlorella* and *Scenedesmus*) in health food production [2.99]. Many species of microalgae have been used as live feed for aquatic organisms and feed for livestock. The main genera for these purposes include the microalgae *Chlorella, Tetraselmis, Spirulina, Nannochloropsis, Nitzchia, Navicula,*

Chaetoceros, Scenedesmus, Haematococcus, and *Crypthecodinium* [2.100] [2.101]. Microalgae, like *Chlorella* spp. and *Dunaliella salina,* are already being used for human consumption and have been classified by the U.S. Food and Drug Administration as "generally safe." *Haematococcus* spp., *Dunaliella* spp., *Arthrospira* spp., and *Schizochytrium* spp. are also included in this classification. The marine diatom *Odontella aurita* from Innovalg (France) has also been approved as a new food since 2002 [2.100].

Microalgae are predicted to be an essential supply feedstock for food, fuels and chemicals in a biobased economy [2.102]. The diatoms can be used to produce biofuels, hydrocarbons, and hydrogen. They can be a clean and renewable energy source because of their high yield and low spatial requirements, when compared to terrestrial plants. Some authors consider microalgae as biodiesel feedstocks for the future. Microalgae have several advantages over other biological sources as a feedstock for biofuels and bioproducts. Algae have a higher ratio of oil production to required cultivated area compared with common bioenergy sources such as corn, soybean, rapeseed, jatropha, and others. Microalgae can also be used for starch production and its subsequent ethanol production through yeast fermentation, achieving efficiencies similar to the best available crops. Another characteristic of algae is that they can double their biomass at an exponential rate, requiring less surface area for the production of biofuels and bioproducts [2.102] [2.103]. It has been reported that after utilization of lipids and starch fractions for bio-oil production, the residual algae cake, which is rich in proteins, is important for producing valuable co-products. Protein (and starch) content can constitute up to 60% of dry weight of algae [2.93]. This residual protein from the biomass can be used for livestock, poultry, and fish feed additives. It has been reported that algae can replace about 5–10% of conventional protein sources in poultry feed. Recently, residual algae cake after lipid extraction has been used in large animal feeding trials. However, high concentrations of nucleic acids in algae can pose challenges for the utilization in animal feed applications [2.104] [2.105].

2.9 Scope of Diatoms in Biorefinery

Diatoms have a silica-based frustule (cell wall), which distinguishes them as a unicellular organism from other types of life [2.106]. The potential of marine diatoms as an untapped living factory for the production of important commodity chemicals is enormous. In recent years, as more and more algae have been discovered and investigated as potential sources of lipids, pigments, and other biomolecules, diatom biorefinery has become increasingly significant [2.107]. The term "biorefinery" was introduced to characterize the cost-effective and environmentally sustainable generation of biofuels as well as high-value co-products from biomasses by integrating bioprocessing with suitable low environmental impact chemical technologies [2.108]. Similar to petro-refineries, which process crude oils to produce gasoline, diesel, wax, kerosene, and other products, biorefineries enable the extraction of several biomolecules, such as lipid, protein, and carbohydrates, from the algal biomass. They may have a positive impact on the demand for food, feed, fiber, medications, and fodder [2.109].

Diatoms receive special attention because they are the dominant category of microalgae in seas. They have received a lot of interest recently as a viable alternate renewable feedstock for the sustainable manufacture of fuels with a carbon base, especially stimulated by their quick growth and high lipid content [2.110]. Similar to plant oils, diatom lipids have the ability to be transformed into a variety of useful compounds. The sustainable production of industrial commodities, such as fuels, bulk/high-value chemicals, and materials, has a lot of potential for marine microalgae [2.88]. In reality, it has been established that the lipid found in diatoms is the primary source of petroleum, and some of their lipid components, such as steranes, norcholestanes, and highly branched isoprenoids, are employed as biomarkers to identify the Earth's petroleum reserve [2.111]. Diatom cultivation won't encounter problems with freshwater since they may be produced in seawater. The transesterification procedure is used to create biodiesel from microalgal oil, which contains triglycerides. Methyl esters of fatty acids (biodiesel) and glycerol are created when the triglycerides and methanol interact in a process known as transesterification or alcoholysis [2.112]. Although vegetable oils used to make biodiesel are now primarily generated from food crops, the problem of competition between food and fuel is brought up. Moreover, the market price of these oil crops has increased due to the rising demand for them, driving up the price of generating biodiesel. To resolve this concern, alternative feedstocks must be produced [2.113].

2.10 Conclusions and Future Prospects

Diatoms stand out among the potential options for the production of biofuel because of their high lipid content (16–77%, with levels exceeding 30% being typical in a number of microalgal species), quick growth, and high lipid production capability [2.114] [2.115]. Utilizing diatoms as a source of biodiesel feedstock has yielded potential benefits as microalgae develop incredibly quickly; their biomass can double in just 24 hours in comparison to terrestrial plants. Moreover, more than 50% of the dry cell weight of the microalgal biomass can be made up of oil and the oil production exceeds that of the oilseed crops in terms of cultivated area. Moreover, microalgae do not compete with land-based crops for resources; by converting CO_2 into biomass, microalgae can potentially lower the atmospheric concentration of CO_2. The biofuels derived from microalgae are henceforth believed to be nontoxic, sulfur-free, and very biodegradable [2.115]. However, there are several critical issues regarding the production of biodiesel from microalgae. Microalgae are significantly harder to harvest and extract oil from than terrestrial oil crops, which increases the expense of producing microalgae-based biodiesel [2.115] [2.116]. Typically, the weight of the dry microalgal biomass in the microalgae culture is about 0.1–1%. Therefore, the dewatering process and microalgal biomass concentration are both expensive and energy-intensive. Another technological barrier to producing microalgal lipid into biodiesel is that due to their tiny size, diatoms cannot be easily removed by traditional mechanical processes used to collect oil from oil crops like oil palm trees or soybeans.

Acknowledgment

The authors acknowledge the support of the Department of Biotechnology, MMEC, Maharishi Markandeshwar (Deemed to be University), Mullana-Ambala, India.

Human and Animal Rights and Informed Consent

This chapter does not contain any experimental studies involving use of animals or humans.

References

[2.1] M. Briggs, Widescale biodiesel production from algae. University of New Hampshire, Physics Department 5 (2004).

[2.2] M.J. Khan, N. Singh, S. Mishra, A. Ahirwar, F. Bast, S. Varjani, B. Schoefs, J. Marchand, K. Rajendran, J.R. Banu, Impact of light on microalgal photosynthetic microbial fuel cells and removal of pollutants by nanoadsorbent biopolymers: updates, challenges and innovations, *Chemosphere* 288 (2022) 132589.

[2.3] T. Ramachandra, D.M. Mahapatra, R. Gordon, Milking diatoms for sustainable energy: biochemical engineering versus gasoline-secreting diatom solar panels, *Industrial & Engineering Chemistry Research* 48(19) (2009) 8769-8788.

[2.4] R. Bhattacharjya, T.K. Marella, A. Tiwari, A. Saxena, P.K. Singh, B. Mishra, Bioprospecting of marine diatoms *Thalassiosira, Skeletonema* and *Chaetoceros* for lipids and other value-added products, *Bioresource Technology* 318 (2020) 124073.

[2.5] G. Saranya, T. Ramachandra, Novel biocatalyst for optimal biodiesel production from diatoms, *Renewable Energy* 153 (2020) 919-934.

[2.6] Q. Hu, M. Sommerfeld, E. Jarvis, M. Ghirardi, M. Posewitz, M. Seibert, A. Darzins, Microalgal triacylglycerols as feedstocks for biofuel production: perspectives and advances, *The plant journal* 54(4) (2008) 621-639.

[2.7] J. Milano, H.C. Ong, H. Masjuki, W. Chong, M.K. Lam, P.K. Loh, V. Vellayan, Microalgae biofuels as an alternative to fossil fuel for power generation, *Renewable and Sustainable Energy Reviews* 58 (2016) 180-197.

[2.8] J.-K. Wang, M. Seibert, Prospects for commercial production of diatoms, *Biotechnology for Biofuels* 10(1) (2017) 1-13.

[2.9] K. Manandhar-Shrestha, M. Hildebrand, Characterization and manipulation of a DGAT2 from the diatom *Thalassiosira pseudonana*: improved TAG accumulation without detriment to growth, and implications for chloroplast TAG accumulation, *Algal research* 12 (2015) 239-248.

[2.10] T. Lebeau, J.-M. Robert, Diatom cultivation and biotechnologically relevant products. Part II: Current and putative products, *Applied microbiology and biotechnology* 60(6) (2003) 624-632.

[2.11] V. Vinayak, R. Gordon, S. Gautam, A. Rai, Discovery of a diatom that oozes oil, *Advanced Science Letters* 20(7-8) (2014) 1256-1267.

[2.12] M.M. Joseph, K. Renjith, G. John, S.M. Nair, N. Chandramohanakumar, Biodiesel prospective of five diatom strains using growth parameters and fatty acid profiles, *Biofuels* 8(1) (2017) 81-89.

[2.13] J. Sheehan, T. Dunahay, J. Benemann, P. Roessler, A look back at the US Department of Energy's aquatic species program: biodiesel from algae, *National Renewable Energy Laboratory* 328 (1998) 1-294.

[2.14] M.D. Guiry, How many species of algae are there?, *Journal of phycology* 48(5) (2012) 1057-1063.

[2.15] M. Branco-Vieira, S. San Martin, C. Agurto, M.A.d. Santos, M.A. Freitas, T.M. Mata, A.A. Martins, N.S. Caetano, Potential of *Phaeodactylum tricornutum* for biodiesel production under natural conditions in Chile, *Energies* 11(1) (2017) 54.

[2.16] X.-l. Li, T.K. Marella, L. Tao, L. Peng, C.-f. Song, L.-l. Dai, A. Tiwari, G. Li, A novel growth method for diatom algae in aquaculture waste water

for natural food development and nutrient removal, *Water Science and Technology* 75(12) (2017) 2777-2783.
[2.17] R. Davis, A. Aden, P.T. Pienkos, Techno-economic analysis of autotrophic microalgae for fuel production, *Applied Energy* 88(10) (2011) 3524-3531.
[2.18] G. Saranya, T. Ramachandra, Scope for biodiesel and bioactive compounds production in the diatom *Nitzschia punctata*, *Fuel* 300 (2021) 120985.
[2.19] K. Chojnacka, F.-J. Marquez-Rocha, Kinetic and stoichiometric relationships of the energy and carbon metabolism in the culture of microalgae, *Biotechnology* 3(1) (2004) 21-34.
[2.20] G. Huang, F. Chen, D. Wei, X. Zhang, G. Chen, Biodiesel production by microalgal biotechnology, *Applied energy* 87(1) (2010) 38-46.
[2.21] A.-y. Liu, C. Wei, L.-l. Zheng, L.-r. Song, Identification of high-lipid producers for biodiesel production from forty-three green algal isolates in China, Progress in Natural Science: *Materials International* 21(4) (2011) 269-276.
[2.22] O. Perez-Garcia, Y. Bashan, Microalgal heterotrophic and mixotrophic culturing for bio-refining: from metabolic routes to techno-economics, *Algal biorefineries* (2015) 61-131.
[2.23] Z. Zhang, D. Sun, T. Wu, Y. Li, Y. Lee, J. Liu, F. Chen, The synergistic energy and carbon metabolism under mixotrophic cultivation reveals the coordination between photosynthesis and aerobic respiration in *Chlorella zofingiensis*, *Algal research* 25 (2017) 109-116.
[2.24] J. Roostaei, Y. Zhang, K. Gopalakrishnan, A.J. Ochocki, Mixotrophic microalgae biofilm: a novel algae cultivation strategy for improved productivity and cost-efficiency of biofuel feedstock production, *Scientific reports* 8(1) (2018) 1-10.
[2.25] O. Perez-Garcia, F.M. Escalante, L.E. De-Bashan, Y. Bashan, Heterotrophic cultures of microalgae: metabolism and potential products, *Water research* 45(1) (2011) 11-36.
[2.26] Z. Shokravi, H. Shokravi, A. Atabani, W.J. Lau, O.H. Chyuan, A.F. Ismail, Impacts of the harvesting process on microalgae fatty acid profiles and lipid yields: Implications for biodiesel production, *Renewable and Sustainable Energy Reviews* 161 (2022) 112410.
[2.27] A. Serrà, R. Artal, J. Garcia-Amoros, E. Gómez, L. Philippe, Circular zero-residue process using microalgae for efficient water decontamination, biofuel production, and carbon dioxide fixation, *Chemical Engineering Journal* 388 (2020) 124278.
[2.28] T. Hülsen, C. Züger, Z.M. Gan, D.J. Batstone, D. Solley, P. Ochre, B. Porter, G. Capson-Tojo, Outdoor demonstration-scale flat plate photobioreactor for resource recovery with purple phototrophic bacteria, *Water Research* 216 (2022) 118327.
[2.29] A. Richmond, Outdoor mass cultures of microalgae, *CRC handbook of microalgal mass culture*, CRC Press, 2017, pp. 285-330.

[2.30] R. Bhagea, V. Bhoyroo, D. Puchooa, Microalgae: the next best alternative to fossil fuels after biomass. A review, *Microbiology Research* 10(1) (2019) 7936.
[2.31] M. Bošnjaković, Biodiesel from algae, *J Mech Eng Autom* 3 (2013) 179-88.
[2.32] W. Haiying, F. Ru, P. Guofeng, A study on lipid production of the mixotrophic microalgae Phaeodactylum tricornutum on various carbon sources, *African Journal of Microbiology Research* 6(5) (2012) 1041-1047.
[2.33] X. Liu, S. Duan, A. Li, N. Xu, Z. Cai, Z. Hu, Effects of organic carbon sources on growth, photosynthesis, and respiration of *Phaeodactylum tricornutum*, *Journal of Applied Phycology* 21(2) (2009) 239-246.
[2.34] G.M. Ceron, F.G. Camacho, A.S. Mirón, J. Sevilla, Y. Chisti, E.M. Grima, Mixotrophic production of marine microalga *Phaeodactylum tricornutum* on various carbon sources, *Journal of microbiology and biotechnology* 16(5) (2006) 689-694.
[2.35] J. Fábregas, J. Arán, E. Morales, T. Lamela, A. Otero, Modification of sterol concentration in marine microalgae, *Phytochemistry* 46(7) (1997) 1189-1191.
[2.36] B. Scholz, G. Liebezeit, Biochemical characterisation and fatty acid profiles of 25 benthic marine diatoms isolated from the Solthörn tidal flat (southern North Sea), *Journal of Applied Phycology* 25(2) (2013) 453-465.
[2.37] C. Delgado, I. Pardo, L. García, Diatom communities as indicators of ecological status in Mediterranean temporary streams (Balearic Islands, Spain), *Ecological indicators* 15(1) (2012) 131-139.
[2.38] F.J. Fields, J.P. Kociolek, An evolutionary perspective on selecting high-lipid-content diatoms (Bacillariophyta), *Journal of Applied Phycology* 27(6) (2015) 2209-2220.
[2.39] S. Hausmann, D.F. Charles, J. Gerritsen, T.J. Belton, A diatom-based biological condition gradient (BCG) approach for assessing impairment and developing nutrient criteria for streams, *Science of the Total Environment* 562 (2016) 914-927.
[2.40] L. Rodolfi, G. Chini Zittelli, N. Bassi, G. Padovani, N. Biondi, G. Bonini, M.R. Tredici, Microalgae for oil: Strain selection, induction of lipid synthesis and outdoor mass cultivation in a low-cost photobioreactor, *Biotechnology and bioengineering* 102(1) (2009) 100-112.
[2.41] G. d'Ippolito, A. Sardo, D. Paris, F.M. Vella, M.G. Adelfi, P. Botte, C. Gallo, A. Fontana, Potential of lipid metabolism in marine diatoms for biofuel production, *Biotechnology for biofuels* 8(1) (2015) 1-10.
[2.42] J.C. Taylor, J. Prygiel, A. Vosloo, P.A. de la Rey, L. van Rensburg, Can diatom-based pollution indices be used for biomonitoring in South Africa? A case study of the Crocodile West and Marico water management area, *Hydrobiologia* 592(1) (2007) 455-464.
[2.43] L. Chen, T. Liu, W. Zhang, X. Chen, J. Wang, Biodiesel production from algae oil high in free fatty acids by two-step catalytic conversion, *Bioresource Technology* 111 (2012) 208-214.

[2.44] X. Tan, Q. Zhang, M.A. Burford, F. Sheldon, S.E. Bunn, Benthic diatom based indices for water quality assessment in two subtropical streams, *Frontiers in microbiology* (2017) 601.

[2.45] Y. Chisti, Biodiesel from microalgae, *Biotechnology advances* 25(3) (2007) 294-306.

[2.46] T.K. Marella, A. Tiwari, Marine diatom *Thalassiosira weissflogii* based biorefinery for co-production of eicosapentaenoic acid and fucoxanthin, *Bioresource technology* 307 (2020) 123245.

[2.47] H. Jiang, K. Gao, Effects of lowering temperature during culture on the production of polyunsaturated fatty acids in the marine diatom *Phaeodactylum tricornutum* (bacillariophyceae) 1, *Journal of Phycology* 40(4) (2004) 651-654.

[2.48] A. Saxena, T.K. Marella, P.K. Singh, A. Tiwari, Indoor mass cultivation of marine diatoms for biodiesel production using induction plasma synthesized nanosilica, *Bioresource Technology* 332 (2021) 125098.

[2.49] M.A. Borowitzka, Microalgae for aquaculture: opportunities and constraints, *Journal of applied phycology* 9(5) (1997) 393-401.

[2.50] K. Yamaguchi, Recent advances in microalgal bioscience in Japan, with special reference to utilization of biomass and metabolites: a review, *Journal of applied phycology* 8(6) (1996) 487-502.

[2.51] A.P. Carvalho, L.A. Meireles, F.X. Malcata, Microalgal reactors: a review of enclosed system designs and performances, *Biotechnology progress* 22(6) (2006) 1490-1506.

[2.52] D. Chaumont, Biotechnology of algal biomass production: a review of systems for outdoor mass culture, *Journal of applied phycology* 5(6) (1993) 593-604.

[2.53] L. Travieso, A. Pellón, F. Benıtez, E. Sánchez, R. Borja, N. O'farrill, P. Weiland, BIOALGA reactor: preliminary studies for heavy metals removal, *Biochemical Engineering Journal* 12(2) (2002) 87-91.

[2.54] A. Scragg, A. Illman, A. Carden, S. Shales, Growth of microalgae with increased calorific values in a tubular bioreactor, *Biomass and Bioenergy* 23(1) (2002) 67-73.

[2.55] Y. Duan, F. Shi, Bioreactor design for algal growth as a sustainable energy source, *Reactor and process design in sustainable energy technology*, Elsevier, 2014, pp. 27-60.

[2.56] M. Borowitzka, Closed algal photobioreactors: design considerations for large-scale systems, *Journal of Marine Biotechnology* 4 (1996) 185.

[2.57] J. Pruvost, J.-F. Cornet, L. Pilon, Large-scale production of algal biomass: photobioreactors, *Algae biotechnology*, Springer, 2016, pp. 41-66.

[2.58] C. Posten, Design principles of photo-bioreactors for cultivation of microalgae, *Engineering in Life Sciences* 9(3) (2009) 165-177.

[2.59] L. Xu, P.J. Weathers, X.R. Xiong, C.Z. Liu, Microalgal bioreactors: challenges and opportunities, *Engineering in Life Sciences* 9(3) (2009) 178-189.

[2.60] J.C.G. Cañedo, G.L.L. Lizárraga, Considerations for photobioreactor design and operation for mass cultivation of microalgae, *Algae-organisms for imminent biotechnology* 14 (2016).
[2.61] E. Hincapie, B.J. Stuart, Design, construction, and validation of an internally lit air-lift photobioreactor for growing algae, *Frontiers in Energy Research* 2 (2015) 65.
[2.62] W. Fu, K. Wichuk, S. Brynjólfsson, Developing diatoms for value-added products: Challenges and opportunities, *New biotechnology* 32(6) (2015) 547-551.
[2.63] N.N. Zulu, K. Zienkiewicz, K. Vollheyde, I. Feussner, Current trends to comprehend lipid metabolism in diatoms, *Progress in lipid research* 70 (2018) 1-16.
[2.64] O. Levitan, J. Dinamarca, G. Hochman, P.G. Falkowski, Diatoms: a fossil fuel of the future, *Trends in biotechnology* 32(3) (2014) 117-124.
[2.65] P. Kuppusamy, I. Soundharrajan, S. Srigopalram, M.M. Yusoff, G.P. Maniam, N. Govindan, K.C. Choi, Potential pharmaceutical and biomedical applications of Diatoms microalgae-An overview, (2017).
[2.66] M. Hildebrand, A.K. Davis, S.R. Smith, J.C. Traller, R. Abbriano, The place of diatoms in the biofuels industry, *Biofuels* 3(2) (2012) 221-240.
[2.67] A.A. Jamali, F. Akbari, M.M. Ghorakhlu, M. de la Guardia, A. Yari Khosroushahi, Applications of diatoms as potential microalgae in nanobiotechnology, BioImpacts: *BI* 2(2) (2012) 83-9.
[2.68] A. Tiwari, T.K. Marella, Potential and application of diatoms for industry-specific wastewater treatment, *Application of Microalgae in Wastewater Treatment* (2019) 321-339.
[2.69] K. Wichuk, S. Brynjólfsson, W. Fu, Biotechnological production of value-added carotenoids from microalgae: Emerging technology and prospects, *Bioengineered* 5(3) (2014) 204-208.
[2.70] R. Dhanker, R. Kumar, A. Tiwari, V. Kumar, Diatoms as a biotechnological resource for the sustainable biofuel production: a state-of-the-art review, *Biotechnology and Genetic Engineering Reviews* 38(1) (2022) 111-131.
[2.71] T. Li, G. Xu, J. Rong, H. Chen, C. He, M. Giordano, Q. Wang, The acclimation of Chlorella to high-level nitrite for potential application in biological NOx removal from industrial flue gases, *Journal of plant physiology* 195 (2016) 73-79.
[2.72] J. Wang, An Absorbent and method of application for the treatment of heavy metal waste water, Chinene Patent: 201410072325.0 (2015).
[2.73] A.L. Ahmad, N.M. Yasin, C. Derek, J. Lim, Microalgae as a sustainable energy source for biodiesel production: a review, *Renewable and sustainable energy reviews* 15(1) (2011) 584-593.
[2.74] S. Merlo, X. Gabarrell Durany, A. Pedroso Tonon, S. Rossi, Marine microalgae contribution to sustainable development, *Water* 13(10) (2021) 1373.

[2.75] A.E. Atabani, A.S. Silitonga, I.A. Badruddin, T. Mahlia, H. Masjuki, S. Mekhilef, A comprehensive review on biodiesel as an alternative energy resource and its characteristics, *Renewable and sustainable energy reviews* 16(4) (2012) 2070-2093.

[2.76] A. Bozarth, U.-G. Maier, S. Zauner, Diatoms in biotechnology: modern tools and applications, *Applied microbiology and biotechnology* 82(2) (2009) 195-201.

[2.77] G. Najafi, B. Ghobadian, T.F. Yusaf, Algae as a sustainable energy source for biofuel production in Iran: A case study, *Renewable and Sustainable Energy Reviews* 15(8) (2011) 3870-3876.

[2.78] S.R. Medipally, F.M. Yusoff, S. Banerjee, M. Shariff, Microalgae as sustainable renewable energy feedstock for biofuel production, *BioMed research international* 2015 (2015).

[2.79] P. Gnanamoorthy, S. Anandhan, V.A. Prabu, Natural nanoporous silica frustules from marine diatom as a biocarrier for drug delivery, *Journal of Porous Materials* 21(5) (2014) 789-796.

[2.80] U. Uthappa, V. Brahmkhatri, G. Sriram, H.-Y. Jung, J. Yu, N. Kurkuri, T.M. Aminabhavi, T. Altalhi, G.M. Neelgund, M.D. Kurkuri, Nature engineered diatom biosilica as drug delivery systems, *Journal of Controlled Release* 281 (2018) 70-83.

[2.81] N. Sharma, D.P. Simon, A.M. Diaz-Garza, E. Fantino, A. Messaabi, F. Meddeb-Mouelhi, H. Germain, I. Desgagné-Penix, Diatoms biotechnology: Various industrial applications for a greener tomorrow, *Frontiers in Marine Science* 8 (2021) 636613.

[2.82] A. Saxena, A. Tiwari, R. Kaushik, H.M.N. Iqbal, R. Parra-Saldívar, Diatoms recovery from wastewater: Overview from an ecological and economic perspective, *Journal of Water Process Engineering* 39 (2021) 101705.

[2.83] W. Li, T. Wang, D.A. Campbell, K. Gao, Ocean acidification interacts with variable light to decrease growth but increase particulate organic nitrogen production in a diatom, *Marine Environmental Research* 160 (2020) 104965.

[2.84] L.K. Pandey, D. Kumar, A. Yadav, J. Rai, J.P. Gaur, Morphological abnormalities in periphytic diatoms as a tool for biomonitoring of heavy metal pollution in a river, *Ecological Indicators* 36 (2014) 272-279.

[2.85] H.K. Lichtenthaler, C. Buschmann, Chlorophylls and carotenoids: Measurement and characterization by UV-VIS spectroscopy, *Current protocols in food analytical chemistry* 1(1) (2001) F4. 3.1-F4. 3.8.

[2.86] A. Gómez-Loredo, J. Benavides, M. Rito-Palomares, Growth kinetics and fucoxanthin production of *Phaeodactylum tricornutum* and *Isochrysis galbana* cultures at different light and agitation conditions, *Journal of Applied Phycology* 28(2) (2016) 849-860.

[2.87] E.B. D'Alessandro, N.R. Antoniosi Filho, Concepts and studies on lipid and pigments of microalgae: A review, *Renewable and Sustainable Energy Reviews* 58 (2016) 832-841.

[2.88] A. Bayu, A. Rachman, D. Noerdjito, M. Putra, W. Widayatno, High-value chemicals from marine diatoms: a biorefinery approach, *IOP Conference Series: Earth and Environmental Science*, IOP Publishing, 2020, p. 012012.
[2.89] M. Gong, A. Bassi, Carotenoids from microalgae: A review of recent developments, *Biotechnology advances* 34(8) (2016) 1396-1412.
[2.90] B. Sajjadi, W.-Y. Chen, A.A.A. Raman, S. Ibrahim, Microalgae lipid and biomass for biofuel production: A comprehensive review on lipid enhancement strategies and their effects on fatty acid composition, *Renewable and Sustainable Energy Reviews* 97 (2018) 200-232.
[2.91] M. Ameen, M.T. Azizan, S. Yusup, A. Ramli, M. Yasir, Catalytic hydrodeoxygenation of triglycerides: An approach to clean diesel fuel production, *Renewable and Sustainable Energy Reviews* 80 (2017) 1072-1088.
[2.92] S. Bellou, M.N. Baeshen, A.M. Elazzazy, D. Aggeli, F. Sayegh, G. Aggelis, Microalgal lipids biochemistry and biotechnological perspectives, *Biotechnology advances* 32(8) (2014) 1476-1493.
[2.93] J. Singh, S. Gu, Commercialization potential of microalgae for biofuels production, *Renewable and sustainable energy reviews* 14(9) (2010) 2596-2610.
[2.94] R. Harun, M. Singh, G.M. Forde, M.K. Danquah, Bioprocess engineering of microalgae to produce a variety of consumer products, *Renewable and sustainable energy reviews* 14(3) (2010) 1037-1047.
[2.95] S. Barua, S. Gogoi, R. Khan, N. Karak, Silicon-based nanomaterials and their polymer nanocomposites, *Nanomaterials and polymer nanocomposites*, Elsevier2019, pp. 261-305.
[2.96] E. De Tommasi, J. Gielis, A. Rogato, Diatom frustule morphogenesis and function: a multidisciplinary survey, *Marine genomics* 35 (2017) 1-18.
[2.97] S. Maher, T. Kumeria, M.S. Aw, D. Losic, Diatom silica for biomedical applications: Recent progress and advances, *Advanced Healthcare Materials* 7(19) (2018) 1800552.
[2.98] X.W. Sun, Y.X. Zhang, D. Losic, Diatom silica, an emerging biomaterial for energy conversion and storage, *Journal of Materials Chemistry A* 5(19) (2017) 8847-8859.
[2.99] E.W. Becker, Micro-algae as a source of protein, *Biotechnology advances* 25(2) (2007) 207-210.
[2.100] M. Van der Spiegel, M. Noordam, H. Van der Fels-Klerx, Safety of novel protein sources (insects, microalgae, seaweed, duckweed, and rapeseed) and legislative aspects for their application in food and feed production, *Comprehensive reviews in food science and food safety* 12(6) (2013) 662-678.
[2.101] H. Furbeyre, J. van Milgen, T. Mener, M. Gloaguen, E. Labussière, Effects of dietary supplementation with freshwater microalgae on growth performance, nutrient digestibility and gut health in weaned piglets, *Animal* 11(2) (2017) 183-192.

[2.102] P.R. Mooij, L.D. de Jongh, M. van Loosdrecht, R. Kleerebezem, Influence of silicate on enrichment of highly productive microalgae from a mixed culture, *Journal of Applied Phycology* 28(3) (2016) 1453-1457.
[2.103] D. de Jesús-Campos, J.A. López-Elías, L.Á. Medina-Juarez, G. Carvallo-Ruiz, D. Fimbres-Olivarria, C. Hayano-Kanashiro, Chemical composition, fatty acid profile and molecular changes derived from nitrogen stress in the diatom Chaetoceros muelleri, *Aquaculture Reports* 16 (2020) 100281.
[2.104] L.Q. Zepka, E. Jacob-Lopes, R. Goldbeck, L.A. Souza-Soares, M.I. Queiroz, Nutritional evaluation of single-cell protein produced by Aphanothece microscopica Nägeli, *Bioresource Technology* 101(18) (2010) 7107-7111.
[2.105] [T. Ramachandra, G. Saranya, Sustainable Bioeconomy prospects of diatom biorefineries in the Indian west coast, *Renewable and Sustainable Energy Reviews* 161 (2022) 112399.
[2.106] [S. Scala, C. Bowler, Molecular insights into the novel aspects of diatom biology, *Cellular and Molecular Life Sciences CMLS* 58(11) (2001) 1666-1673.
[2.107] T.K. Marella, I.Y. López-Pacheco, R. Parra-Saldívar, S. Dixit, A. Tiwari, Wealth from waste: Diatoms as tools for phycoremediation of wastewater and for obtaining value from the biomass, *Science of the Total Environment* 724 (2020) 137960.
[2.108] S.N. Naik, V.V. Goud, P.K. Rout, A.K. Dalai, Production of first and second generation biofuels: a comprehensive review, *Renewable and sustainable energy reviews* 14(2) (2010) 578-597.
[2.109] M.H. Eppink, G. Olivieri, H. Reith, C.v.d. Berg, M.J. Barbosa, R.H. Wijffels, From current algae products to future biorefinery practices: a review, *Biorefineries* (2017) 99-123.
[2.110] V.B. Borugadda, V.V. Goud, Biodiesel production from renewable feedstocks: Status and opportunities, *Renewable and Sustainable Energy Reviews* 16(7) (2012) 4763-4784.
[2.111] T. Obata, A.R. Fernie, A. Nunes-Nesi, The central carbon and energy metabolism of marine diatoms, *Metabolites* 3(2) (2013) 325-346.
[2.112] G. Pokoo-Aikins, A. Nadim, M.M. El-Halwagi, V. Mahalec, Design and analysis of biodiesel production from algae grown through carbon sequestration, *Clean Technologies and Environmental Policy* 12(3) (2010) 239-254.
[2.113] S.A. Khan, M.Z. Hussain, S. Prasad, U. Banerjee, Prospects of biodiesel production from microalgae in India, *Renewable and sustainable energy reviews* 13(9) (2009) 2361-2372.
[2.114] Y. Sharma, B. Singh, Development of biodiesel: current scenario, *Renewable and sustainable energy reviews* 13(6-7) (2009) 1646-1651.

[2.115] H.-W. Yen, I.-C. Hu, C.-Y. Chen, S.-H. Ho, D.-J. Lee, J.-S. Chang, Microalgae-based biorefinery–from biofuels to natural products, *Bioresource technology* 135 (2013) 166-174.

[2.116] T. Suganya, M. Varman, H. Masjuki, S. Renganathan, Macroalgae and microalgae as a potential source for commercial applications along with biofuels production: a biorefinery approach, *Renewable and Sustainable Energy Reviews* 55 (2016) 909-941.

[2.117] Kuczynska, P., Jemiola-Rzeminska, M. and Strzalka, K., 2015. Photosynthetic pigments in diatoms. *Marine drugs*, 13(9), pp.5847-5881.

[2.118] Hu, Q., Sommerfeld, M., Jarvis, E., Ghirardi, M., Posewitz, M., Seibert, M. and Darzins, A., 2008. Microalgal triacylglycerols as feedstocks for biofuel production: perspectives and advances. *The plant journal*, 54(4), pp.621-639.

3

Biochemical Compounds in *Phaeodactylum tricornutum*

Vandana Sirotiya and Vandana Vinayak*

Diatom Nano Engineering and Metabolism Laboratory (DNM), School of Applied Sciences, Dr. Harisingh Gour Central University Sagar, Madhya Pradesh, India

Abstract

A suite of stress factors can trigger the rapid biosynthesis of biochemical compounds like lipid, pigments, carbohydrates, etc., in marine diatom *Phaeodactylum tricornutum*. Stress factors such as high light intensity, nutrient supplementation, alkaline pH, and nitrogen depletion can potentially affect the natural metabolic processes and induce the accumulation of fucoxanthin and polyunsaturated fatty acids (PUFAs). *P. tricornutum* has a robust metabolism with the efficiency of an immense array of metabolic fluxes. Characterization of fundamental metabolic pathways, possible regulatory mechanism of central metabolism and enzymes for higher and rapid biosynthesis of biochemical compounds have vital flux roles. Study of these metabolic pathways suggest that these stress factors significantly influence the performance for the biosynthesis and accumulation of a high number of value-added compounds. Demand for natural compounds and their applications in various industries lead to optimization of biosynthesis pathways of biochemical compounds with low-cost stress chemicals as well as other factors for exponential cell growth to obtain a great amount of biomass.

***Keywords*:** Biocompounds, diatom, fucoxanthin, lipids, stress

**Corresponding author*: kapilvinayak@gmail.com

Vandana Vinayak and Richard Gordon (eds.) Diatom Cultivation for Biofuel, Food and High-Value Products, (51–78) © 2025 Scrivener Publishing LLC

3.1 Introduction

Both marine and fresh water diatoms are subjected to constant environmental stress. Diatom cells develop lipid droplets by intense remodeling, which is induced by numerous chemical and physical stress conditions such as nutrient deficiency, high temperature, high light, exposure to nitric oxide, hydrogen peroxide etc. [3.76]. Uncertainty exists regarding the logic behind development as a component of general stress reactions [3.68]. Furthermore, *Phaeodactylum tricornutum* frequently occurs in three distinct morphotypes: fusiform, oval, and triradiate, making it a great model for studying cellular processes involved in morphological transition [3.85].

Diatom nutrient deficiencies, including a lack of nitrogen and phosphate, cause neutral lipids to build up [3.107]. The tricarboxylic acid (TCA) cycle provides the majority of the reducing equivalents and carbon skeletons that are necessary for cells to assimilate nitrogen into proteins in order to maximize their ability to grow. However, as nitrogen availability decreases, so does the sink for TCA cycle metabolites, and acetyl-CoA, the cycle's carbon source, can be diverted toward fatty acid (FA) biosynthesis. Because of this, under nitrogen stress, storage lipids rise while cellular protein content declines [3.57]. Diverse strategies and tactics have been developed to fully use the potential of diatom as a feedstock, including the development of diatom-based "cell factories" and the production of bio-based commodities like biofuels and value-added goods like carotenoids [3.106]. They account for over half of primary production and serve as the foundation for marine food webs, as well as a variety of products derived from marine diatoms such as pigments and polyunsaturated fatty acids (PUFAs) [3.47]. Table 3.1 illustrates the different biochemical compounds from diatoms, their yield, their functional use and types of photobioreactors used.

Many photosynthetic pigments—such as chlorophyll *a* (*Chl a*), chlorophyll *c* (*Chl c*), and fucoxanthin—have been found in diatoms and are thought to play an important part in light harvesting photosystems [3.20].

Fucoxanthin content in diatoms is, however, comparatively higher than that of chlorophyll [3.65]. Fucoxanthin, a photosynthetic pigment that belongs to the family of xanthophylls, and is responsible for the diatoms distinctive golden-brown color [3.81]. According to multiple recent studies, fucoxanthin possesses a variety of biological qualities, such as antioxidant, antibacterial, anti-obesity and anticancer activities, that make it useful in a number of biomedical applications [3.71]. Some green algae and diatoms may store carbon and energy in the form of lipids called triacylglycerides, which has re-energized the possibility of using algal oil for the

Table 3.1 Biochemical compounds from diatom their yield, use and type of photobioreactors used.

Product	Product type	Product yield/ productivity reported	Functional end-use	Photobioreactors	Reference
Lipids	EPA	3% DW; 56 mg/L/d	Nutraceutical; cardiovascular health; precursor for prostaglandin-3, thromboxane-3, and leukotriene-5 eicosanoids	Outdoor chemostat split-cylinder airlift PBR (50–60 l), Almeria, Spain; photoautotrophic/ mixotrophic with glycerol	[3.82]
	DHA	0.64% DW	Nutraceutical; primary structural component of the human brain, cerebral cortex, skin, and retina	Indoor horizontal fence PBR (550 l); photoautotrophic	[3.39]
		0.26% DW		Indoor raceway (1250 l); photoautotrophic	
		0.92% DW		Indoor flasks; photoautotrophic	[3.100]
	ARA	1.89% DW	Nutraceutical; prostaglandin precursor	Indoor flasks; photoautotrophic	

(Continued)

Table 3.1 Biochemical compounds from diatom their yield, use and type of photobioreactors used. (*Continued*)

Product	Product type	Product yield/ productivity reported	Functional end-use	Photobioreactors	Reference
Lipids	TAG	58.5 mg/L/day, 45% DW	Biodiesel	Green wall panel III (≤40 l); outdoor; photoautotrophic	[3.78]
	Brassicosterol		Decreased risk of coronary heart disease; anti-inflammatory activities	Lab PBR (≤1 l); photoautotrophic	[3.24]
Carbohydrates	Chrysolaminarin	14% DW; 94 mg/L/day	Antioxidant	Indoor flat-plate PBR (50 l); photoautotrophic	[3.33]
Terpenoids	Fucoxanthin	0.7% DW; 4.7 mg/L/day	Antioxidant, anti-obesity, anticancer, anti-inflammator	Indoor flat-plate PBR (50 l); photoautotrophic	

(*Continued*)

Table 3.1 Biochemical compounds from diatom their yield, use and type of photobioreactors used. (*Continued*)

Product	Product type	Product yield/ productivity reported	Functional end-use	Photobioreactors	Reference
Pharmaceutical Products	Lupeol	0.01% DW	Antiprotozoal, antimicrobial, antitumor, chemopreventative	Lab Algem PBR (≤1 l); photoautotrophic	[3.24]
		0.0013% DW		Indoor horizontal fence PBR (550 l); photoautotrophic	
	Betulin	Detectable levels	Antitumor	Lab Algem PBR (≤1 l); photoautotrophic	
Heterologous compounds and proteins	Polyhydroxybutyrate (PHB)	10.6% DW	Bioplastics	Indoor flasks (≤1 l); photoautotrophic	[3.40]
	Human IgGαHBsAg: antibody against hepatitis B virus surface protein	0.0021% DW; 8.7% TSP	Monoclonal antibody	Indoor flasks (≤1 l); photoautotrophic	[3.41]

(*Continued*)

Table 3.1 Biochemical compounds from diatom their yield, use and type of photobioreactors used. (*Continued*)

Product	Product type	Product yield/ productivity reported	Functional end-use	Photobioreactors	Reference
Antibodies	IgG1/kappa Ab CL4mAb: antibody to hepatitis B virus surface protein lacking the ER retention signal (DDEL) at the C-termini of both antibody chains	2.5 mg/L (secreted)	Monoclonal antibody	Indoor flasks (≤1 l); photoautotrophic	[3.42]
	Monoclonal IgG antibodies against the nucleoprotein of Marburg virus (close relative of Ebola virus	2 mg/L (secreted)	Monoclonal antibody	Indoor flasks (≤1L); photoautotrophic	[3.43]
Whole cell	Biomass	25.4 g/L, 1.7 g/L/ day	Aquaculture/animal feed	Outdoor split-cylinder airlift PBR (60 l); mixotrophic (0.1 M glycerol)	[3.17] [3.82]

generation of liquid fuels in addition to photoautotrophic growth (i.e., carbon fixation by sunlight) [3.75]. The 3% of the U.S. arable agricultural land may be used to produce microalgal biodiesel, which could provide 50% of the nation's liquid transportation fuel requirements. Numerous studies have attempted to increase the synthesis of microalgal lipids using biogeochemical and genetic methods [3.22].

We are aware of many elements of carbon absorption in diatoms, but for many eukaryotic photoautotrophs under various growth circumstances, we do not know the direct effects or the contribution of inorganic carbon to lipid formation [3.17]. The preponderance of methods has seen some degree of success, but much more research is needed to completely understand how to increase lipid synthesis in microalgae in an efficient and cost-effective manner [3.69]. In plants, animal products, and lipid-rich foods, bioactive substances are additional nutritious components that naturally occur in tiny amounts [3.79]. Several clinical and experimental studies are being conducted to find and extract these chemicals from diverse plant and animal sources, as well as to evaluate their health effect due to their qualities. Plants are the primary source of the majority of bioactive substances, including pigments, phenolic compounds, fatty acids, and lipids [3.63]. The pigments found in diatoms play an important role in photoprotection and have antioxidant capabilities [3.109]. They consist of cutting-edge metabolic pathways, which have been proven to be great sources of bioactive compounds for usage in a variety of sectors, including food, cosmetics, medicine, pharmaceuticals, bioremediation, and nanotechnology [3.25]. It is believed that the phenolic compounds in diatoms are a natural source of antioxidative substances with beneficial health benefits and aid in avoiding disease brought on by oxidative stress [3.12]. Table 3.2 describes the accumulation of biochemical compounds under stress conditions.

The polyunsaturated fatty acids eicosapentaenoic acid (EPA), docosapentaenoic acid (DHA), myristic acid, palmitic acid, and palmitoleic acid are particularly abundant in diatoms [3.58]. Unicellular phytoplanktonic organisms, which are passive drifters in the water column and frequently suffer abrupt changes in their surrounding environment, dominate marine photosynthesis. One of the most numerous and diverse families of photosynthetic organisms is the diatom family [3.93]. They are especially well-suited to thriving in extremely dynamic conditions, such as choppy coastal waters, upwelling zones, and polar seas [3.90]. Such algae's potential for adaptation shows that they possess complex systems for perceiving and reacting quickly to environmental changes [3.28]. In support of this idea, the genome sequence information of representative diatom model species like *Thalassiosira pseudonana* and *Phaeodactylum tricornutum*, as well as the availability of transcriptomic and proteomic data in various species

Table 3.2 Accumulation of biochemical compounds under stress conditions.

Sr. no.	Stress	Quantity of stress	Impact on biogenic compounds	Reference
1	Nitrogen stress	Initial nitrogen concentration decreases from 49.4 mg/L to 1.24 mg/L	EPA decreases 75%, PUFA decreased 74%, but total fatty acid increased 241%.	[3.8]
2	Light stress	Light intensity increases from 50 μmol/m²/s to 150 μmol/m²/s	EPA no differences, DHA increased 12%, total fatty acid no differences.	[3.74]
3	Dark stress	Dark (after 4 days dark treatment)	Lipid increased 230% by glycolysis and fatty acid synthesis.	[3.7]
4	Temperature stress	Temperature decreases from 25 °C to 15 °C	EPA increased 18%, DHA increased 59%, but total fatty acid decreased 32%.	[3.74]
5	CO_2 Stress	Higher CO_2 level (0.15%)	EPA increased 73%, as well as lipid content.	[3.103]

exposed to various stimuli and stresses, have highlighted the existence of some diatom-specific adaptive strategies, pointing to molecular regulators of responses to environmental change.

The main source of polysaccharides in diatoms is intracellular food storage polymers, followed by the external mucilaginous matrix, also referred to as exopolysaccharides (EPSs) [3.94], and finally the polysaccharides from the frustule or diatom cell wall, particularly sulfated polysaccharides. There is evidence that diatom polysaccharides offer a variety of positive health benefits, including anti-inflammatory, anticancer, and immunostimulant properties. Chrysolaminarin, a material used for food preservation, is known to stop colon cancer cells from proliferating. According to studies, the dietary additive chrysolaminarin from the *Chaetoceros muelleri* has an immunostimulatory effect [3.9] [3.91]. Diatoms feature unusual gene sets associated with nitrogen metabolism, including a full urea cycle, which may be employed as a temporary energy storage system or as a sink for photorespiration [3.35]. The metabolism between plastids and mitochondria in diatoms finally achieve photosynthetic optimization [3.32].

Due to their very effective photoprotective systems, diatoms' ecological dominance also depends on their ability to deal with light stressors. They have a high ability to release surplus light energy as heat through high energy quenching, which may be seen by the non-photochemical quenching of *Chl a* fluorescence [3.55]. Key elements of the energy quenching procedure in diatoms have been identified as the LHCX1 protein, a member of the light-harvesting protein family, and the xanthophyll diatoxanthin pigment, which is produced from the de-epoxidation of diadinoxanthin during illumination [3.15]. Similar to the light harvesting complex stress-related proteins of green algae and mosses, *P. tricornutum* cells with unregulated LHCX1 expression exhibit a considerably reduced non-photochemical quenching capability and a lower fitness, indicating a vital function for this protein in light acclimation [3.88].

Fucoxanthin, a kind of xanthophyll pigment, is the main carotenoid in *P. tricornutum* and functions as an accessory pigment in chloroplasts alongside chlorophyll *a* and *c*, creating a complex with the chlorophyll *a/c* binding protein [3.14]. The brown color of diatoms is caused by a high concentration of fucoxanthin, which has absorption spectra at blue-green and yellow-green portions of light spectra [3.80]. Fucoxanthin also has protective benefits on a number of organs, including the blood vessels, bones, liver, skin and eyes [3.36] [3.108].

The byproduct of Vitamin B3 i.e niacin metabolism is trigonelline. It is made by *P. tricornutum* and has several therapeutic potentials for cardiovascular illnesses, cancer, diabetes, hyperlipidaemia, and disorders of the central nervous system [3.67].

3.2 Biochemical Compounds

3.2.1 Pigments

Natural pigments found in diatoms include light-harvesting chlorophyll *a*, chlorophyll *c*, fucoxanthin (Fx), and photoprotective carotene, diadinoxanthin (Ddx), diatoxanthin (Dtx), zeaxanthin (Zx), antheraxanthin (Ax), and violaxanthin (Vx). Isoprenoid precursor molecules dimethylallyl diphosphate (DMAPP) and its isomer isopentyl pyrophosphate (IPP) are the source of carotenoids [3.98]. (Fx) is the main constituent of the fucoxanthin-chlorophyll protein (FCP) complex, which plays a important role in carrying out the light-harvesting function in diatoms. On the offset, diadinoxanthin (Ddx) and diatoxanthin (Dtx) aid in the non-photochemical quenching and heat dissipation [3.14]. Fucoxanthin is an allelic carotenoid that may be isolated from brown edible seaweeds, diatoms, and other plants. Potential natural sources of fucoxanthin include the marine microalgae diatoms, which may grow quickly and perform aggressively in regulated bioreactors [3.51]. Due to their numerous beneficial characteristics, including their antioxidant, anticancer, anti-inflammatory, anti-obesity, and neuroprotective activities, carotenoids are highly valued in the nutraceutical and pharmaceutical industries [3.4]. The primary diatom-specific carotenoids and their quantity changes depending on growth circumstances that are controlled by light involvement in the xanthophylls cycle [3.81]. Strong antioxidant qualities are displayed by zeaxanthin (Zx), antheraxanthin (Ax), and violaxanthin (Vx), and a vital marine metabolite also has a positive impact on health [3.52]. Since these pigments have allenic bonds, acetyl, and epoxy groups, diadinoxanthin (Ddx) and diatoxanthin (Dtx) exhibit strong bioactivities and function as chemicals that promote health. The key structural changes in pigments like carotenoids include the presence of C-7 allenic carbon, two hydroxyl groups, a carbonyl group, 5, 6-monoepoxide, and an acetyl group in the end side ring of fucoxanthin [3.54]. The allenic bond is what gives fucoxanthin its strong antioxidant capacity [3.1]. Fucoxanthin possesses six oxygen atoms, which makes it more vulnerable to radicals in an anoxic environment. It has a remarkable allenic bond and 5,6-monoepoxide, which contributes to its distinctive structure. Fucoxanthin has a unique molecular structure that is related to neoxanthin and peridinin and has great biological activity. Fucoxanthin was the first allenic carotenoid found in sea forests. Fucoxanthin is a golden brown, orange-colored pigment, associated with other chlorophylls (*a* and *c*) and carotenoids found in brown seaweeds and diatoms [3.21]. Fucoxanthin makes up 1–2.5% of the dry cell weight of diatoms and is a vital part of the fucoxanthin chlorophyll protein complex. By capturing light energy, FCP plays a crucial part in photosynthesis in diatoms [3.97]. The primary carotenoid of *P. tricornutum* is fucoxanthin, a kind of xanthophyll pigment. It functions as an accessory pigment in chloroplasts alongside chlorophyll *a* and *c*, creating a complex with the chlorophyll *a/c* binding proteins (FCPs) [3.14]. Fucoxanthin alone has high antioxidant properties and has huge therapeutic applications with its positive effect on healing inflammation, diabetes, cardiac health, etc. [3.83].

3.2.2 Proteins

Diatom peptides, also known as natural peptides, have been demonstrated to have less adverse effects on immune system activity than synthetic supplements, which may affect the body's defensive processes and cause cell stress [3.84]. An increasingly viable source of bioactive peptides are microalgae [3.9]. The most investigated diatoms as a source of active polypeptides include *Navicula* species, *Nitzschia* species, and *Phaeodactylum* species. An excellent source of peptides, *Navicula incerta* has a protein level between 46.5 and 51.7%. Additionally, two bioactive peptides from *Navicula* sp., Pro-Gly-Trp-Asn Gln-Trp-Phe-Leu and Val-Glu-Val-Leu-Pro-Pro-Ala-Glu-Leu, were also discovered. In HepG2/CYP2E1 cells, significant antioxidant capabilities have been demonstrated [3.29]. *In vitro* and in cell cultures, bioactive proteins and hydrolysates are shown to have strong antioxidant activities. The suppression of the angiotensin I-converting enzyme (ACE), which is crucial in the control of hypertension, is another *in vitro* assay that has been investigated. Angiotensin-I, a powerful vasoconstrictor peptide, may be converted to Angiotensin-II by hydrolysates, regulating blood pressure [3.56].

3.2.2.1 Hydrolysates

The method of acquisition of hydrosylates from diatoms may involve extraction using chemicals or enzymes, followed by purifying techniques. Free amino acids and peptides with various chain lengths make up the complex combination of free amino acids seen in protein hydrolysates [3.101]. The degree of hydrolysis (DH), a general metric that represents the proportion of peptide bonds in a protein that have been broken, defines the hydrolysis level of the hydrolysates. It is generally accepted that the higher the DH content in hydrolysates, the greater is their biological activity. Antioxidant properties have been linked to hydrolysates. We still don't fully understand the precise processes through which the hydrolysates display antioxidant action. However, various aspects of their chemical makeup, like their hydrophobicity, low molecular weight, and amino acid sequences have been favorably connected to the antioxidant activity. An indole/imidazole/pyrrolidine ring, together with a steric structure at the C- and N-terminal region, is the key property associated with this action. There is evidence that the N-terminal area is less important for antioxidant efficacy and radical scavenging than the C-terminal region [3.9].

3.2.2.2 *Chlorophyll Proteins: 10-Hydroxy-Phaeophorbide A and Phaeophorbide A*

In addition to Phaeophorbide A, 10-Hydroxyphaeophorbide A is a byproduct of the degradation of chlorophyll. Because of the amino groups in their structure, these molecules are referred to as chlorophyll proteins [3.110]. Different biological properties linked to antioxidant, anti-inflammatory, anticancer, and antibacterial activities have been shown for these compounds. The anti-tuberculosis action of these chlorophyll proteins from *Skeletonema costatum* is one of the most alluring current development uses [3.111].

3.2.3 Carbohydrates

The amount of carbohydrates present in the algae may be influenced by a number of environmental conditions, including temperature, light exposure, nutrients, and salinity. The composition of monosaccharides in diatom exopolysaccharides (EPSs) varies greatly. Diatoms change their EPS machinery depending on the environmental conditions [3.48]. It has been demonstrated that EPSs are heteropolysaccharides. EPS in diatoms are good source of sulfated glucoronomannans and chitin fibres as well. A typical (1,6) ramified (1,3) polyglucan food storage molecule in diatoms is chrysolaminarin [3.38] [3.112].

3.2.3.1 *Chrysolaminarin*

Chrysolaminarin is a polysaccharide in marine diatoms, that stores the most energy. In terms of structure, it has a backbone made up of 11:1 (-1,3)-glucans and (-1,6)-glucans, respectively [3.60]. The buildup of chrysolaminarin increases with the amount of nutrients in the medium and can reach up to 33% dry weight. Chrysolaminarin has a number of qualities that have been linked to health as discussed in earlier sections.

3.2.3.2 *Exopolysaccharides (EPSs)*

Extracellular mucilage made of significant amounts of carbohydrates is produced by diatoms. This matrix is referred to as EPSs and may also contain other extracellular polymeric compounds [3.104] [3.113]. Like other chemicals in algae, EPS has a variety of components that are influenced by many conditions [3.89]. EPS have adsorption properties and its been seen that diatoms embellish themselves on steel, glass and plastics resulting in biofouling and marine litter [3.113]. The various components of EPS give the EPS its shape, which ranges from mucilaginous capsules to very stiff fibers [3.30]. EPS is difficult to extract and purify because of its complicated and diverse makeup. Soluble EPS are the most popular ones. In addition, diatom EPSs feature methyl-pentoses like rhamnose involved in adhesion qualities and have a medicinal use as a bioderived glue to seal wounded tissues [3.19].

3.2.3.3 Sulfated Polysaccharides

Diatoms have sulfated polysaccharides (SPs) in their cell walls. The amount and shape of sulphur compounds, which are involved in frustule development, fluctuate according to the environmental circumstances. The diatom may adapt to changes more quickly when there is an increase in uronic acid, sulfate, and fucose under stressful circumstances [3.11]. Given that their sulfate group (-SO_3H) confers them with antibacterial, antiviral, and antioxidant properties, SPs have received the greatest attention from researchers studying the biological activity of diatom polymers [3.10].

3.2.4 Lipids

As a replacement for fossil fuel, neutral lipids found in microalgae are easily converted to biodiesel and have drawn a lot of interest. Diatoms grow quickly and gather a lot of neutral lipids that can be turned into biodiesel, up to 20% of its total dry weight under typical, unstressed growing circumstances [3.92]. Previous studies have shown that a variety of abiotic stressors, including lack of nitrogen, phosphorus, salt, and light, increase the quantity of neutral lipids in an organism. UV-C (200–280 nm) radiation causes high levels of fatty acid accumulation in the microalga *Tetraselmis* sp. under stress, leading to the accumulation of neutral lipids, which may be used as a marker for cells under stress [3.70]. One of the most effective methods for increasing lipid cell content in many microalgae species, including *Phaeodactylum tricornutum*, was nutrient deprivation, nitrogen (-N) in particular [3.16]. It has been proposed that a mixture of ideal growth circumstances and growth-limited conditions might boost biodiesel production, despite the fact that N starvation and other environmental stresses reduce algal growth rate and total lipid output [3.95]. Genetic engineering might also be used to overexpress crucial genes involved in fatty acid synthesis, which could increase the cost-effectiveness of biofuel production. This is despite the fact that research has demonstrated that overexpressing a type-2 diacylglycerol acyltransferase in a genetically modified strain of a marine diatom can accelerate lipid biosynthesis [3.49] [3.53].

In comparison to other microalgal species, diatoms are very effective at synthesizing lipids, and therefore hold promise for potential future advancements in the generation of biodiesel. For instance, the well-known marine diatom *P. tricornutum* can collect up to 61.4% of lipid per dry weight of the algae [3.96]. In a study conducted on three different microalgae in which *Amphiprora* sp. was cultured under normal culture condition and the other two microalgae (*Pleurosigma* sp. and *Amphora* sp.) cultured in SiO-NP medium, the study showed that the lipid production of *Amphiprora* sp. was higher (41.48%) than *Pleurosigma* sp. (27.3%) and *Amphora* sp. (22.49%) [3.46]. In research conducted on *Chaetoceros muelleri* grown under air 10% CO_2 it was observed that total lipid (43.40% dry weight) and neutral lipid contents were all obtained with 10% CO_2 aeration [3.99]. Earlier studies have shown that *Gyrosigma* sp. when cultivated in palm oil wastewater showed lipid production of 24.1 mg L^{-1} day^{-1} [3.37].

A further benefit of using marine diatoms for biofuel generation is that, unlike freshwater algae, they may benefit from an endless supply of seawater [3.3]. Our society's infrastructure has been heavily reliant on petroleum-based goods since the industrial revolution for fuels [3.59]. Petroleum has been used for one hundred years thanks to a balanced relationship between supply and demand. However, the rising energy consumption has produced a situation where demand may start to outpace supply. It is obvious that renewable biofuels (such as ethanol, butanol, H_2, CH_4 and biodiesel) are required to help replace petroleum reliance in the United States and across the world given the enormous inflow of carbon into the atmosphere caused by the burning and consumption of petroleum-based goods. To increase the viability and usability of renewable energy sources that utilize direct phototrophic CO_2 fixation into liquid biofuels, both basic and applied research is required, even though current technology may be used to assist, avoid and revert current environmental trends [3.61].

3.2.5 Fatty Acids

Omega-3 fatty acids, which comprise DHA and EPA, are polyunsaturated fatty acids with a number of health advantages. These fatty acids are linked to long-term, healthy aging. Certain fatty acids are produced, such as omega-3, which is important for cardiovascular health, prenatal development, and preventing Alzheimer's disease [3.27]. Humans need proper doses of omega-3 fatty acids in a certain quantity through fish and fish products. Additionally abundant in lipids, diatoms include certain long-chain polyunsaturated fatty acids like EPA that are good for human cardiovascular health. Additionally, L-arginine, RNA, and EPA are combined with DHA in fish oil preparation to lessen infections. The fatty acids that are present in diatom species in quite high concentrations include EPA and DHA. It has been studied that in *Chaetoceros muelleri* that there is >90% predominance of C14-C18 fatty acids and >80% that for monosaturated fatty acids at same CO_2 concentrations [3.99]. In a similar study conducted on marine diatom *Gyrosigma* sp. cultivated in inorganic nutrient dissolved in palm oil mill effluent showed a significant amount of the following fatty acids: palmitic acid (48.6%), eicosapentaenoic acid (10.6%), myristic acid (8.1%), stearic acid (8.0%), linoleic acid (7.5%), oleic acid (6.4%), and linolenic acid (5.8%) [3.37]. A particular form of fatty acid called DHA is present in aquatic fish such as tuna, halibut, sea blubber, etc. Inside the human body, DHA can be converted to EPA. As a result of their role in lowering blood cholesterol, which is a major risk factor for coronary heart disease, EPAs are known to have preventive effects against a number of chronic and acute disorders [3.66].

3.2.5.1 Omega-3 Fatty Acids

When an acid has at least two double bonds in its structure, it is referred to as a polyunsaturated fatty acid (PUFA) [3.64]. The two main categories of PUFAs are omega-3 (n3) and omega-6 (n6), while there are other classifications as well. Alpha-linolenic acid and three long-chain PUFAs (EPA, DPA, and DHA) are found in N-3 PUFAs, whereas linoleic and arachidonic acids are found in N-6 PUFAs [3.5]. Dietary fibers having a lengthy list of health advantages which include omega-3 long-chain PUFA, EPA, and DHA. Both the anti-inflammatory activities and the viscosity of cell membranes are dependent on Omega 3 fatty acids in our body [3.50]. Being a large part of cell membranes, DHA is widely distributed in the brain and retina. In recent research, it was shown that EPA and DHA function as precursors of many metabolites, powerful lipid mediators that are essential for the prevention and treatment of a number of disorders [3.62]. Marine microalgae generate PUFA and EPA, which are employed in food and nutraceutical products. The diatom *P. tricornutum* has been researched in the lab and outdoors. Due to their increase in EPA content, lipid formation following nutritional stress, and bulk culture. *P. tricornutum*, generates an acceptable amount of EPA but has very little DHA. Due to its fast cultivation, simplicity in genetic modification, and availability of a whole genome sequence, this diatom serves as an excellent industrial source in the synthesis of EPA. Besides, EPA, this diatom accumulates polar lipids, particularly monogalactosyldiacylglycerol (MGDG), a galactolipid [3.34].

3.3 Demand, Valorization and Biotechnological Applications

When addressing microalgal development from a biotechnological angle, essential criteria to be taken into account include light intensity and its spectral properties [3.31]. Even if using artificial light is more expensive than using sunshine, doing so can be profitable if high-value microalgal biomasses are generated. Light-emitting diodes (LED) are a relatively new and cost-free technology compared to conventional fluorescent light tubes, and they offer the great advantage of selecting nearly monochromatic lights at various wavelengths [3.26].

Diatoms can potentially biosynthesize chemicals for a wide range of industrial and commercial uses because of their peculiar physiology and specific metabolic pathways. Fucoxanthin is the most abundant isoprenoid in diatoms [3.6]. The global market for carotenoids reached $1.23 billion USD in 2015, and by 2022, it was anticipated to reach $1.81 billion USD. Carotenoids derived from natural sources have a high economic value. Diatoms synthesize a variety of sterols, including those found in plants. Nutraceutical and bioactive characteristics of sterols have been established. By 2013, the phytosterol market was predicted to be worth $300 million USD. The global sterols market is increasing at a rate of 7–9% per year, with $935 million USD predicted by 2022 [3.73]. This chapter highlighted the widespread applications of diatoms and their physiologically active byproducts, such as fucoxanthin and omega-3 fatty acids, as well as those with unique structural and functional traits. Many types of surface changes based on diatom structure and purification techniques are studied in order to provide a reliable drug delivery method [3.45]. Diatoms are the only organism that can be used to transport drugs because they are easily accessible, have a sizable surface area, are biocompatible, are nontoxic, have a significant cost-effectiveness ratio, are environmentally benign, and have a systematic porosity structure that makes surface modification simple [3.102]. Fucoxanthin has a wide range of bioactivities, including those that are anti-inflammatory, anti-obesity, anti-diabetic, antioxidant, hepatoprotective, and skin protective. Additionally, omega-3 fatty acids have bioactivities that support fetal development, guard against CVDs and other diseases like Parkinson's and Alzheimer's, among others [3.44]. By year 2018, the worldwide fatty acids market was worth $4.31 billion USD, and it is predicted to grow to 91.6 billion by 2026. The polyphenols found in diatoms have been linked to a variety of biological functions. The market for polyphenols was valued at $1.28 billion USD in 2018; from 2019 to 2025, it is predicted to grow by 7.2%. With a projected value of $1.79 billion USD in 2016, the worldwide market for green natural colors is predicted to reach $2.97 billion USD by 2025. The FDA approves diatom-derived derivatives of chlorophyll and their intended uses in the food sector as natural green hues [3.105].

Diatoms produce nutrient-rich oil for human use, offering a possible option to satisfy the demand of future population growth [3.2]. Renewable energy is especially intriguing because it can produce bioenergy, which can help us get past the limitations of fossil fuels [3.87]. The development of an economical culture technique to extract the most in a short amount of time is another important feature in diatom research. Future algae companies may concentrate on either the contents of long-chained PUFAs as a source of health food or on the short-chain fatty acids for the manufacture of biofuel due to *P. tricornutum's* rapid growth and high lipid content [3.23]. The frustule, which has several biotechnological uses, is the most distinctive feature of diatom morphology [3.2]. Variations in light intensity (100–300 mol phot m^2 s^{-1}) and spectrum (five monochromatic LEDs: blue, green-yellow, red-orange, red, and a complete visible spectrum white LED) were shown to be associated with changes in the characteristics of the frustule of *Coscinodiscus granii* (valve thickness, Si concentration, frustule diameter, foramen number, size and density). It was shown that these changes had an effect on the frustule's photonic characteristics. Different growth rates were caused by light wavelengths simultaneously, with white, blue, and red LEDs supporting the greatest growth at both light intensities studied [3.86]. The EPA concentration in *P. tricornutum* was shown to be affected by light intensity in a recent study, with light intensities between 60 and 100 mol phot m^2 s^{-1} being the optimum for promoting EPA accumulation in comparison to higher intensities (250–750 mol phot m^2 s^{-1}). The production of fucoxanthin in the marine diatom *P. tricornutum,* cultivated in nutrient-complete culture medium, was examined in relation to both light intensity and other environmental growth variables [3.77]. It is noticed that there was some relationship between light intensity and the fucoxanthin levels. Furthermore, it was discovered that in the same diatom, acclimating to high light treatments included activating the xanthophyll-cycle photoprotective mechanism without changing the quantity of fucoxanthin

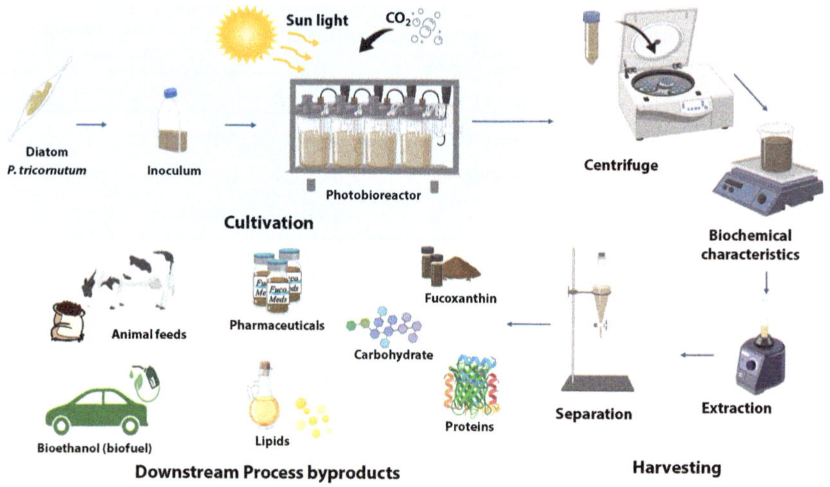

Figure 3.1 Extraction of proteins, valorization of residual biomass for the manufacture of biofuels and biosilica, and biorefinery from *P. tricornutum* biomass for the production of high-value chemicals, with fucoxanthin as the principal product. The overview of each upstream and downstream process is displayed using rectangles.

[3.5]. Blue light is thought to be crucial for *P. tricornutum's* photoacclimation to high light intensities, but it also promotes the accumulation of carotenoids [3.5]. However, other studies suggest that blue light strongly restricts the photoprotective processes in the coastal diatom *Skeletonema marinoi*, likely increasing the biochemical energy available for growth. In addition, *S. marinoi* was exposed to a sinusoidal blue light treatment with a peak at midday and researchers discovered that this light regimen was ideal for both growth and the production of primary metabolites (lipids, carotenoids and proteins) [3.5]. Extraction of proteins, valorization of residual biomass for the manufacture of biofuels and biosilica, and biorefinery from *P. tricornutum* biomass are used for the production of high-value chemicals, with fucoxanthin as the principal product [3.13]. The overview of each upstream and downstream process is displayed in Figure 3.1.

3.4 Conclusion

P. tricornutum has remarkable metabolic versatility, enabling the production of a diverse range of bioactive compounds that are widely used across various industrial applications. These compounds include lipids, essential for biofuels; pigments like fucoxanthin, which have antioxidant properties for use in aquaculture, nutraceuticals, cosmetics, and pharmaceuticals; carbohydrates used in food additives and polyunsaturated fatty acids (PUFAs), valued for their health benefits in functional foods and pharmaceutical formulations. The natural origin and beneficial properties of these bioactive molecules make them increasingly attractive to industries seeking sustainable alternatives. Moving forward, continued research and optimization efforts are crucial to enhancing production yields and scalability, thereby maximizing the potential of *P. tricornutum* as a valuable resource for meeting the evolving global demands.

References

[3.1] Aghajanpour, M., Nazer, M.R., Obeidavi, Z., Akbari, M., Ezati, P. and Kor, N.M. (2017) Functional foods and their role in cancer prevention and health promotion: a comprehensive review. *American journal of cancer research* 7(4), 740.

[3.2] Ahirwar, A., Kesharwani, K., Deka, R., Muthukumar, S., Khan, M.J., Rai, A., Vinayak, V., Varjani, S., Joshi, K.B. and Morjaria, S. (2022) Microalgal drugs: A promising therapeutic reserve for the future. *Journal of Biotechnology* 349, 32-46.

[3.3] Ahirwar, A., Das, S., Das, S., Yang, Y.-H., Bhatia, S.K., Vinayak, V. and Ghangrekar, M.M. (2023) Photosynthetic microbial fuel cell for bioenergy and valuable production: A review of circular bio-economy approach. *Algal Research* 70, 102973.

[3.4] Ahirwar, A., Rai, A., Sirotiya, V., Khandelwal, P., Singh, G., Jhadav, D., Harish, Marchand, J., Schoefs, B. and Varjani, S. (2024) Fermentation of algal biomass for its nutritional value: perspectives and revolutions in the food industry. *Environmental Technology Reviews* 13(1), 1-28.

[3.5] Amjad Khan, W., Chun-Mei, H., Khan, N., Iqbal, A., Lyu, S.-W. and Shah, F. (2017) Bioengineered plants can be a useful source of omega-3 fatty acids. *BioMed research international* 2017.

[3.6] Athanasakoglou, A. and Kampranis, S.C. (2019) Diatom isoprenoids: Advances and biotechnological potential. *Biotechnology advances* 37(8), 107417.

[3.7] Bai, X., Song, H., Lavoie, M., Zhu, K., Su, Y., Ye, H., Chen, S., Fu, Z. and Qian, H. (2016) Proteomic analyses bring new insights into the effect of a dark stress on lipid biosynthesis in *Phaeodactylum tricornutum*. *Scientific reports* 6(1), 1-10.

[3.8] Baldisserotto, C., Sabia, A., Ferroni, L. and Pancaldi, S. (2019) Biological aspects and biotechnological potential of marine diatoms in relation to

different light regimens. *World Journal of Microbiology and Biotechnology* 35(2), 1-9.
[3.9] Barkia, I., Saari, N. and Manning, S.R. (2019) Microalgae for high-value products towards human health and nutrition. *Marine drugs* 17(5), 304.
[3.10] Barzkar, N., Jahromi, S.T., Poorsaheli, H.B. and Vianello, F. (2019) Metabolites from marine microorganisms, micro, and macroalgae: Immense scope for pharmacology. *Marine drugs* 17(8), 464.
[3.11] Bedi, N., Srivastava, D.K., Srivastava, A., Mahapatra, S., Dkhar, D.S., Chandra, P. and Srivastava, A. (2022) Marine biological macromolecules as matrix material for biosensor fabrication. *Biotechnology and Bioengineering* 119(8), 2046-2063.
[3.12] Begum, R., Howlader, S., Mamun-Or-Rashid, A., Rafiquzzaman, S., Ashraf, G.M., Albadrani, G.M., Sayed, A.A., Peluso, I., Abdel-Daim, M.M. and Uddin, M. (2021) Antioxidant and Signal-Modulating Effects of Brown Seaweed-Derived Compounds against Oxidative Stress-Associated Pathology. *Oxidative Medicine and Cellular Longevity* 2021.
[3.13] Branco-Vieira, M., San Martin, S., Agurto, C., Freitas, M.A., Martins, A.A., Mata, T.M. and Caetano, N.S. (2020) Biotechnological potential of *Phaeodactylum tricornutum* for biorefinery processes. *Fuel* 268, 117357.
[3.14] Büchel, C. (2020) Light harvesting complexes in chlorophyll c-containing algae. *Biochimica et Biophysica Acta (BBA)-Bioenergetics* 1861(4), 148027.
[3.15] Buck, J.M., Kroth, P.G. and Lepetit, B. (2021) Identification of sequence motifs in Lhcx proteins that confer qE-based photoprotection in the diatom *Phaeodactylum tricornutum*. *The Plant Journal* 108(6), 1721-1734.
[3.16] Burch, A.R. and Franz, A.K. (2016) Combined nitrogen limitation and hydrogen peroxide treatment enhances neutral lipid accumulation in the marine diatom Phaeodactylum tricornutum. *Bioresource technology* 219, 559-565.
[3.17] Butler, T., Kapoore, R.V. and Vaidyanathan, S. (2020) *Phaeodactylum tricornutum*: a diatom cell factory. *Trends in biotechnology* 38(6), 606-622.
[3.18] Caballero, M.A., Jallet, D., Shi, L., Rithner, C., Zhang, Y. and Peers, G. (2016) Quantification of chrysolaminarin from the model diatom Phaeodactylum tricornutum. *Algal Research* 20, 180-188.
[3.19] Cheah, Y.T. and Chan, D.J.C. (2022) A methodological review on the characterization of microalgal biofilm and its extracellular polymeric substances. *Journal of Applied Microbiology* 132(5), 3490-3514.
[3.20] Chen, M., Blankenship, R.E., Croce, R., van Grondelle, R., van Amerongen, H. and van Stokkum, I. (2018) Pigments: General properties and biosynthesis. In: *Light Harvesting in Photosynthesis*. CRC Press: 3-20.
[3.21] Chesalin, M., Al-Ghassani, S., Ryabushko, V., Bobko, N., Gureeva, E. and Nekhoroshev, M. (2017) Fucoxanthin Content and Chemical Composition in Brown Seaweed, *Nizamuddinia zanardinii* (Phaeophyta)

Collected from Mirbat, Southern Oman (The Arabian Sea). *International Journal on Algae* 19(3).

[3.22] Correa, D.F., Beyer, H.L., Possingham, H.P., Thomas-Hall, S.R. and Schenk, P.M. (2019) Global mapping of cost-effective microalgal biofuel production areas with minimal environmental impact. *GCB Bioenergy* 11(8), 914-929.

[3.23] Cui, Y., Thomas-Hall, S.R. and Schenk, P.M. (2019) Phaeodactylum tricornutum microalgae as a rich source of omega-3 oil: Progress in lipid induction techniques towards industry adoption. *Food chemistry* 297, 124937.

[3.24] D'Adamo, S., Schiano di Visconte, G., Lowe, G., Szaub-Newton, J., Beacham, T., Landels, A., Allen, M.J., Spicer, A. and Matthijs, M. (2019) Engineering the unicellular alga *Phaeodactylum tricornutum* for high-value plant triterpenoid production. *Plant biotechnology journal* 17(1), 75-87.

[3.25] de la Luz Cadiz-Gurrea, M., del Carmen Villegas-Aguilar, M., Leyva-Jiménez, F.J., Pimentel-Moral, S., Fernández-Ochoa, Á., Alañón, M.E. and Segura-Carretero, A. (2020) Revalorization of bioactive compounds from tropical fruit by-products and industrial applications by means of sustainable approaches. *Food research international* 138, 109786.

[3.26] Twidell, J. (2021). Renewable energy resources. Routledge.

[3.27] Djuricic, I. and Calder, P.C. (2021) Beneficial outcomes of omega-6 and omega-3 polyunsaturated fatty acids on human health: an update for 2021. *Nutrients* 13(7), 2421.

[3.28] Duan, Q. and Lee, J. (2020) Fast-developing machine learning support complex system research in environmental chemistry. *New Journal of Chemistry* 44(4), 1179-1184.

[3.29] Ejike, C.E., Collins, S.A., Balasuriya, N., Swanson, A.K., Mason, B. and Udenigwe, C.C. (2017) Prospects of microalgae proteins in producing peptide-based functional foods for promoting cardiovascular health. *Trends in Food Science & Technology* 59, 30-36.

[3.30] El-Sheekh, M. and Abomohra, A.E.-F. (2021) *Handbook of algal biofuels: aspects of cultivation, conversion, and biorefinery*. Elsevier.

[3.31] Fuente, D., Keller, J., Conejero, J.A., Rögner, M., Rexroth, S. and Urchueguía, J.F. (2017) Light distribution and spectral composition within cultures of micro-algae: quantitative modelling of the light field in photobioreactors. *Algal research* 23, 166-177.

[3.32] Gain, G., Vega de Luna, F., Cordoba, J., Perez, E., Degand, H., Morsomme, P., Thiry, M., Baurain, D., Pierangelini, M. and Cardol, P. (2021) Trophic state alters the mechanism whereby energetic coupling between photosynthesis and respiration occurs in *Euglena gracilis*. *New Phytologist* 232(4), 1603-1617.

[3.33] Gao, B., Chen, A., Zhang, W., Li, A. and Zhang, C. (2017) Co-production of lipids, eicosapentaenoic acid, fucoxanthin, and chrysolaminarin by

Phaeodactylum tricornutum cultured in a flat-plate photobioreactor under varying nitrogen conditions. *Journal of Ocean University of China* 16(5), 916-924.

[3.34] Ghosh, A., Khanra, S., Mondal, M., Halder, G., Tiwari, O., Saini, S., Bhowmick, T.K. and Gayen, K. (2016) Progress toward isolation of strains and genetically engineered strains of microalgae for production of biofuel and other value added chemicals: A review. *Energy Conversion and Management* 113, 104-118.

[3.35] Gomes-Dias, J.S., Teixeira, J.A. and Rocha, C.M. (2022) Recent Advances in the Valorization of Algae Polysaccharides for Food and Nutraceutical Applications: a Review on the Role of Green Processing Technologies. *Food and Bioprocess Technology*, 1-29.

[3.36] Gupta, A.K., Seth, K., Maheshwari, K., Baroliya, P.K., Meena, M., Kumar, A. and Vinayak, V. (2021) Biosynthesis and extraction of high-value carotenoid from algae. *Frontiers in Bioscience-Landmark* 26(6), 171-190.

[3.37] Govindan, N., Maniam, G.P., Yusoff, M.M., Chatsungnoen, T., Ramaraj, R. and Chisti, Y. (2020) Statistical optimization of lipid production by the diatom *Gyrosigma* sp. grown in industrial wastewater. *Journal of Applied Phycology* 32(1), 375-387.

[3.38] Gügi, B., Le Costaouec, T., Burel, C., Lerouge, P., Helbert, W. and Bardor, M. (2015) Diatom-specific oligosaccharide and polysaccharide structures help to unravel biosynthetic capabilities in diatoms. *Marine Drugs* 13(9), 5993-6018.

[3.39] Hamilton, M.L., Warwick, J., Terry, A., Allen, M.J., Napier, J.A. and Sayanova, O. (2015) Towards the industrial production of omega-3 long chain polyunsaturated fatty acids from a genetically modified diatom *Phaeodactylum tricornutum. PloS One* 10(12), e0144054.

[3.40] Hempel, F., Bozarth, A.S., Lindenkamp, N., Klingl, A., Zauner, S., Linne, U., Steinbüchel, A. and Maier, U.G. (2011) Microalgae as bioreactors for bioplastic production. *Microbial Cell Factories* 10(1), 1-6.

[3.41] Hempel, F., Lau, J., Klingl, A. and Maier, U.G. (2011) Algae as protein factories: expression of a human antibody and the respective antigen in the diatom *Phaeodactylum tricornutum. PloS One* 6(12), e28424.

[3.42] Hempel, F. and Maier, U.G. (2012) An engineered diatom acting like a plasma cell secreting human IgG antibodies with high efficiency. *Microbial cell factories* 11(1), 1-6.

[3.43] Hempel, F., Maurer, M., Brockmann, B., Mayer, C., Biedenkopf, N., Kelterbaum, A., Becker, S. and Maier, U.G. (2017) From hybridomas to a robust microalgal-based production platform: molecular design of a diatom secreting monoclonal antibodies directed against the Marburg virus nucleoprotein. *Microbial cell factories* 16(1), 1-10.

[3.44] Hosseini, S.F., Rezaei, M. and McClements, D.J. (2022) Bioactive functional ingredients from aquatic origin: a review of recent progress in marine-derived nutraceuticals. *Critical Reviews in Food Science and Nutrition* 62(5), 1242-1269.

[3.45] Hussein, H.A. and Abdullah, M.A. (2020) Anticancer compounds derived from marine diatoms. *Marine Drugs* 18(7), 356.

[3.46] Jayakumar, S., Bhuyar, P., Pugazhendhi, A., Rahim, M.H.A., Maniam, G.P. and Govindan, N. (2021) Effects of light intensity and nutrients on the lipid content of marine microalga (diatom) *Amphiprora* sp. for promising biodiesel production. *Science of the Total Environment* 768, 145471.

[3.47] Jónasdóttir, S.H. (2019) Fatty acid profiles and production in marine phytoplankton. *Marine drugs* 17(3), 151.

[3.48] Khan, M.J., Ahirwar, A., Schoefs, B., Pugazhendhi, A., Varjani, S., Rajendran, K., Bhatia, S.K., Saratale, G.D., Saratale, R.G. and Vinayak, V. (2021a) Insights into diatom microalgal farming for treatment of wastewater and pretreatment of algal cells by ultrasonication for value creation. *Environmental research* 201, 111550.

[3.49] Khan, M.J., Bawra, N., Verma, A., Kumar, V., Pugazhendhi, A., Joshi, K.B. and Vinayak, V. (2021b) Cultivation of diatom *Pinnularia saprophila* for lipid production: A comparison of methods for harvesting the lipid from the cells. *Bioresource technology* 319, 124129.

[3.50] Keservani, R.K., Sharma, A.K. and Kesharwani, R.K. (2020) *Nutraceuticals and Dietary Supplements: Applications in Health Improvement and Disease Management*. CRC Press.

[3.51] Khan, M.I., Shin, J.H. and Kim, J.D. (2018) The promising future of microalgae: current status, challenges, and optimization of a sustainable and renewable industry for biofuels, feed, and other products. *Microbial cell factories* 17(1), 1-21.

[3.52] Kumar, V., Kashyap, M., Gautam, S., Shukla, P., Joshi, K.B. and Vinayak, V. (2018) Fast Fourier infrared spectroscopy to characterize the biochemical composition in diatoms. *Journal of biosciences* 43(4), 717-729.

[3.53] Klaitong, P., Fa-Aroonsawat, S. and Chungjatupornchai, W. (2017) Accelerated triacylglycerol production and altered fatty acid composition in oleaginous microalga *Neochloris oleoabundans* by overexpression of diacylglycerol acyltransferase 2. *Microbial Cell Factories* 16(1), 1-10.

[3.54] Kuczynska, P., Jemiola-Rzeminska, M. and Strzalka, K. (2015) Photosynthetic pigments in diatoms. *Marine drugs* 13(9), 5847-5881.

[3.55] Lacour, T., Lariviere, J., Ferland, J., Bruyant, F., Lavaud, J. and Babin, M. (2018) The role of sustained photoprotective non-photochemical quenching in low temperature and high light acclimation in the bloom-forming arctic diatom *Thalassiosira gravida*. *Frontiers in Marine Science* 5, 354.

[3.56] Laosam, P., Panpipat, W., Chaijan, M., Roytrakul, S., Charoenlappanit, S., Panya, A., Phonsatta, N., Cheong, L.-Z. and Yusakul, G. (2022) Molecular Structures and *In Vitro* Bioactivities of Enzymatically Produced Porcine

Placenta Peptides Fractionated by Ultrafiltration. *Food and Bioprocess Technology* 15(3), 669-682.

[3.57] Levitan, O., Dinamarca, J., Zelzion, E., Lun, D.S., Guerra, L.T., Kim, M.K., Kim, J., Van Mooy, B.A., Bhattacharya, D. and Falkowski, P.G. (2015) Remodeling of intermediate metabolism in the diatom *Phaeodactylum tricornutum* under nitrogen stress. *Proceedings of the National Academy of Sciences* 112(2), 412-417.

[3.58] Mai, T.D., Lee-Chang, K.J., Jameson, I.D., Hoang, T., Cai, N.B.A. and Pham, H.Q. (2021) Fatty acid profiles of selected microalgae used as live feeds for shrimp postlarvae in Vietnam. *Aquaculture Journal* 1(1), 26-38.

[3.59] Manikandan, S., Subbaiya, R., Biruntha, M., Krishnan, R.Y., Muthusamy, G. and Karmegam, N. (2022) Recent development patterns, utilization and prospective of biofuel production: Emerging nanotechnological intervention for environmental sustainability–A review. *Fuel* 314, 122757.

[3.60] Marella, T.K., Bhattacharjya, R. and Tiwari, A. (2021) Impact of organic carbon acquisition on growth and functional biomolecule production in diatoms. *Microbial Cell Factories* 20(1), 1-13.

[3.61] Mathushika, J. and Gomes, C. (2021) Development of microalgae-based biofuels as a viable green energy source: challenges and future perspectives. *Biointerface Res Appl Chem* 12, 3849-3882.

[3.62] Matos, Â.P. (2017) The impact of microalgae in food science and technology. *Journal of the American Oil Chemists' Society* 94(11), 1333-1350.

[3.63] McClements, D.J. and Grossmann, L. (2021) A brief review of the science behind the design of healthy and sustainable plant-based foods. *NPJ science of food* 5(1), 1-10.

[3.64] Mourya, M., Khan, M.J., Ahirwar, A., Schoefs, B., Marchand, J., Rai, A., Varjani, S., Rajendran, K., Banu, J.R. and Vinayak, V. (2022) Latest trends and developments in microalgae as potential source for biofuels: The case of diatoms. *Fuel* 314, 122738.

[3.65] Nagao, R., Ueno, Y., Akita, F., Suzuki, T., Dohmae, N., Akimoto, S. and Shen, J.-R. (2019) Biochemical characterization of photosystem I complexes having different subunit compositions of fucoxanthin chlorophyll a/c-binding proteins in the diatom *Chaetoceros gracilis*. *Photosynthesis research* 140(2), 141-149.

[3.66] Nigam, D., Yadav, R. and Tiwari, U. (2018) Omega-3 fatty acids and its role in human health. In: *Functional food and human health*. Springer: 173-198.

[3.67] Olasehinde, T.A., Olaniran, A.O. and Okoh, A.I. (2017) Therapeutic potentials of microalgae in the treatment of Alzheimer's disease. *Molecules* 22(3), 480.

[3.68] Ovide, C., Kiefer-Meyer, M.-C., Bérard, C., Vergne, N., Lecroq, T., Plasson, C., Burel, C., Bernard, S., Driouich, A. and Lerouge, P. (2018) Comparative in depth RNA sequencing of *P. tricornutum*'s morphotypes reveals specific features of the oval morphotype. *Scientific reports* 8(1), 1-19.

[3.69] Pal, P., Chew, K.W., Yen, H.-W., Lim, J.W., Lam, M.K. and Show, P.L. (2019) Cultivation of oily microalgae for the production of third-generation biofuels. *Sustainability* 11(19), 5424.

[3.70] Paliwal, C., Mitra, M., Bhayani, K., Bharadwaj, S.V., Ghosh, T., Dubey, S. and Mishra, S. (2017) Abiotic stresses as tools for metabolites in microalgae. *Bioresource technology* 244, 1216-1226.

[3.71] Polat, S., Trif, M., Rusu, A., Šimat, V., Čagalj, M., Alak, G., Meral, R., Özogul, Y., Polat, A. and Özogul, F. (2021) Recent advances in industrial applications of seaweeds. *Critical Reviews in Food Science and Nutrition*, 1-30.

[3.72] Wang, S., Liu, J., Li, Z., Wei, X., Luo, J., Tang, J., Zhang, Y., Wang, P., Wang, D., Wang, X. and Hu, X., (2024) Preparation and properties of hydrogen storage materials of porous carbon composite diatom skeleton. *International Journal of Hydrogen Energy*, 64, pp.773-780.

[3.73] Putnik, P., Bursać Kovačević, D., Režek Jambrak, A., Barba, F.J., Cravotto, G., Binello, A., Lorenzo, J.M. and Shpigelman, A. (2017) Innovative "green" and novel strategies for the extraction of bioactive added value compounds from citrus wastes—A review. *Molecules* 22(5), 680.

[3.74] Qiao, H., Cong, C., Sun, C., Li, B., Wang, J. and Zhang, L. (2016) Effect of culture conditions on growth, fatty acid composition and DHA/EPA ratio of Phaeodactylum tricornutum. *Aquaculture* 452, 311-317.

[3.75] Rai, A., Khan, M.J., Ahirwar, A., Deka, R., Singh, N., Schoefs, B., Marchand, J., Varjani, S. and Vinayak, V. (2022) Hydrogen economy and storage by nanoporous microalgae diatom: special emphasis on designing photobioreactors. *International Journal of Hydrogen Energy*.

[3.76] Rai, V., Karthikaichamy, A., Das, D., Noronha, S., Wangikar, P.P. and Srivastava, S. (2016) Multi-omics frontiers in algal research: techniques and progress to explore biofuels in the postgenomics world. *OMICS: A Journal of Integrative Biology* 20(7), 387-399.

[3.77] Remmers, I.M., Martens, D.E., Wijffels, R.H. and Lamers, P.P. (2017) Dynamics of triacylglycerol and EPA production in *Phaeodactylum tricornutum* under nitrogen starvation at different light intensities. *PloS one* 12(4), e0175630.

[3.78] Rodolfi, L., Biondi, N., Guccione, A., Bassi, N., D'Ottavio, M., Arganaraz, G. and Tredici, M.R. (2017) Oil and eicosapentaenoic acid production by the diatom Phaeodactylum tricornutum cultivated outdoors in Green Wall Panel (GWP®) reactors. *Biotechnology and bioengineering* 114(10), 2204-2210.

[3.79] Roncero, J.M., Álvarez-Ortí, M., Pardo-Giménez, A., Rabadán, A. and Pardo, J.E. (2020) Review about non-lipid components and minor fat-soluble bioactive compounds of almond kernel. *Foods* 9(11), 1646.

[3.80] Sajjadi, B., Chen, W.-Y., Raman, A.A.A. and Ibrahim, S. (2018) Microalgae lipid and biomass for biofuel production: A comprehensive review on lipid enhancement strategies and their effects on fatty acid composition. *Renewable and Sustainable Energy Reviews* 97, 200-232.

[3.81] Seth, K., Kumar, A., Rastogi, R.P., Meena, M. and Vinayak, V. (2021) Bioprospecting of fucoxanthin from diatoms—Challenges and perspectives. *Algal Research* 60, 102475.

[3.82] Sevilla, J.F., Cerón García, M., Sánchez Mirón, A., Belarbi, E.H., Camacho, F.G. and Grima, E.M. (2004) Pilot-plant-scale outdoor mixotrophic cultures of *Phaeodactylum tricornutum* using glycerol in vertical bubble column and airlift photobioreactors: studies in fed-batch mode. *Biotechnology progress* 20(3), 728-736.

[3.83] Sil, A. and Dasgupta, U. (2021) Nutraceutical and Cosmeceutical Potential of Seaweed Derived Fucoxanthin. *Indian Journal of Pharmaceutical Sciences* 83(4), 618-633.

[3.84] Skjånes, K., Aesoy, R., Herfindal, L. and Skomedal, H. (2021) Bioactive peptides from microalgae: Focus on anti-cancer and immunomodulating activity. *Physiologia Plantarum* 173(2), 612-623.

[3.85] Song, Z., Lye, G.J. and Parker, B.M. (2020) Morphological and biochemical changes in *Phaeodactylum tricornutum* triggered by culture media: Implications for industrial exploitation. *Algal Research* 47, 101822.

[3.86] Su, Y., Lundholm, N., Friis, S.M. and Ellegaard, M. (2015) Implications for photonic applications of diatom growth and frustule nanostructure changes in response to different light wavelengths. *Nano Research* 8(7), 2363-2372.

[3.87] Suganya, T., Varman, M., Masjuki, H. and Renganathan, S. (2016) Macroalgae and microalgae as a potential source for commercial applications along with biofuels production: a biorefinery approach. *Renewable and Sustainable Energy Reviews* 55, 909-941.

[3.88] Taddei, L., Chukhutsina, V.U., Lepetit, B., Stella, G.R., Bassi, R., van Amerongen, H., Bouly, J.-P., Jaubert, M., Finazzi, G. and Falciatore, A. (2018) Dynamic changes between two LHCX-related energy quenching sites control diatom photoacclimation. *Plant Physiology* 177(3), 953-965.

[3.89] Tang, C.-C., Wang, R., Wang, T.-Y., He, Z.-W., Tian, Y. and Wang, X.C. (2021) Characteristic identification of extracellular polymeric substances and sludge flocs affected by microalgae in microalgal-bacteria aggregates treating wastewater. *Journal of Water Process Engineering* 44, 102418.

[3.90] Tréguer, P., Bowler, C., Moriceau, B., Dutkiewicz, S., Gehlen, M., Aumont, O., Bittner, L., Dugdale, R., Finkel, Z. and Iudicone, D. (2018) Influence of diatom diversity on the ocean biological carbon pump. *Nature Geoscience* 11(1), 27-37.

[3.91] Miranda-Arizmendi, V., Fimbres-Olivarria, D., Miranda-Baeza, A., Martinez-Robinson, K., Rascon-Chu, A., De Anda-Flores, Y., Lizardi-Mendoza, J., Mendez-Encinas, M.A., Brown-Bojorquez, F.,

Canett-Romero, R. and Carvajal-Millan, E., (2022) Sulfated polysaccharides from *Chaetoceros muelleri*: macromolecular characteristics and bioactive properties. *Biology*, 11(10), p.1476.
[3.92] Veeramuthu, A. and Ngamcharussrivichai, C. (2020) Potential of microalgal biodiesel: Challenges and applications. *Renewable Energy-Technologies and Applications.*
[3.93] Vernet, M., Geibert, W., Hoppema, M., Brown, P.J., Haas, C., Hellmer, H., Jokat, W., Jullion, L., Mazloff, M. and Bakker, D. (2019) The Weddell Gyre, Southern Ocean: present knowledge and future challenges. *Reviews of Geophysics* 57(3), 623-708.
[3.94] Vinayak, V., Khan, M.J. and Joshi, K.B. (2023) Nanoengineering diatoms in microfluidic lab on chip devices. In: *Applications of Multifunctional Nanomaterials.* Elsevier: pp. 95-124.
[3.95] Vinayak, V., Bhaskar, P., Pandey, L.K. and Khan, M.J. (2022) Diatoms: the Living Jewels and their Biodiversity, Phycosphere and Associated Phenotypic Plasticity: A Lesson to Learn from the Current Pandemic of Coronavirus. In: *Biodiversity in India: Status, Issues and Challenges.* Springer: 385-429.
[3.96] Vinayak, V., Manoylov, K.M., Gateau, H., Blanckaert, V., Hérault, J., Pencréac'h, G., Marchand, J., Gordon, R. and Schoefs, B. (2015) Diatom milking: a review and new approaches. *Marine drugs* 13(5), 2629-2665.
[3.97] Wang, L.-J., Fan, Y., Parsons, R.L., Hu, G.-R., Zhang, P.-Y. and Li, F.-L. (2018) A rapid method for the determination of fucoxanthin in diatom. *Marine drugs* 16(1), 33.
[3.98] Wang, W., Yu, L.-J., Xu, C., Tomizaki, T., Zhao, S., Umena, Y., Chen, X., Qin, X., Xin, Y. and Suga, M. (2019) Structural basis for blue-green light harvesting and energy dissipation in diatoms. *Science* 363(6427), eaav0365.
[3.99] Wang, X.-W., Liang, J.-R., Luo, C.-S., Chen, C.-P. and Gao, Y.-H. (2014) Biomass, total lipid production, and fatty acid composition of the marine diatom *Chaetoceros muelleri* in response to different CO2 levels. *Bioresource technology* 161, 124-130.
[3.100] Wang, X., Liu, Y.-H., Wei, W., Zhou, X., Yuan, W., Balamurugan, S., Hao, T.-B., Yang, W.-D., Liu, J.-S. and Li, H.-Y. (2017) Enrichment of long-chain polyunsaturated fatty acids by coordinated expression of multiple metabolic nodes in the oleaginous microalga *Phaeodactylum tricornutum. Journal of agricultural and food chemistry* 65(35), 7713-7720.
[3.101] Wen, C., Zhang, J., Zhang, H., Duan, Y. and Ma, H. (2020) Plant protein-derived antioxidant peptides: Isolation, identification, mechanism of action and application in food systems: A review. *Trends in Food Science & Technology* 105, 308-322.
[3.102] Uthappa, U.T., Brahmkhatri, V., Sriram, G., Jung, H.Y., Yu, J., Kurkuri, N., Aminabhavi, T.M., Altalhi, T., Neelgund, G.M. and Kurkuri, M.D., (2018)

Nature engineered diatom biosilica as drug delivery systems. *Journal of Controlled Release*, 281, pp.70-83.

[3.103] Wu, S., Huang, A., Zhang, B., Huan, L., Zhao, P., Lin, A. and Wang, G. (2015) Enzyme activity highlights the importance of the oxidative pentose phosphate pathway in lipid accumulation and growth of *Phaeodactylum tricornutum* under CO2 concentration. *Biotechnology for biofuels* 8(1), 1-11.

[3.104] Yadav, G., Mishra, A., Ghosh, P., Sindhu, R., Vinayak, V. and Pugazhendhi, A. (2021) Technical, economic and environmental feasibility of resource recovery technologies from wastewater. *Science of The Total Environment* 796, 149022.

[3.105] Yi, Z. (2018) Biotechnological approaches to enhance fucoxanthin production in a model diatom *Phaeodactylum tricornutum.*

[3.106] Yi, Z., Xu, M., Magnusdottir, M., Zhang, Y., Brynjolfsson, S. and Fu, W. (2015) Photo-oxidative stress-driven mutagenesis and adaptive evolution on the marine diatom *Phaeodactylum tricornutum* for enhanced carotenoid accumulation. *Marine drugs* 13(10), 6138-6151.

[3.107] Yodsuwan, N., Sawayama, S. and Sirisansaneeyakul, S. (2017) Effect of nitrogen concentration on growth, lipid production and fatty acid profiles of the marine diatom *Phaeodactylum tricornutum. Agriculture and Natural Resources* 51(3), 190-197.

[3.108] Zarekarizi, A., Hoffmann, L. and Burritt, D. (2019) Approaches for the sustainable production of fucoxanthin, a xanthophyll with potential health benefits. *Journal of Applied Phycology* 31(1), 281-299.

[3.109] Rai, A., Sirotiya, V., Mourya, M., Khan, M. J., Ahirwar, A., Sharma, A. K., ... & Vinayak, V. (2022). Sustainable treatment of dye wastewater by recycling microalgal and diatom biogenic materials: Biorefinery perspectives. *Chemosphere*, 305, 135371.

[3.110] Lauritano, C., Martín, J., de la Cruz, M., Reyes, F., Romano, G. and Ianora, A., (2018) First identification of marine diatoms with anti-tuberculosis activity. *Scientific Reports*, 8(1), p.2284.

[3.111] Rojas, V., Rivas, L., Cárdenas, C. and Guzmán, F., 2020. Cyanobacteria and eukaryotic microalgae as emerging sources of antibacterial peptides. *Molecules*, 25(24), p.5804.

[3.112] Størseth, T.R., Hansen, K., Reitan, K.I. and Skjermo, J., (2005) Structural characterization of β-D-(1→ 3)-glucans from different growth phases of the marine diatoms *Chaetoceros müller*i and *Thalassiosira weissflogii. Carbohydrate Research*, 340(6), pp.1159-1164.

[3.113] Khan, M.J., Singh, R., Shewani, K., Shukla, P., Bhaskar, P.V., Joshi, K.B. and Vinayak, V., (2020). Exopolysaccharides directed embellishment of diatoms triggered on plastics and other marine litter. *Scientific Reports*, 10(1), p.18448.

Part II
HIGH-VALUE PRODUCTS

4

Diatoms: A Natural Resource of High-Valued Products and their Future Prospective

Khushboo Kesharwani[1‡], Shruti Sharma[1‡], Aanand Kautu[1‡], Satyendra Kumar Tripathi[1], Vikas Kumar[2,3*] and Khashti Ballabh Joshi[1†]

[1]*Department of Chemistry, School of Chemical Science and Technology, Dr. Harisingh Gour Vishwavidyalaya (A Central University), Madhya Pradesh, India*
[2]*Department of Chemistry, Government College, Khimlasa, Sagar, Madhya Pradesh, India*
[3]*Department of Organic Chemistry, Indian Institute of Science, Bangalore, India*

Abstract

Diatoms are photosynthetic unicellular microalgae, known as nature's bio-nanoreactors, that produce high-value products such as lipids, sterols, isoprenoids, and toxins with use in apoptotic, fertility-controlling, and cancer drugs. From species to genera, they contain several biosynthetic metabolites with varied levels of polyunsaturated fatty acids, with applications in various pharmacological, food, and biofuel drug industries. The structural and spectroscopic profile of kerogen reveals that the organic compounds produced by the diatoms are more closely analogous to petroleum. It is known that crude oil or petroleum is derived from natural sources, such as decomposed organic matter produced by various organisms, and gets deposited as sediments in sedimentary basins from the geological past. Diatoms possess a uniform array of small pores on their surface made up of silica, known as frustules, that are applied for the oozing of diatom oil. Scientists are focusing on regenerative approaches for the production and clean harvesting of diatom oil to be used as an additive for natural oil under the scientific brand term Diafuel™. It has attracted huge scientific interest toward understanding the

**Corresponding author*: vikasnamdeo1389@gmail.com
†*Corresponding author*: kbjoshi77@gmail.com
‡These authors contributed equally.

Vandana Vinayak and Richard Gordon (eds.) Diatom Cultivation for Biofuel, Food and High-Value Products, (81–114) © 2025 Scrivener Publishing LLC

machinery behind the production of such high-value-added products by diatoms and developing techniques to mimic nature's nanomachines.

Keywords: Diatoms, bio-nanoreactors, diafuel, crude oil, petroleum, nanomachines

4.1 Introduction

The post-COVID period has led to the depletion of world oil reserves. This, along with the continuous rise and fall of oil prices, together with increased greenhouse gases and the predicted peril of climate change, have generated renewed interest in using biomass-derived liquid fuels as an alternative and renewable feedstock for energy production [4.45] [4.34] [4.24] [4.112]. Biofuels provide potential prospects to encounter the global energy demand, satisfying carbon-neutral solutions, and assisting carbon dioxide (CO_2) sequestration from the atmosphere [4.124]. Oceans cover 70% of the Earth's surface. The photosynthetic organisms living within the photic zone of the ocean are accountable for around one-half of global productivity, although they constitute only a small share of the Earth's green biomass. The most prosperous organisms are believed to be photosynthetic prokaryotes and a class of eukaryotic unicellular algae known as diatoms [4.32].

Diatoms are unicellular microscopic eukaryotic algae which mainly belong to the Bacillariophyceae family, involving 15,000 or more species living either in sea water, in fresh water, in moist habitats and soil or brackish water [4.46]. The class of diatoms is mainly represented by two hundred different genera and six hundred species; their thallus is unicellular, and, somehow, they possess bilateral or radial symmetry [4.98]. The cell wall structure is generally called a unique frustule which is made up of two halves that enclose the cytoplasmic content, which is extremely resistant to putrefaction, highly complex in shape as well as silicified skeleton. They require nitrogen, phosphate, and silica to build their cell wall, which itself has various applications related to several fields. In this silicified skeleton, a rough surface is found as the outer wall, which is transparent and is termed "transparent opaline silica," allowing the entrance of light as well as playing a major role in the diffusion and excretion of various materials.

With the help of chlorophyll and some other key nutrients, they collect energy from sunlight and convert it into various high-value-added chemicals via photosynthesis [4.152] that are useful for fuels, foods, and bio-chemical industries. Diatoms generally produce oil when exposed to sunlight as they flow in a very nutrient-rich environment (water solutions with several metals or minerals responsible for their growth) [4.149] and their culture involves various ambient conditions. Their usage is cost-effective because it involves neither costly culture media nor expensive instruments. They consist of a high percentage of fatty acids and carotenoids, and are responsible for 41–50% of CO_2 fixation in oceans [4.102]. But due to their thick silica wall, it becomes difficult to extract these useful chemicals from them. Therefore, treatment of diatoms with different chemical entities, such as a peptide, metal nanoparticles, and various metal ions, or applying physical stress, may provide helping hands to obtain all possible value-added materials from diatoms in a nondestructive manner [4.44]. Physical stress, such as applying a low-intensity electrical field, mechanical stress, or piezoelectric force, may cause a light fracture in the cell wall of diatoms instead of complete cell destruction, by which oil can be collected and fractured cells can recover when placed into nutrient-rich culture media [4.99].

Various species of diatoms possess different levels of organic content in them which can be distinguished by their FTIR spectra. This information is useful for carrying out the removal of these organics from them by depositing some organic moieties like peptides, and inorganic material such as metal ions and metal nanoparticles on their surface [4.62]. This deposition over the diatom surface takes place by physical adsorption via non-covalent interactions between the inorganic and organic moieties, which may enhance diatom oil production [4.117]. Owing to many secondary metabolites which are pertinent to health, chemical, and pharmaceutical industries obtained from diatoms, they cannot be considered as ordinary algae products, but are known as high-valued molecules (HVMs) [4.120]. It is also pertinent to mention here that these compounds are the vital source of many prospective applications in cosmetics, which utilize these intracellularly synthesized fatty acids and amino acids by diatoms, in pharmaceuticals, and as power-harvesting devices [4.61]. Apart from the production of HVMs, diatoms have attracted huge scientific attention due to their hierarchical and symmetrical pore structures that are used to test the resolution of an optical microscope [4.79] [4.151]. Therefore, 3D frustules of diatoms provide a big platform for a scientist to elaborate the application of diatoms on 3D engineered nanostructured materials [4.61].

As discussed above, diatoms produce oil as HVMs; therefore, it is important to discuss the significance of diatom oil and its contribution to crude oil production [4.57]. Crude oil or petroleum is a complex mixture of petroleum originating from the deposition of natural sources as secondary metabolites which provide high-value organic matter along with the sediments in sedimentary basins from the geological past [4.118]. There are various kinds of living organisms that contribute to the production of hydrocarbon as a normal part of their existence; lipids produce petroleum, and algae-like diatoms are one of the principal contributors. Modern geologists have not accepted the abiogenic origin of petroleum because most living organisms are considered to yield a large percentage of organic compounds more closely analogous to petroleum that is identified through the chemical structure of kerogen [4.118]. Brongersma-Sanders highlighted the areas of upwelled nutrient-rich ocean waters, in which there is a notable development of abundant plankton, so it seems both geological and geochemical aquatic plant and animal life are accountable as source material for the bulk of our hydrocarbon reserves [4.14]. According to recent genetic and sedimentary evidence, diatoms originated in the early Jurassic period, 185 million years ago [4.14]. Medlin *et al.* suggested that after the end-Permian extinction, many marine niches were opened, and therefore the origin of these diatoms might be related to it [4.58] [4.89].

Diatoms comprise the bulk of the ocean phytoplankton, so attributing crude oil to diatoms is the pivotal line of reasoning of geologists. This view should be tempered by likening the geological record of diatoms to the estimated age of the oil deposits. There are two opinions: according to the first one, the preceded diatoms did not originate from them, and crude oil might have been produced by prehistoric organisms. The fatty acid profile matches that of diatoms [4.89] and the presence of 24-norcholestane biomarkers confirmed the diatoms as the source of crude oil [4.143].

Fluorescence spectroscopic studies and composition profile of crude oil confirmed the presence of approximate 60–70% saturated fatty acids, which get converted into condensed lipids during early diagenesis [4.108] [4.123]. Many researchers have suggested diatoms as major contributors to crude oil deposits [4.4] [4.23] [4.29] [4.31] [4.69] [4.134]; however, the overall contribution of diatoms has not been estimated in worldwide petroleum reserves. We must make a total guesstimate of worldwide petroleum deposits continuing naturally contributed by diatoms because, as per the reports, approximately half of crude oil has been utilized.

The production of oil droplets by diatoms is linked to the maintenance of dynamic availability of nutrients and photon flux density of diatom territories, which is the most required capability to uphold metabolic homeostasis. To fulfill this requirement, storage lipids that are confiscated lipid droplets (LDs) get turned into lipid droplets. In these lipid droplets, neutral lipid core is surrounded by a polar lipid monolayer and these LDs behave as a carbon and electron sink and work as a shield to maintain intracellular environment from the accumulation of hydrophobic compounds.

Therefore, this machinery works on the interconnection between other intracellular systems which include photosynthesis and autophagy [4.70]. It was proposed that the function of oil droplets in diatoms is to provide the buoyancy and counter the weight of the dense silica shell (for Si/C mole ratios up to 0.81144); the presence of oil in cells may also be responsible for the reduction in specific gravity of diatoms. Oil accumulation in diatoms has been accountable for providing a food reserve and helping in keeping them afloat while waiting for the return of conditions favorable for multiplication [4.94] [4.127].

4.1.1 Diatom Morphology

Diatoms possess different types of morphology depending on their size and shape. They are generally of size 2–200 μm. They can be differentiated by the shape of their frustules that are of two types: centric and pennate. Pennate diatoms possess bilateral symmetry in which each of their valves have openings with the slits along the raphes. Their shells are elongated parallel to these raphes that generate cell movement through streams along the raphes. Central diatoms possess upper and lower valves as epitheca and hypotheca and are radially symmetric. Epitheca and hypotheca consists of a valve and girdle band which can easily slide underneath each other and expand to increase cell content over the diatom's progression. Centric diatoms demonstrate a variety of shapes and sizes, and presence of spines, which depends on the axis from which the shell extends.

4.1.2 General Features of Oil Bodies in Diatoms

A) Size, number, and morphology of diatom oil bodies

It is known that oil bodies are an organelle in which neutral lipids are accumulated in response to nitrogen, phosphorous, and silicon nutrient deprivation [4.70]. The size and number of oil bodies were measured by the staining method with lipophilic fluorescence dyes such as Nile Red or BODIPY 505/515, although these features vary according to the lipid accumulation phase. From these techniques, the amount of oil to be accumulated and already accumulated oil can be estimated. The oleaginous diatom *Fistulifera solaris* possesses two oil bodies at the polar regions of the cell, which suggests that there might be diverse mechanisms of oil localizations among diatoms, which are indeed species dependent. In general, oil bodies are found spherical in shape, however, their shape may vary according to the shape of cell and other cell organelles that restrict the confinement of

Table 4.1 General features of oil bodies in diatoms.

Order	Species	Size of oil bodies (μm)	Number of oil bodies
Pennales	*Phaeodactylum tricornutum*	0.4–2.6	1–5
	Fistulifera solaris	1.3–5.6	2
	Pseudo-nitzschia multiseries	0.7–1.4	6–15
	Nitzschia closterium	0.8–1.3	3–4
Centrales	*Chaetoceros calcitrans*	1.0–2.0	3–4
	Thalassiosira pseudonana	0.3–1.5	3–5
	Cyclotella cryptica	1.1–3.1	1–10
	Skeletonema costatum	1.1–2.5	3–5

oil droplets in various shapes and sizes (Table 4.1) [4.82] [4.93] [4.103] [4.128] [4.138] [4.147].

B) Productivity and lipid content in diatoms

Lipid productivity of microalga cultured in large-scale bioreactors is directly correlated with the size and number of oil bodies at the microscale phenotypic level [4.134]. The selection of auspicious candidates to produce biofuel plays a key role in increasing the efficiency of lipid productivity. Biomass and lipid productivity may vary in the same species depending on the cultural conditions.

Among the reported diatom species, *F. solaris* possesses the highest rate of lipid production in which biomass productivity has been found as 299 mg $l^{-1}d^{-1}$, lipid productivity has been observed as 160 mg $l^{-1}d^{-1}$ and lipid content was observed as 53.5% [4.75]. Another species of diatom *Amphora* were also reported for their high lipid productivity, which is 345 mg $l^{-1}d^{-1}$, 63 mg $l^{-1}d^{-1}$, and 71 mg $l^{-1}d^{-1}$; however, no further investigation was carried out to confirm the lipid productivity by these diatoms [4.43] [4.115]. Lipid productivity of another diatom, *Chaetoceros gracilis,* was observed as 2210 mg $l^{-1}d^{-1}$ [4.31], but recently the total TAG productivity under optimal conditions was reported as 27.2 mg $l^{-1}d^{-1}$ [4.133]. This difference could be due to the cultural changes, the use of different strains in each study, and the biomass productivity of *Amphora* and *C. gracilis* not being reported in the same unit (mg $l^{-1}d^{-1}$).

C) Fatty acid composition

Fatty acid composition describes the lipid quantity, which is highly affected by the physiological and culture conditions [4.50]. There are several diatom species, including *P. tricornutum, Thalassiosira pseudonana* [4.121] [4.133] [4.150], *F. solaris* [4.60], *Nistzchia closterium* [4.126] and *C. gracilis* [4.133], in which between the different oil accumulation phases variations of fatty acid composition were investigated. These studies delivered perceptions of lipid metabolism during adaptation to environmental changes. The most common fatty acids are C14:0, C16:0, C16:1, and polyunsaturated fatty acids, eicosapentaenoic acid. They are much different than the fatty acids that are found in green algae which contain a significant amount of C18 fatty acids [4.50]. The composition of fatty acids determines the quality of biodiesel which estimates the potential of microalgae for biodiesel production [4.60]. Figure 4.1 shows the diatom lipid classification and chemical composition of fatty acids [4.50] [4.60] [4.110].

American Society for Testing and Materials ASTM D6751 and EN 14214 in Europe, established a standard for the determination of the quality of biodiesel, which also recognized the Japanese Industrial Standard K 2390:2008 and the Automotive fuels – Fatty acid methyl ester (FAME) as blend stock specification. As per these standards, the PUFA content should be low in good quality diatom oil, therefore, C16:0 and C16:1 is preferentially used for biofuel applications.

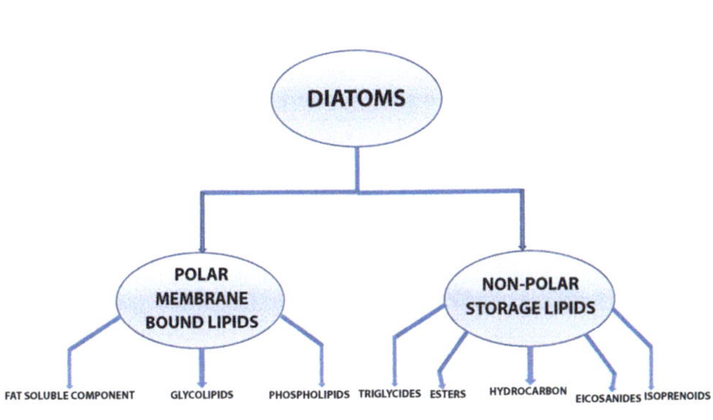

Figure 4.1 Diatom lipid classification [4.50] [4.60] [4.110].

Globally, diatoms play a major role in the ecological system as they are the producers within the food chain. During their rapid growth, huge amounts of nutrients and trace elements are absorbed from the surface of the water layer, especially silicon to form their outer shells or skeleton, and many other metals, such as zinc, are used physiologically. Generally, diatoms are very much sensitive to various environmental variables, which include salinity, nutrient contents, and pH factors. The diatoms' frustules, which occur naturally, are a source of nanomaterials. Several new facets and various potential or special applications of diatoms have been reported, here we summarize the facets of biological and environmental applications

Table 4.2 Applications of diatoms in different fields [4.16] [4.25] [4.26] [4.67] [4.76] [4.100] [4.101].

Uses	Properties
Nanotechnology	➢ Reproducibility of the three-dimensional structures ➢ Possibility of genetic engineering and low cost of production ➢ Intricate pore sizes which can be modified ➢ Ability to self-replicate
Biosensing	➢ Micron-sized and homogeneously spaced with striae ➢ Optical properties ➢ Prospect of cheaply creating thousands of channels on a single silicon chip ➢ Low-cost and naturally available material
Immunoisolation, immunodiagnostics, and immunosensors	➢ High sensitivity and option to chemically modify the surface to attach bioactive molecules ➢ Filtration and encapsulation properties of diatom frustules. Evades complements of the immune system
Filtration and water purification	➢ Filters microorganisms ➢ Homogeneous permeability and fixed pore size ➢ Cost-effective ➢ U.S. Environmental Protection Agency (EPA) approved
Drug delivery	➢ Uniform nanoscale pore structure ➢ Chemically inert and biocompatible ➢ Sustained release of drugs ➢ Nontoxic

of diatoms under one umbrella, which include biomineralization, biosilicification, drug delivery vehicles, optical and hybrid immune-biosensors, aquaculture feeds, immunodiagnostics, metabolites, and the use of diatoms as nanomaterials and biofuel (Table 4.2) [4.110].

4.2 Biosilicification and Silicification as a Crucial Application in Bone Repair

The outer cell wall or rough skeleton of the diatoms is made up of amorphous hydrated silica $SiO_2.nH_2O$ [4.1]. It is formed by the immobilization of silicic acid in the process generally known as biosilicification. *In vivo*, diatoms can be prepared at ambient conditions such as temperature, pressure, nutrients, etc. Biosilicification takes place at 7.0 pH. Biosilica of diatoms is used as an absorptive carrier in many more fields such as the chemical and pharmaceutical industries. The structure of biosilica diatom is quite like natural bone, which is formed by biomineralization. Due to aging and poor physical activity, bone disorders are very common in the population nowadays. To treat bone defects, a bone graft is prominently used as a cure. The diatom frustules contain chemical compositions strongly associated with each other made up of both inorganic and organic components. The cell wall of these diatoms contains proteins and long-chain polyamines (LCPA) which strengthen the silicon skeleton [4.59] [4.66].

Biosilica of diatoms has a structure that looks like hierarchical pore-like bone formed by biomineralization. The connective tissues as tendons and bone have the presence of silicon, therefore, this element itself plays an important role in bone formation. Since the biosilica of diatoms can also be used for bone formation as they are known to be silicon donors [4.68]. To suppress osteoclast activity, various particles are used as supplements. However, their toxicity effect must be controlled by pre-processing of raw material, dose, size, shape, etc. The mechanism of silica condensation in diatoms has been found by the zwitterion interaction, self-assembly via electrostatic interaction of the zwitterionic molecules, silaffins with the presence of positive charges (amines) and negative charges (phosphates). Moreover, the sea depth and temperature could influence frustule morphology by several factors such as the presence of gaseous nitrogen and water pressure during morphogenesis [4.39] [4.142].

4.3 Effect of Metals as a Therapeutic Application on Diatom Frustules

Diatoms are typically primary producers in an ecosystem and are photoautotrophic. They require various trace elements, minerals, and vitamins from the environment for metabolic functions as well as a growth medium. These trace elements require the production of the enzyme; for example, zinc is used for various physiological roles and antibacterial properties while magnesium is found in chlorophyll as a central metal cofactor [4.11] [4.33]. In addition to zinc and magnesium, some of the very crucial elements used for protein production in the electron transport system are nitrogen and phosphorous. Ecological study also shows that diatoms can remove heavy metals (such as cadmium and lead) and other contaminants from the water environment [4.40] [4.41]. In the 1970s, it was observed that various materials other than silica could be incorporated into the siliceous skeleton of diatoms. Martin and Knauer performed some studies that investigated the presence of trace elements in both the organic (cell) and inorganic silica frustule [4.87]. Thus, the result shows that aluminum, iron, and titanium were mainly present in the organic frustule of diatoms. After several of these trials found that diatom frustules can be incorporated by several metal ions, a whole new area for possibilities with sources of applications was unlocked successfully. Therefore, the study of chemistry of diatom frustules has now shown that they have enough potential to form many functionalized microcontainers and biomaterials which can be tailored to an area for specific target applications. However, we are still unaware of any studies dealing with silica cell membrane of diatoms with metal ions, and as of yet there is no conclusion regarding the study dealing with the exact mechanism(s) for how other materials (such as metals) are bought into silica frustules that the *in-vivo* feeding of diatoms requires; in diatoms, growth media metal salt can be added in the form of hydroxides but it alters the growth media pH, which can cause cell death [4.91]. Ideally, nontoxic salts of the desired concentration and high affinity can be added. There must be balancing feasibility in biocompatibility, solubility, utilization, and toxicity. The salt must be dissolved in diatoms growth media so that it can be absorbed by the diatom's cell. This must be done very carefully while keeping two key concerns in mind: solubility and toxicity.

4.4 Successful Deposition of Metals on Diatom Frustules

4.4.1 Germanium

The first study investigates the use of germanium metal in the form of germanium dioxide in an attempt to inhibit the growth of diatoms that infect the other cultures based on the theory that it is quite like a competitive inhibitor and therefore disrupts the frustule formation of the diatoms. The first case reported which incorporated it into their frustules shows that germanium, in the form of transistors, isotopically labeled germanic acid, $^{68}Ge(OH)_4$, was able to be incorporated by the diatoms successfully [4.7]. The incorporation of germanium may reduce the pore size, without affecting the morphology of the frustule [4.83].

4.4.2 Titanium

This substitution or incorporation of titanium has had the most interest as this metal is mostly substituted for semiconductor TiO_2 nanostructures and biological study. Metabolic insertion of titanium at low enough concentration in diatom frustules is found until all of the silica gets consumed by the diatom's cells, the metal itself limiting the species in the growth media or cultures. Titanium nanoparticles are used simply to understand the photocatalytic activity better and to understand the biology of diatoms [4.19] [4.48] [4.64] [4.65] [4.71] [4.72].

4.4.3 Calcium

Calcium-doped biosilica supports, by the *in vivo* feeding of algae with $CaCl_2$ salts, have potential application in bone regeneration. This is done by simply adding the metal in the form of soluble salt in a growth medium of diatoms, after which the metal ion is taken up by normal metabolic pathways by the diatom cells while the metal is incorporated in both the organic cell and inorganic frustule. This method of incorporation is very simple, easy, environmentally and cost-effective too. This method can be confirmed by FTIR (Fourier-transform infrared spectroscopy) techniques. The natural form of silicon in diatoms is in the form of silica (SiO_2); the other metal oxides may be incorporated into the diatom's frustules by the same mechanisms, which is known as biosilicification. But as calcium doesn't have the chemical composition of binary oxides, it must follow a different pathway, as they play a major role that leads to more potential applications in biomedical engineering, particularly bone tissue repair [4.77].

4.4.4 Strontium

The mechanism by which strontium affects bone is the same as calcium metal; its interaction with bone shows similarities with calcium as strontium is allowed to have interactions with the osteoblast-mediated process. Proteins act as a catalyst to fabricate complex porous structures. The most specific function of strontium is calcium sensing in osteoblast, which leads to a decrease in bone resorption [4.22]. So, the silica frustule is very beneficial even in its natural state or also as therapeutic metal therapeutic (magnesium and strontium), which could be a huge leap forward for its use as a source for bone medicine, BTE (bone tissue engineering), etc., to overcome the problems faced by populations. Finally, diatoms can easily be processed with several metal frustules as they don't require very hazardous chemicals or energy [4.15].

4.5 Biomedical and Environmental Applications

In the field of nanomedicine, diatoms can be employed as electric biosensors and microcarriers for drug delivery systems, etc.; however, their major shortcomings are toxicity, biodegradability, and impact on the environment [4.83]. In biotechnological tools, there are some effective advances that are used to characterize the diatom frustules for optoelectronics fabrication. The incorporation or uptake of various elements into the frustule of diatoms like *Stephanodiscus hantzschii*, *Thalassiosira pseudonana*, etc., can cause a change in the pore size, shape, and other characteristics, which have been studied for several functions for photonic device applications to help in biosensing. Manufacturing of these elements on incorporation with diatom frustules can lead to several nanotechnological applications as well as overall durability.

4.5.1 Biomedical Applications

Diatomite or diatomaceous earth (DE) is generated from fossil diatoms and plays a very crucial role as targeted drug delivery cells [4.125]. They have recently been proven successful yet their production requires toxic metals and is quite costly. Drug release has two steps: the first is rapid release (over 6 h is attributed to the surface deposited drug), and the second is slow and sustained release with zero order kinetics. These functionalized micro-algae drug-loaded nanomaterials specifically target and kill tumor cells and deliver targeted poorly water-soluble chemotherapeutic drugs to cancer cells. *Thalassiosira pseudonana* is genetically engineered to display an IgG-binding domain of protein G on the biosilica surface, enabling the attachment of cell-targeting antibodies [4.78] [4.97]. Significant drawbacks are that it requires time-consuming, toxic chemicals (hydrofluoric acid and silanes) and is costly [4.116].

In terms of possible diatom-loaded drug delivery, oral administration is the most common and convenient way for delivery so that the drug could be easily degraded through digestion in the GI tract. Diatomite was recognized as safe by the FDA (U.S. Food and Drug Administration). When compared to the other routes of delivery, injections, such as intravenous (IV) or intramuscular (IM) administrations, it offers the advantage of a targeted effect that can be safely administrated [4.6] [4.23]. Diatoms are also used as food additives which can be easily administrated through the oral cavity route: their size can be decreased in a controllable manner. The poor degradability of mesoporous silica is one of the drawbacks in the use of raw diatoms. Retaining good biodegradability, porosity, and structural morphology (during conversion of diatom silica to pure silica) is necessary for drug delivery and medical applications.

4.5.2 Environmental Applications

Alarming water pollution is toxic and leads to a sharp decline in the species diversity of the ecosystem. Diatoms are golden brown microalga, exist in different shapes and sizes of 2–200 μm, and are responsible for fixing 25% of the atmospheric CO_2 [4.113]. Diatoms are arranged in beautiful ring-walled nanostructures [4.3] [4.5] [4.8]. Besides being found in different water bodies, they are unicellular eukaryotic microalgae that convert atmospheric CO_2 to carbohydrates, thus forming various biomolecules such as proteins, lipids, fatty acids, carotenoids, etc. [4.141]. They are not only a high source of naturally occurring valued products with several nanotechnological advantages but can also indicate water quality. Almost 70% of the ecosystem contains water. Thus, they play a very crucial role as an indicator for different kinds of water pollution of organic pollutants, metals, dyes, pesticides, hydrocarbons, drugs, etc., in aquatic ecosystems

[4.104] [4.153]. Since the main reason behind groundwater contamination is industrialization and urbanization, their disposal results in the release of nutrients. So, there are many technologies for treating wastewater and recovering sources from it [4.73]. Furthermore, heavy metals, such as Cu, Cd, Tl, Hg, As, Zn, Ni, Cr, and Pb, in wastewater are found in traces as products not only harmful to humans but also to other biota [4.96]. Heavy metal resistance is shown by some species like *Cyclotella cryptica, Skeletonema, costatum, Cylindrotheca fusiformis, Phaeodactylum tricornutum*, and *Thalassiosira pseudonana*, because of biotoxicity which is the rich source of natural microporous silica [4.129] [4.130]. It is observed that diatoms have more sulfur-containing groups in the cell wall, higher specific surface area and cadmium binding capacity in cells. Diatom nanoshells in conjugation with several ligands have been shown to absorb a greater amount of toxic metals in salinity as the marine diatoms have a stronger affinity for the adsorption of Cd^{2+} ions [4.109]. More Si is found as polysilicon acid when the salinity decreases; therefore, diatom frustules are considered as a key factor for change in silicon content for absorbing the Cd on its surface [4.56] [4.145]. Each cell wall (frustule) of diatoms contains tiny holes of silica from which they absorb nutrients and get rid of waste. Water filtration and purification by diatoms (microalga) is a remediation that is also approved by the U.S. Environmental Protection Agency (EPA).

4.6 Deposition of Different Metal Nanoparticles for Various Applications

The photoautotrophic, eukaryotic, unicellular organisms known as diatoms have gained the attention of researchers in the biosynthesis of well-defined, advanced, and eco-friendly metallic nanoparticles in the area of nanotechnology and its application. The nanoparticles provide massive advantages in antibacterial activity, filtering waste and chemical compounds; in bone repair, catalyst, drug targeting, delivery, and release; and with their additional potential to combine diagnosis and therapy, have emerged as one of the major tools in nanomedicine. To modify the chemical composition of frustules, various approaches are used such as the sol-gel process, chemical bath deposition, and atomic layer deposition. Since cytotoxicity and degradation of nanoparticles remain a major problem, improvement in biocompatibility obviously is a main concern of future research. Nanoparticle-mediated drugs will have a great potential effect on society since it will drastically improve patient's healthcare by reducing the severity of disease and detecting pathogens earlier, resulting in improved clinical outcomes for the patient. Nanoparticles can be easily administrated by oral, nasal, parenteral, and ocular routes [4.55]. Nanoparticles for directing the drugs to specific target cells and specific ligands can be attached to their surface. Nanoparticles improve the stability and therapeutics index while reducing toxic effects on bioavailability. Several types of nanoparticles are used for drug delivery systems (DDS) such as:

- Metal-based nanoparticles – Au, Ag, Cd-Se
- Lipid-based nanoparticles – Liposomes and neosome
- Polymer-based nanoparticles – Dendrimer, micelle.
- Biological nanoparticles – BSA based NPs.

4.6.1 Iron Oxide Nanoparticles

In the United States, cancer remains one of the leading causes of death. The primary challenge in cancer therapy is especially to deliver therapeutics to the tumor's site without affecting the normal tissues [4.12] [4.119]. The advanced advantages of nanotechnology are a great opportunity as per the current limitations of chemotherapy. One promising approach in this field is magnetic delivery. By applying the external magnetic field, drugs can be targeted to tumors by tagging drug carriers in nanoparticles (NPs). In diatoms, small molecules can be co-loaded as iron oxide nanoparticles (IONP). Many efforts have been given to tumor-targeting motif particles on diatoms' surfaces. Iron oxide nanoparticles can be used *in vivo* as a carrier in magnetic drug delivery for cancer diagnosis and therapy [4.30] [4.80]. For cocktail cancer therapy there is also a possibility to load a combination of therapeutics. Through physical adsorption, small molecules were loaded onto diatoms while with their large pore size, they can hold a vast range of functionalities and a rapid release of drugs [4.122] [4.148]. In future studies, it is important to enrich small-size diatoms to afford better pharmacokinetic properties. Hence, to improve drug efficiency as a therapeutic application there is an urgent need to focus on diatoms as microcontainers.

4.6.2 Silver Nanoparticles

Among metal particles, silver nanoparticles (AgNPs) have increasingly become an interesting area of research for a number of reasons, including their bactericidal effects, catalysts, nanomedicine, biological labeling, sensing [4.88], solar cell surface coatings, electronics, staining pigments for glasses and ceramics, and antimicrobial agents [4.146]. In biomedical applications, AgNPs are mostly preferred as antimicrobial agents due to their low toxicity against antibiotic-resistant bacteria and effectiveness in very low concentrations. For the biosynthesis of AgNPs many biological resources such as actinomycetes, fungus, bacteria, plant extracts, and photosynthetic algae are exploited [4.84–4.86] [4.107]. Diatom frustules decorated with AgNPs between 10 and 20 nm in diameter, are accessible at room temperature using glucose as a reducing agent under basic conditions and without the need for linking agents. Diatom culture *Amphora* sp. (IMMTCC-46) is denoted as *Amphora*-46. *Amphora* sp. is reported for the first time for a light-dependent biosynthesis of SNPs using its aqueous extract and $AgNO_3$ [4.20]. The synthesized SNPs are spherical with an average particle size of 20–25 nm. Fucoxanthin, a light-sensitive photosynthetic pigment, was found to be responsible for the reduction of Ag^+ ion to Ag^0. Since this biologically synthesized metal nanoparticle shows bactericidal activity against gram (+) and gram (-) bacteria, it can have a lot of applications in the biomedical field. *Amphora* sp. shows the feasibility of biotechnology's important future in green technology [4.54].

Several physio-chemical methods have been proposed for the synthesis of AgNP, such as sol-gel, chemical, photochemical reduction, electrochemical method, laser ablation, ion sputtering, etc., but they all suffer from several drawbacks such as requiring toxic and expensive chemicals, having a complex synthetic route, and releasing byproducts hazardous to health, thus leading to high risk. In contrast, biological synthesis is in high demand since they are nontoxic, eco-friendly, biocompatible, and commercially feasible [4.9] [4.27]. Currently, naturally derived agents like plant extracts, algae, fungi, bacteria, yeast, enzymes, biopolymers, etc., are widely used for the synthesis of nanoparticles owing to their excellent reducing and capping behavior. Diatom algae are the most enduring system for the green synthesis of nanoparticles, be it metallic nanoparticles, nanostructured polymers, or a diverse array of new nanomaterials [4.90]. Diatoms are a reservoir of metabolites with diverse applications and AgNP from diatoms holds immense therapeutic potential against pathogenic microbes owing to their silica frustules.

Frustules of diatoms decorated with silver nanoparticles between 10–20 nm, are reachable at room temperature under basic conditions in which glucose is used as a reducing agent and without the need for a linking agent. The process is sensitive to both pH and the time elapsed after plasma cleaning prior to decoration [4.137]. Nanosized particles have unique physical and chemical properties by which they are becoming a progressively important material for device technology with applications in catalysis, biological labeling, nanomedicine, optoelectronics, data storage, composite materials, and chemical sensing [4.81]. There is a need for novel strategies to construct new-generation nanoscale devices, and to organize nanoparticles into controlled structures that are motivating us to use new approaches, including using organic monolayers, co-colloidal suspensions, chemical crosslinking crystallization, and using oligonucleotides to self-assemble inorganic nanoparticles into ordered arrays. All of these require the use of a linking agent or stabilizer, which can affect the catalytic and antimicrobial activity, for example, as well as the toxicology of the relative materials [4.88] [4.114].

Silver nanoparticles (AgNPs) are interesting for several reasons, including their bactericidal effects and antimicrobial activity, and have been used as a preservative, mixed into plastics, and in the medical field to treat burns and infections [4.81]. They are used as quantitative real-time biosensing platforms with the help of localized surface plasmon resonance (LSPR), which is given by arrays of silver nanoparticles [4.49]. So, frustules can be used as a scaffold for the formation of ordered structures of AgNPs without the need for the use of linking agents and stabilizers. The frustule provides a cost-effective way for unlocking highly ordered complex 3-dimensional structures that have potential applications in, for example, gas sensing, drug delivery, providing catalytic surface and support, photonic crystal, and large molecule separation [4.10] [4.53]. Functionalization of frustules with more technologically relevant materials, such as gold, iron oxide, titania, or silver, is a challenge but some work has shown the ability to coat frustules with gold and titania and with titania metabolically inserted into the frustules during growth and development of the diatoms [4.104] [4.135] [4.136].

Toster *et al.* report a method of decorating the silica frustules of diatoms with AgNPs of approximately 10–20 nm in size without the use of any linking agents [4.81], in which mined frustules were cleaned by plasma treatment, followed by immediate immersion in a silver nitrate solution. Sodium hydroxide was added to facilitate the reduction process through the formation of silver oxide particles bound to the surface, which are then readily reduced by glucose. The resulting decoration of frustules is analyzed by SEM, TEM microscopic methods, and nanoparticles characterized by XRD [4.13]. In conclusion, the authors have developed a simple yet highly effective method for decorating diatom frustules with AgNPs 10–20 nm in size, which are devoid of surfactants. This involves the reduction of *in situ* generated silver oxide particles attached to the surface, which is effective at room temperature using only glucose as a reducing agent with no requirement for any linking agents between the silica frustule and silver nanoparticles.

Kumar *et al.* worked on peptide fabricated on diatom frustules and proposed conditions in which to synthesize (*in situ*) silver nanoparticles (AgNPs) of size between 30–42 nm in diameter under physiological conditions. The formation of AgNPs was attained by a photoinduced reduction process using UV-B rays from an electric lamp. The kinetics of NPs synthesis via photoreduction is highly sensitive and depends upon peptide-diatom interactions, which modify the rate of synthesis. Thus, an alternative and effortless synthesis of such AgNPs-peptide embedded diatom devices could be utilized as high-efficiency dye-synthesized solar cells (DSSC) in the progress of renewable and sustainable energy fields.

4.6.3 Gold Nanoparticles

Diatom-mediated nanosynthesis is advantageous over chemical-physical methods because of its simple protocol, easy mass-scale production of diatoms, biocompatibility, and no use of toxic chemicals [4.17]. For the very first time in a study, the potential of diatom microalgae in nanobiotechnology and microbial control has been put forward. Currently, a major trend in nanotechnology is to find an eco-friendly nanosynthesis protocol. This is of particular interest in view of the fact that diatoms are a huge hub of nanoparticles and can enter into the food chain in the ecosystem. It was observed that gold nanoparticles (AuNPs) can be synthesized and accumulated by diatoms while maintaining cell viability [4.111]. Other studies also show that diatom-fabricated AuNPs-antibiotics complex showed promising inhibitory action against pathogenic bacteria. Nano-antibiotics complex on pathogens will also be an interesting research path in nanosynthesis with future applications in medicine, science, engineering, among others. Gold nanoparticles have been biosynthesized with an aqueous solution of $HAuCl_4$ under laboratory conditions. The interaction of diatoms with an aqueous salt of $HAuCl_4$ (tetrachloroaurate) forms the precipitation of gold nanoparticles [4.18]. The $HAuCl_4$ reduction by diatoms was found to be a low-cost and effective method for the preparation of nanocomposites. The described method is very simple and environmentally friendly compared to other chemical methods which use toxic reagents.

4.6.4 Titanium Nanoparticle

Among various nanoparticles, titanium dioxide (TiO_2) nanoparticles have been widely used in industry, agriculture, household goods, and consumer goods and products, and not just limited to sunscreens, cosmetics, electronics, food packing, and food products. These products can be produced in large quantities due to the whiteness and opacity of nanoparticles [4.38]. Biogenic titanium nanoparticles can be prepared on the surface of the diatom frustule by incorporation of metal, and high production of TiO_2-deposited diatoms is required to utilize the photocatalytic activity. Titanium-incorporated diatom (silica-titanium frustules) have 7.2% titanium [4.74]. The authors revealed that *Calliostoma schroederi* and *Navicula sp.* are good candidates for wide used for laboratory purposes but also aquaculture [4.35].

The hybrid nanosponge is a device that helps in the clean separation and collection of oil in diatom solar panels (DSP) as well as desorbs oil on exposure to UV light. El-Roz *et al.* fabricated TiO_2 Luffa sponge as a green photocatalyst for the purification of oil and water. The TiO_2 simultaneously helps in the absorption and desorption of oil on exposure to UV light because of its photocatalytic property, which may be used in photobioreactors [4.57]. The fabrication of hydrophobic polydimethylsiloxane (PDMS)-Luffa nanosponge doped with TiO_2 nanoparticles is used for the cost-effective separation of oil from a diatom. This hybrid nanosponge exhibited a 3D microcellular porous structure which allows light to pass through it and gives great binding ability with inorganic materials [4.21] [4.57]. PDMS-Luffa-TiO_2 nanosponges show an efficient selective oil adsorption capacity. The porous PDMS and PDMS-Luffa-TiO_2 nanosponges absorbed the amount of oil selectively and left the water [4.61]. This absorbed oil gets deposited inside the porous structure of PDMS. It has been supposed that the nanosponge floats over the water surfaces after the absorption of oil and can be easily removed without any leakage. The absorbing capacity of PDMS-Luffa-TiO_2 nanosponge is higher than PDMS only due to the presence of Luffa fibers and TiO_2 nanoparticles, as it exhibits a larger surface area and more porous structure. There is no residual oil detected after absorption, validating the excellent selective oil separation capacity of the sponges [4.57].

4.6.5 Magnetite Nanoparticles

It was reported that magnetite-doped bacteria (MTB) were found to have Fe_3O_4 (magnetite) or Fe_3S_4 (greigite) nanoparticles called magnetosomes which aligned along the direction of the magnetic field [4.28]. These magnetite nanoparticles are decorated with bacteria by the adsorption process. Magnetite-doped (MT) diatoms can be separated in a photobioreactor from oozed oil in the presence of a magnetic field and can be recycled for further production [4.63]. The diatoms have interaction ability with inorganic materials such as metal nanoparticles, in which metal nanoparticles get deposited into frustules of diatoms. These deposited metal nanoparticles regulate the oozing of oil from diatoms. Due to their thermoplastic effect, different metal nanoparticles like gold, silver, magnetite, and titania nanoparticles may also lead to the opening of frustules and facilitate the oozing of oil from diatoms. The diatom *Diadesmis confervaceae* has the self-oozing property of oil under lab conditions. It has 289 amino acids and amino acid sequences 187–188 have metal binding sites, which helps in the adsorption of magnetite nanoparticles on binding sites in the chloroplast as well as on the hydroxyl groups of SiOH molecules on the silica frustules of this diatom. Incubation time also plays an important role when magnetite nanoparticles interact with diatoms. The adsorption of magnetite on the surface of SiOH molecules on the silica frustules of diatom have an amine group form Fe-SiOH-NH cluster. These magnetite nanoparticles are deposited onto the surface of a diatom. The prolonged incubation (2–3 days) of metal nanoparticles with diatom results in great deposition as well as the disappearance of pores and roughness over the surface taking place. The interaction and growth of clustered magnetite nanoparticles can be confirmed by topographical analysis of metal-doped diatoms. The diatom *D. confervaceae* oozes and separates oil from the rest of the cellular biomass in both living and dead cells. Therefore, the doping of diatoms with magnetite will facilitate clean and easy separation of oil from diatoms [4.36] [4.63] [4.139].

4.7 Interaction of Diatoms with Peptides and Their Plausible Applications

A biotinylated dipeptide having two tryptophan units that self-assemble into vesicle-type morphology with plausible application in drug delivery vehicles was studied for its interaction with diatoms [4.61] [4.106]. The distinct feature of this peptide is to attract several moieties towards them when found in closer proximity because of the presence of noncovalent interaction sides in them. Independently, diatoms are the primary producers of phytoplanktons and can help with about 40% of oceanic primary

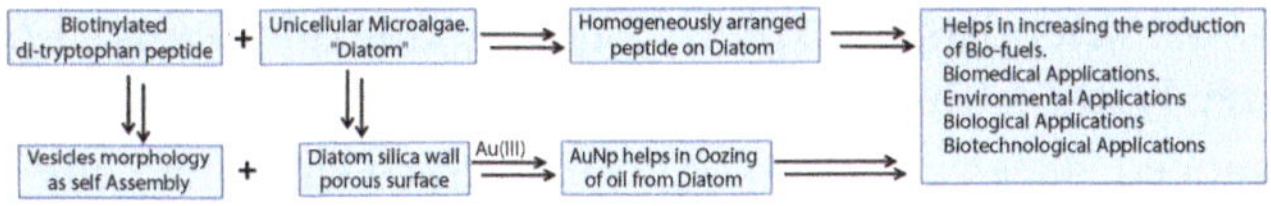

Figure 4.2 Flowchart depicting plausible applications of peptide-loaded diatoms [4.61].

productivity [4.37], and are also considered major producers of biofuel oil, producing about 30% of crude oil [4.140]. Diatom oil is also considered to be the major source of renewable energy as it contains polyunsaturated fatty acids. But due to its thick silica walls, it requires high energy input to produce this oil, which makes it difficult to obtain such natural energy for various purposes. The said peptide has been reported for its capability of photoinduced synthesis of silver and gold nanoparticles into its vesicular structure; therefore, this biotinylated ditryptophanyl short peptide may help to overcome this difficult situation of production and clean harvesting of biofuels or oils from diatoms [4.61] [4.140]. The surface provided by the diatom frustules was just the indication of the formation of a hierarchical nanoarray of peptide vesicles. It is again pertinent to mention here that the diatoms are reportedly very stable at normal physiological conditions [4.140], therefore, they can be used for various biologically and chemically applied mechanisms and have potential use for many applications (Figure 4.2) [4.2] [4.51] [4.52] [4.83] [4.131] [4.132] [4.144].

4.8 Diafuel: A Diatom Application with the Most Potential

Rapid industrialization and globalization have severely impacted the ecosystem and the problem can only be solved by developing technologies to decrease our dependency on conventional energy resources and finding renewable energy resources [4.92]. The unicellular microalgae diatoms can be promising candidates to produce biofuel which can replace non-renewable sources of energy. Geoscientists have proved that the algal lipids work as biomarkers which remain stable for millions of years and are the major components of petroleum. The highly branched isoprenoid alkenes (C28 and C29 steranes) are found in large quantities [4.42] [4.47] [4.95]; however, the physiological conditions, such as light, stress on diatom cells, and supply of nutrients, may alter the chemical composition of the oil. Knowledge of the potential of diatoms and optimization of fatty acids and biomass increase the economic value of diatoms and can help environmental scientists to explore this very unexploited area of the biofuel industry [4.105]. Therefore, a greater focus should be placed on developing advanced genetic technology to produce diatoms cells as biofuel machinery for future use.

4.9 Conclusion

This chapter discussed the potential of diatoms to produce diafuel as an alternative or additive to natural oil to fill the increasing demand for fuel for daily use. Scientific approaches for the regenerative production of diatoms and oozing of oil by using technological advancement were explained, which can provide new ideas for scientific inventions for diafuel production. The chapter summarized how the use of different metal ions and metal nanoparticles can help to activate these nature's nanomachines to use their maximum capacity as bio-nanoreactors to produce good quality secondary metabolites to be used as high value-added products.

Acknowledgments

KK and AK thank DHSGU-Sagar for the university fellowship. SS and SKT thank SERB-DST-India and CSIR-UGC India for project associate-I and senior research fellowship, respectively. VK thanks the Department of Higher Education, Madhya Pradesh and Dr. D. S. Kothari Foundation for the postdoctoral fellowship. This work is supported by SERB-DST-India project no. CRG/2021/002526 to KBJ. Dr. Vandana Vinayak, Department of Forensic Science and Criminology, Dr. Harisingh Gour Vishwavidyalaya (A Central University) is greatly acknowledged for her useful discussion and necessary support during manuscript preparation.

References

[4.1] Abramson, L., Wirick, S., Lee, C., Jacobsen, C., Brandes, J. A., The use of soft X-ray spectromicroscopy to investigate the distribution and composition of organic matter in a diatom frustule and a biomimetic analog. *Deep Sea Research Part II: Topical Studies in Oceanography*, 56(18), 1369, 2009.

[4.2] Agarwal, P., Gupta, R., Agarwal, N., Advances in synthesis and applications of microalgal nanoparticles for wastewater treatment. *Journal of Nanotechnology*, 2019.

[4.3] Ahirwar, A., Meignen, G., Khan, M., Khan, N., Rai, A., Schoefs, B., Vinayak, V., Nanotechnological approaches to disrupt the rigid cell walled microalgae grown in wastewater for value-added biocompounds: commercial applications, challenges, and breakthrough. *Biomass Conversion and Biorefinery*, 1, 2021.

[4.4] Anderson, F. M., Origin of California petroleum. *Bulletin of the Geological Society of America*, 37(4), 585, 1926.

[4.5] Arif, M., Bai, Y., Usman, M., Jalalah, M., Harraz, F. A., Al-Assiri, M. S., Zhang, C., Highest accumulated microalgal lipids (polar and non-polar) for biodiesel production with advanced wastewater treatment: role of lipidomics. *Bioresource Technology*, 298, 122299, 2020.

[4.6] Aw, M. S., Bariana, M., Yu, Y., Addai-Mensah, J., Losic, D., Surface-functionalized diatom microcapsules for drug delivery of water-insoluble drugs. *Journal of Biomaterials Applications*, 28(2), 163, 2013.

[4.7] Azam, F., Hemmingsen, B. B., Volcani, B. E., Germanium incorporation into the silica of diatom cell walls. *Archiv für Mikrobiologie*, 92(1), 11, 1973.

[4.8] Bach, L. T., Riebesell, U., Sett, S., Febiri, S., Rzepka, P., Schulz, K. G., An approach for particle sinking velocity measurements in the 3–400 μm size range and considerations on the effect of temperature on sinking rates. *Marine Biology*, 159(8), 1853, 2012.

[4.9] Baltazar-Encarnación, E., Escárcega-González, C. E., Vasto-Anzaldo, X. G., Cantú-Cárdenas, M. E., Morones-Ramírez, J. R., Silver nanoparticles synthesized through green methods using *Escherichia coli* top 10 (Ec-Ts) growth culture medium exhibit antimicrobial properties against nongrowing bacterial strains. *Journal of Nanomaterials*, 2019.

[4.10] Blanco, A., Chomski, E., Grabtchak, S., Ibisate, M., John, S., Leonard, S. W., van Driel, H. M., Large-scale synthesis of a silicon photonic crystal with a complete three-dimensional bandgap near 1.5 micrometres. *Nature*, 405(6785), 437, 2000.

[4.11] Blankenship, R. E., Molecular mechanisms of photosynthesis. *John Wiley & Sons*, 2021.

[4.12] Brannon-Peppas, L., Blanchette, J. O., Nanoparticle and targeted systems for cancer therapy. *Advanced Drug Delivery Reviews*, 56(11), 1649, 2004.

[4.13] Briceño, S., Chavez-Chico, E. A., González, G., Diatoms decorated with gold nanoparticles by *In-situ* and *Ex-situ* methods for *in vitro* gentamicin release. *Materials Science and Engineering*: C, 123, 112018, 2021.

[4.14] Brongersma-Sanders, M., On conditions favouring the preservation of chlorophyll in marine sediments. In 3rd World Petroleum Congress. *OnePetro*, 1951.

[4.15] Brown, E. M., Is the calcium receptor a molecular target for the actions of strontium on bone?. *Osteoporosis International*, 14(3), 25, 2003.

[4.16] Boom, R. C. J. A., Sol, C. J., Salimans, M. M., Jansen, C. L., Wertheim-van Dillen, P. M., Van der Noordaa, J. P. M. E., Rapid and simple method for purification of nucleic acids. *Journal of Clinical Microbiology*, 28(3), 495, 1990.

[4.17] Borase, H. P., Patil, C. D., Suryawanshi, R. K., Koli, S. H., Mohite, B. V., Benelli, G., Patil, S. V., Mechanistic approach for fabrication of gold nanoparticles by *Nitzschia* diatom and their antibacterial activity. *Bioprocess and Biosystems Engineering*, 40(10), 1437, 2017.

[4.18] Castro, L., Blázquez, M. L., González, F. G., Ballester, A., Mechanism and applications of metal nanoparticles prepared by bio-mediated process. *Reviews in Advanced Sciences and Engineering*, 3(3), 199, 2014.

[4.19] Chauton, M. S., Skolem, L., Olsen, L. M., Vullum, P. E., Walmsley, J., Vadstein, O., Titanium uptake and incorporation into silica nanostructures by the diatom *Pinnularia* sp. (Bacillariophyceae). *Journal of Applied Phycology*, 27(2), 777, 2015.

[4.20] Chetia, L., Kalita, D., Ahmed, G. A., Synthesis of Ag nanoparticles using diatom cells for ammonia sensing. *Sensing and Bio-sensing Research*, 16, 55, 2017.

[4.21] Choi, S. J., Kwon, T. H., Im, H., Moon, D. I., Baek, D. J., Seol, M. L., Choi, Y. K., A polydimethylsiloxane (PDMS) sponge for the selective absorption of oil from water. *ACS Applied Materials & Interfaces*, 3(12), 4552, 2011.

[4.22] Coulombe, J., Faure, H., Robin, B., Ruat, M., *In vitro* effects of strontium ranelate on the extracellular calcium-sensing receptor. *Biochemical and Biophysical Research Communications*, 323(4), 1184, 2004.

[4.23] Dallas H. G., An early reference to the theory that diatoms are the source of bituminous substances. *AAPG Bulletin*, 12(5), 555, 1928.

[4.24] Demirbas, A., Demirbas, M. F., Algae energy: algae as a new source of biodiesel. *Springer Science & Business Media*, 2010.

[4.25] De Stefano, L., Rendina, I., De Stefano, M., Bismuto, A., Maddalena, P., Marine diatoms as optical chemical sensors. *Applied Physics Letters*, 87(23), 233902, 2005.

[4.26] De Stefano, L., Maddalena, P., Moretti, L., Rea, I., Rendina, I., De Tommasi, E., De Stefano, M., Nano-biosilica from marine diatoms: A brand new material for photonic applications. *Superlattices and Microstructures*, 46(1-2), 84, 2009.

[4.27] Durán, N., Marcato, P. D., Durán, M., Yadav, A., Gade, A., Rai, M., Mechanistic aspects in the biogenic synthesis of extracellular metal nanoparticles by peptides, bacteria, fungi, and plants. *Applied Microbiology and Biotechnology*, 90(5), 1609, 2011.

[4.28] Ebrahiminezhad, A., Varma, V., Yang, S., Berenjian, A., Magnetic immobilization of *Bacillus subtilis* natto cells for menaquinone-7 fermentation. *Applied Microbiology and Biotechnology*, 100(1), 173, 2016.

[4.29] Ehrenberg, C. G., Ueber die, DysodilgenannteMineralspecies, als ein Product aus Infusorienschalen. *Annalen der Physik*, 124(12), 573, 1839.

[4.30] Fang, J., Nakamura, H., Maeda, H., The EPR effect: unique features of tumor blood vessels for drug delivery, factors involved, and limitations and augmentation of the effect. *Advanced Drug Delivery Reviews*, 63(3), 136, 2011.

[4.31] Feyzullayev, A. A., Guliyev, I. S., Tagiyev, M. F., Source potential of the Mesozoic–Cenozoic rocks in the South Caspian Basin and their role in forming the oil accumulations in the Lower Pliocene reservoirs. *Petroleum Geoscience*, 7(4), 409, 2001.

[4.32] Field, C. B., Behrenfeld, M. J., Randerson, J. T., Falkowski, P., Primary production of the biosphere: integrating terrestrial and oceanic components. *Science*, 281(5374), 237, 1998.

[4.33] Finkle, B. J., Appleman, D., The effect of magnesium concentration on chlorophyll and catalase development in *Chlorella*. *Plant Physiology*, 28(4), 652, 1953.

[4.34] Fortman, J. L., Chhabra, S., Mukhopadhyay, A., Chou, H., Lee, T. S., Steen, E., Keasling, J. D., Biofuel alternatives to ethanol: pumping the microbial well. *Trends in Biotechnology*, 26(7), 375, 2008.

[4.35] French, R. A., Jacobson, A. R., Kim, B., Isley, S. L., Penn, R. L., Baveye, P. C., Influence of ionic strength, pH, and cation valence on aggregation kinetics of titanium dioxide nanoparticles. *Environmental Science & Technology*, 43(5), 1354, 2009.

[4.36] Ftouni, K., Lakiss, L., Thomas, S., Daturi, M., Fernandez, C., Bazin, P., El-Roz, M., TiO2/Zeolite Bifunctional (Photo) catalysts for a selective conversion of methanol to dimethoxymethane: on the role of brønsted acidity. *The Journal of Physical Chemistry C*, 122(51), 29359, 2018.

[4.37] Gao, K., Helbling, E. W., Häder, D. P., Hutchins, D. A., Responses of marine primary producers to interactions between ocean acidification, solar radiation, and warming. *Marine Ecology Progress Series*, 470, 167, 2012.

[4.38] Gautam, S., Kashyap, M., Gupta, S., Kumar, V., Schoefs, B., Gordon, R., Vinayak, V., Metabolic engineering of TiO_2 nanoparticles in *Nitzschia palea* to form diatom nanotubes: an ingredient for solar cells to produce electricity and biofuel. *RSC Advances*, 6(99), 97276, 2016.

[4.39] Gehlen, M., Beck, L., Calas, G., Flank, A. M., Van Bennekom, A. J., Van Beusekom, J. E. E., Unraveling the atomic structure of biogenic silica: evidence of the structural association of Al and Si in diatom frustules. *Geochimica et Cosmochimica Acta*, 66(9), 1601, 2002.

[4.40] Gélabert, A., Pokrovsky, O. S., Schott, J., Boudou, A., Feurtet-Mazel, A., Cadmium and lead interaction with diatom surfaces: a combined thermodynamic and kinetic approach. *Geochimica et Cosmochimica Acta*, 71(15), 3698, 2007.

[4.41] Gélabert, A., Pokrovsky, O. S., Reguant, C., Schott, J., Boudou, A., A surface complexation model for cadmium and lead adsorption onto diatom surface. *Journal of Geochemical Exploration*, 88(1-3), 110, 2006.

[4.42] Giri, T., Goutam, U., Arya, A., Gautam, S., Effect of Nutrients on Diatom Growth: A Review. *Trends in Sciences*, 19(2), 1752, 2022.

[4.43] Griffiths, M. J., Harrison, S. T., Lipid productivity as a key characteristic for choosing algal species for biodiesel production. *Journal of Applied Phycology*, 21(5), 493, 2009.

[4.44] Grouneva, I., Jakob, T., Wilhelm, C., Goss, R., The regulation of xanthophyll cycle activity and of non-photochemical fluorescence quenching by two alternative electron flows in the diatoms *Phaeodactylum tricornutum* and *Cyclotella meneghiniana*. *Biochimica et Biophysica Acta (BBA) - Bioenergetics*, 1787(7), 929, 2009.

[4.45] Guliyev, I. S., Feizulayev, A. A., Huseynov, D. A., Isotope geochemistry of oils from fields and mud volcanoes in the South Caspian Basin, Azerbaijan. *Petroleum Geoscience*, 7(2), 201, 2001.

[4.46] Guiry M.D., How many species of algae are there?.*J. Phycol.*, 48, 1057, 2012.

[4.47] Hildebrand, M., Davis, A. K., Smith, S. R., Traller, J. C., Abbriano, R., The place of diatoms in the biofuels industry. *Biofuels*, 3(2), 221, 2012.

[4.48] Hildebrand, M., Frigeri, L. G., Davis, A. K., synchronized growth of *Thalassiosira pseudonana* (Bacillariophyceae) provides novel insights into cell-wall synthesis processes in relation to the cell cycle 1. *Journal of Phycology*, 43(4), 730, 2007.

[4.49] Holland, B. T., Blanford, C. F., Do, T., Stein, A., Synthesis of highly ordered, three-dimensional, macroporous structures of amorphous or crystalline inorganic oxides, phosphates, and hybrid composites. *Chemistry of Materials*, 11(3), 795, 1999.

[4.50] Hu, Q., Sommerfeld, M., Jarvis, E., Ghirardi, M., Posewitz, M., Seibert, M., Darzins, A., Microalgal triacylglycerols as feedstocks for biofuel production: perspectives and advances. *The Plant Journal*, 54(4), 621, 2008.

[4.51] Ingle, A. P., Duran, N., Rai, M., Bioactivity, mechanism of action, and cytotoxicity of copper-based nanoparticles: a review. *Applied Microbiology and Biotechnology*, 98(3), 1001, 2014.

[4.52] Jeffryes, C., Gutu, T., Jiao, J., Rorrer, G. L., Metabolic insertion of nanostructured TiO2 into the patterned biosilica of the diatom *Pinnularia* sp. by a two-stage bioreactor cultivation process. *ACS Nano*, 2(10), 2103, 2008.

[4.53] Jeffryes, C., Campbell, J., Li, H., Jiao, J., Rorrer, G., The potential of diatom nanobiotechnology for applications in solar cells, batteries, and electroluminescent devices. *Energy & Environmental Science*, 4(10), 3930, 2011.

[4.54] Jena, J., Pradhan, N., Dash, B. P., Panda, P. K., Mishra, B. K., Pigment mediated biogenic synthesis of silver nanoparticles using diatom *Amphora* sp. and its antimicrobial activity. *Journal of Saudi Chemical Society*, 19(6), 661, 2015.

[4.55] Jibowu, T., A review on nanoporous metals. *Front Nanosci Nanotech*, 2(4), 165, 2016.

[4.56] Khan, M. J., Rai, A., Ahirwar, A., Sirotiya, V., Mourya, M., Mishra, S., Vinayak, V., Diatom microalgae as smart nanocontainers for biosensing wastewater pollutants: recent trends and innovations. *Bioengineered*, 12(2), 9531, 2021.

[4.57] Khan, M. J., Singh, R., Joshi, K. B., Vinayak, V., TiO_2 doped polydimethylsiloxane (PDMS) and *Luffa cylindrica* based photocatalytic nanosponge to absorb and desorb oil in diatom solar panels. *RSC Advances*, 9(39), 22410, 2019.

[4.58] Kooistra, W. H., Medlin, L. K., Evolution of the diatoms (Bacillariophyta): IV. A reconstruction of their age from small subunit rRNA coding regions and the fossil record. *Molecular Phylogenetics and Evolution*, 6(3), 391, 1996.

[4.59] Kröger, N., Deutzmann, R., Bergsdorf, C., Sumper, M., Species-specific polyamines from diatoms control silica morphology. *Proceedings of the National Academy of Sciences*, 97(26), 14133, 2000.

[4.60] Knothe, G., Steidley, K. R., Kinematic viscosity of biodiesel fuel components and related compounds. *Influence of compound structure and comparison to petrodiesel fuel components. Fuel*, 84(9), 1059, 2005.

[4.61] Kumar, V., Gupta, S., Rathod, A., Vinayak, V., Joshi, K. B., Biomimetic fabrication of biotinylated peptide nanostructures upon diatom scaffold; a plausible model for sustainable energy. *RSC Advances*, 6(77), 73692, 2016.

[4.62] Kumar, V., Kashyap, M., Gautam, S., Shukla, P., Joshi, K. B., Vinayak, V., Fast Fourier infrared spectroscopy to characterize the biochemical composition in diatoms. *Journal of Biosciences*, 43(4), 717, 2018.

[4.63] Kumar, V., Singh, R., Thakur, S., Joshi, K. B., Vinayak, V., Doping of magnetite nanoparticles facilitates clean harvesting of diatom oil as biofuel for sustainable energy. *Materials Research Express*, 5(4), 045503, 2018.

[4.64] Lang, Y., Del Monte, F., Collins, L., Rodriguez, B. J., Thompson, K., Dockery, P., Pandit, A., Functionalization of the living diatom *Thalassiosira weissflogii* with thiol moieties. *Nature Communications*, 4(1), 1, 2013.

[4.65] Lang, Y., Monte, F. D., Rodriguez, B. J., Dockery, P., Finn, D. P., Pandit, A., Integration of TiO2 into the diatom *Thalassiosira weissflogii* during frustule synthesis. *Scientific Reports*, 3(1), 1, 2013.

[4.66] Lechner, C. C., Becker, C. F., Silaffins in silica biomineralization and biomimetic silica precipitation. *Marine Drugs*, 13(8), 5297, 2015.

[4.67] Lettieri, S., Setaro, A., De Stefano, L., De Stefano, M., Maddalena, P., The gas-detection properties of light-emitting diatoms. *Advanced Functional Materials*, 18(8), 1257, 2008.

[4.68] Le, T. D. H., Bonani, W., Speranza, G., Sglavo, V., Ceccato, R., Maniglio, D., Migliaresi, C., Processing and characterization of diatom nanoparticles and microparticles as potential source of silicon for bone tissue engineering. *Materials Science and Engineering: C*, 59, 471, 2016.

[4.69] Levorsen, A. I., Berry, F. A, *Geology of Petroleum* (Vol. 724). San Francisco: WH Freeman, 1967.

[4.70] Leyland, B., Boussiba, S., Khozin-Goldberg, I., A review of diatom lipid droplets. *Biology*, 9(2), 38, 2020.

[4.71] Li, H., Jeffryes, C., Gutu, T., Jiao, J., Rorrer, G. L., Peptide-mediated deposition of nanostructured TiO2 into the periodic structure of diatom biosilica and its integration into the fabrication of a dye-sensitized solar cell device. *MRS Online Proceedings Library (OPL)*, 1189, 2009.

[4.72] Li, J., Han, J., Sun, Q., Wang, Y., Mu, Y., Zhang, K., Feng, C., Biosynthetic calcium-doped biosilica with multiple hemostatic properties for hemorrhage control. *Journal of Materials Chemistry B*, 6(47), 7834, 2018.

[4.73] Li, K., Liu, Q., Fang, F., Luo, R., Lu, Q., Zhou, W., Ruan, R., Microalgae-based wastewater treatment for nutrients recovery: A review. *Bioresource Technology*, 291, 121934, 2019.

[4.74] Li, Y., Zhang, C., Hu, Z., Selective removal of pharmaceuticals and personal care products from water by titanium incorporated hierarchical diatoms in the presence of natural organic matter. *Water Research*, 189, 116628, 2021.

[4.75] Liang, Y., Maeda, Y., Yoshino, T., Matsumoto, M., Tanaka, T., Profiling of fatty acid methyl esters from the oleaginous diatom *Fistulifera* sp. strain JPCC DA0580 under nutrition-sufficient and-deficient conditions. *Journal of Applied Phycology*, 26(6), 2295, 2014.

[4.76] Lin, X., Tirichine, L., Bowler, C., Protocol: Chromatin immunoprecipitation (ChIP) methodology to investigate histone modifications in two model diatom species. *Plant Methods*, 8(1), 1, 2012.

[4.77] Lo Presti, M., Vona, D., Ragni, R., Cicco, S. R., Farinola, G. M., Perspectives on applications of nanomaterials from shelled plankton. *MRS Communications*, 11(3), 213, 2021.

[4.78] Losic, D., Mitchell, J. G., Voelcker, N. H., Diatomaceous lessons in nanotechnology and advanced materials. *Advanced Materials*, 21(29), 2947, 2009.

[4.79] Losic, D., Short, K., Mitchell, J. G., Lal, R., Voelcker, N. H., AFM nanoindentations of diatom biosilica surfaces. *Langmuir*, 23(9), 5014, 2007.

[4.80] Lu, C. W., Hung, Y., Hsiao, J. K., Yao, M., Chung, T. H., Lin, Y. S., Chen, Y. C., Bifunctional magnetic silica nanoparticles for highly efficient human stem cell labeling. *Nano Letters*, 7(1), 149, 2007.

[4.81] Lu, W., Senapati, D., Wang, S., Tovmachenko, O., Singh, A. K., Yu, H., Ray, P. C., Effect of surface coating on the toxicity of silver nanomaterials on human skin keratinocytes. *Chemical Physics Letters*, 487(1-3), 92, 2010.
[4.82] Maeda, Y., Nojima, D., Yoshino, T., Tanaka, T., Structure and properties of oil bodies in diatoms. Philosophical Transactions of the Royal Society B: *Biological Sciences*, 372(1728), 20160408, 2017.
[4.83] Maher, S., Kumeria, T., Aw, M. S., Losic, D., Diatom silica for biomedical applications: Recent progress and advances. *Advanced Healthcare Materials*, 7(19), 1800552, 2018.
[4.84] Marella, T. K., Saxena, A., Tiwari, A., Diatom mediated heavy metal remediation: A review. *Bioresource Technology*, 305, 123068, 2020.
[4.85] Marella, T. K., López-Pacheco, I. Y., Parra-Saldívar, R., Dixit, S., Tiwari, A., Wealth from waste: Diatoms as tools for phycoremediation of wastewater and for obtaining value from the biomass. *Science of the Total Environment*, 724, 137960, 2020.
[4.86] Marella, T. K., Tiwari, A., Marine diatom *Thalassiosira weissflogii* based biorefinery for co-production of eicosapentaenoic acid and fucoxanthin. *Bioresource Technology*, 307, 123245, 2020.
[4.87] Martin, J. H., Knauer, G. A., The elemental composition of plankton. *Geochimica et Cosmochimica Acta*, 37(7), 1639, 1973.
[4.88] McFarland, A. D., Van Duyne, R. P., Single silver nanoparticles as real-time optical sensors with zeptomole sensitivity. *Nano Letters*, 3(8), 1057, 2003.
[4.89] Medlin, L. K., Kooistra, W. H., Sims, P., Wellbrock, U., Is the origin of the diatoms related to the end-Permian mass extinction?. *Nova Hedwigia*, 1-11, 1997.
[4.90] Mishra, B., Saxena, A., Tiwari, A., Biosynthesis of silver nanoparticles from marine diatoms *Chaetoceros* sp., *Skeletonema* sp., *Thalassiosira* sp., and their antibacterial study. *Biotechnology Reports*, 28, e00571, 2020.
[4.91] Morel, F. M., Kinetics of nutrient uptake and growth in phytoplankton 1. *Journal of Phycology*, 23(1), 137, 1987.
[4.92] Nagarajan, D., Varjani, S., Lee, D. J., Chang, J. S., Sustainable aquaculture and animal feed from microalgae–Nutritive value and techno-functional components. *Renewable and Sustainable Energy Reviews*, 150, 111549, 2021.
[4.93] Nakano, V., Padilla, G., do Valle Marques, M., Avila-Campos, M. J., Plasmid-related β-lactamase production in *Bacteroides fragilis* strains. *Research in Microbiology*, 155(10), 843, 2004.
[4.94] Nelson, D. M., Tréguer, P., Brzezinski, M. A., Leynaert, A., Quéguiner, B., Production and dissolution of biogenic silica in the ocean: revised global estimates, comparison with regional data and relationship to biogenic sedimentation. *Global Biogeochemical Cycles*, 9(3), 359, 1995.

[4.95] Nguyen, T. T., Bui, X. T., Ngo, H. H., Nguyen, K. Q., Nguyen, H. H., Némery, J., ... Varjani, S., Nutrient recovery and microalgae biomass production from urine by membrane photobioreactor at low biomass retention times. *Science of the Total Environment*, 785, 147423, 2021.

[4.96] Olteanu, M., Baraitaru, A., Panait, A. M., Dumitru, D., Boboc, M., Deák, G., Advanced SiO2 Composite Materials for Heavy Metal Removal from Wastewater. *Water, Air, & Soil Pollution*, 230(8), 1, 2019.

[4.97] Poulsen, N., Berne, C., Spain, J., Kroeger, N., Silica immobilization of an enzyme through genetic engineering of the diatom *Thalassiosira pseudonana*. *Angewandte Chemie International Edition*, 46(11), 1843, 2007.

[4.98] Rangaswami, G., Mahadevan, A., Diseases of crop plants in India. *PHI Learning Pvt. Ltd.*, 1998.

[4.99] Ranjith K. R., Hanumantha Rao, P., Arumugam, M., Lipid extraction methods from microalgae: a comprehensive review. *Frontiers in Energy Research*, 2, 61, 2015.

[4.100] Rea, I., Terracciano, M., De Stefano, L., Synthetic vs natural: diatoms bioderived porous materials for the next generation of healthcare Nanodevices. *Advanced Healthcare Materials*, 6(3), 1601125, 2017.

[4.101] Rea, I., Terracciano, M., Chandrasekaran, S., Voelcker, N. H., Dardano, P., Martucci, N. M., De Stefano, L. (2016). Bioengineered silicon diatoms: Adding photonic features to a nanostructured semiconductive material for biomolecular sensing. *Nanoscale Research Letters*, 11(1), 1, 2016.

[4.102] Reep, P., Green, M. P., U.S. Patent Application No. 13/207,298, 2012.

[4.103] Richardson, J. W., Johnson, M. D., Zhang, X., Zemke, P., Chen, W., Hu, Q., A financial assessment of two alternative cultivation systems and their contributions to algae biofuel economic viability. *Algal Research*, 4, 96, 2014.

[4.104] Rogato, A., De Tommasi, E., Physical, chemical, and genetic techniques for diatom frustule modification: Applications in nanotechnology. *Applied Sciences*, 10(23), 8738, 2020.

[4.105] Round, F. E., Crawford, R. M., Mann, D. G., Diatoms: biology and morphology of the genera. Cambridge University Press, 1990.

[4.106] Ruggiero, I., Terracciano, M., Martucci, N. M., De Stefano, L., Migliaccio, N., Tatè, R., Rea, I., Diatomite silica nanoparticles for drug delivery. *Nanoscale Research Letters*, 9(1), 1, 2014.

[4.107] Sánchez, C., Cristóbal, G., Bueno, G., Diatom identification including life cycle stages through morphological and texture descriptors. *PeerJ*, 7, e6770, 2019.

[4.108] Sanei, H., Goodarzi, F., Relationship between organic matter and mercury in recent lake sediment: the physical–geochemical aspects. *Applied Geochemistry*, 21(11), 1900, 2006.

[4.109] Sbihi, K., Cherifi, O., Bertrand, M., El Gharmali, A., Biosorption of metals (Cd, Cu and Zn) by the freshwater diatom *Planothidium lanceolatum*: a laboratory study. *Diatom Research*, 29(1), 55, 2014.

[4.110] Scala, S., Bowler, C., Molecular insights into the novel aspects of diatom biology. *Cellular and Molecular Life Sciences CMLS*, 58(11), 1666, 2001.

[4.111] Schröfel, A., Kratošová, G., Bohunická, M., Dobročka, E., Vávra, I., Biosynthesis of gold nanoparticles using diatoms—silica-gold and EPS-gold bionanocomposite formation. *Journal of Nanoparticle Research*, 13(8), 3207, 2011.

[4.112] Scott, S. A., Davey, M. P., Dennis, J. S., Horst, I., Howe, C. J., Lea-Smith, D. J., Smith, A. G., Biodiesel from algae: challenges and prospects. *Current opinion in biotechnology*, 21(3), 277, 2010.

[4.113] Seth, K., Kumar, A., Rastogi, R. P., Meena, M., Vinayak, V., Bioprospecting of fucoxanthin from diatoms—Challenges and perspectives. *Algal Research*, 60, 102475, 2021.

[4.114] Sharma, V. K., Yngard, R. A., Lin, Y., Silver nanoparticles: green synthesis and their antimicrobial activities. *Advances in Colloid and Interface Science*, 145(1-2), 83, 2009.

[4.115] Sheehan, J., Dunahay, T., Benemann, J., Roessler, P., A look back at the US Department of Energy's aquatic species program: biodiesel from algae. *National Renewable Energy Laboratory*, 328, 1, 1998.

[4.116] Sheppard, V. C., Scheffel, A., Poulsen, N., Kröger, N., Live diatom silica immobilization of multimeric and redox-active enzymes. *Applied and Environmental Microbiology*, 78(1), 211, 2012.

[4.117] Shi, J., Jiang, Y., Wang, X., Wu, H., Yang, D., Pan, F., Jiang, Z., Design and synthesis of organic–inorganic hybrid capsules for biotechnological applications. *Chemical Society Reviews*, 43(15), 5192, 2014.

[4.118] Shukla, S. K., Mohan, R., The contribution of diatoms to worldwide crude oil deposits. *In The Science of Algal Fuels, Springer, Dordrecht*, 355, 2012.

[4.119] Siegel, R., Naishadham, D., Jemal, A., Cancer statistics for hispanics/latinos, 2012. *CA: A Cancer Journal for Clinicians*, 62(5), 283, 2012.

[4.120] Singh, S., Verma, D. K., Thakur, M., Tripathy, S., Patel, A. R., Shah, N.,Aguilar, C. N., Supercritical fluid extraction (SCFE) as green extraction technology for high-value metabolites of algae, its potential trends in food and human health. *Food Research International*, 150, 110746, 2021.

[4.121] Smith, S. R., Glé, C., Abbriano, R. M., Traller, J. C., Davis, A., Trentacoste, E. Hildebrand, M., Transcript level coordination of carbon pathways during silicon starvation-induced lipid accumulation in the diatom *Thalassiosira pseudonana. New Phytologist*, 210(3), 890, 2016.

[4.122] Sprynskyy, M., Pomastowski, P., Hornowska, M., Król, A., Rafińska, K., Buszewski, B., Naturally organic functionalized 3D biosilica from diatom microalgae. *Materials & Design*, 132, 22, 2017.

[4.123] Stasiuk, L. D., Sanei, H., Characterization of diatom-derived lipids and chlorophyll within Holocene laminites, Saanich Inlet, British Columbia, using conventional and laser scanning fluorescence microscopy. *Organic Geochemistry*, 32(12), 1417, 2001.

[4.124] Stephenson, P. G., Moore, C. M., Terry, M. J., Zubkov, M. V., Bibby, T. S., Improving photosynthesis for algal biofuels: toward a green revolution, *Trends in Biotechnology*, 29(12), 615, 2011.

[4.125] Sumper, M., Brunner, E., Learning from diatoms: nature's tools for the production of nanostructured silica. *Advanced Functional Materials*, 16(1), 17, 2006.

[4.126] Su, X., Xu, J., Yan, X., Zhao, P., Chen, J., Zhou, C., Li, S., Lipidomic changes during different growth stages of *Nitzschia closterium f. minutissima*. *Metabolomics*, 9(2), 300, 2013.

[4.127] Sverdrup, H. U., Johnson, M. W., Fleming, R. H., *The Oceans: Their Physics, Chemistry, and General Biology* (Vol. 1087). New York: Prentice-Hall, 1942.

[4.128] Tanaka, T., Maeda, Y., Veluchamy, A., Tanaka, M., Abida, H., Maréchal, E., Fujibuchi, W., Oil accumulation by the oleaginous diatom *Fistulifera solaris* as revealed by the genome and transcriptome. *The Plant Cell*, 27(1), 162, 2015.

[4.129] Terracciano, M., De Stefano, L., Rea, I., Diatoms green nanotechnology for biosilica-based drug delivery systems. *Pharmaceutics*, 10(4), 242, 2018.

[4.130] Terracciano, M., Rea, I., Stefano, L. D., Rendina, I., Oliviero, G., Nici, F., Borbone, N., Synthesis of mixed-sequence oligonucleotides on mesoporous silicon: Chemical strategies and material stability. *Nanoscale Research Letters*, 9(1), 1, 2014.

[4.131] Ting, C. S., Owens, T. G., Photochemical and nonphotochemical fluorescence quenching processes in the diatom *Phaeodactylum tricornutum*. *Plant Physiology*, 101(4), 1323, 1993.

[4.132] Tiwari, A., Melchor-Martínez, E. M., Saxena, A., Kapoor, N., Singh, K. J., Saldarriaga-Hernández, S., Iqbal, H. M., Therapeutic attributes and applied aspects of biological macromolecules (polypeptides, fucoxanthin, sterols, fatty acids, polysaccharides, and polyphenols) from diatoms—a review. *International Journal of Biological Macromolecules*, 171, 398, 2021.

[4.133] Tokushima, H., Inoue-Kashino, N., Nakazato, Y., Masuda, A., Ifuku, K., Kashino, Y., Advantageous characteristics of the diatom *Chaetoceros gracilis* as a sustainable biofuel producer. *Biotechnology for Biofuels*, 9(1), 1-19, 2016.

[4.134] Tolman, C. F., Biogenesis of hydrocarbons by diatoms. *Economic Geology*, 22(5), 454, 1927.

[4.135] Toster, J., Harnagea, C., Iyer, K. S., Rosei, F., Raston, C. L., Controlling anatase coating of diatom frustules by varying the binding layer. *CrystEngComm*, 14(10), 3446, 2012.

[4.136] Toster, J., Iyer, K. S., Burtovyy, R., Burgess, S. S., Luzinov, I. A., Raston, C. L., Regiospecific assembly of gold nanoparticles around the pores of diatoms: toward three-dimensional nanoarrays. *Journal of the American Chemical Society*, 131(24), 8356, 2009.

[4.137] Toster, J., Zhou, Q. L., Smith, N. M., Iyer, K. S., Rosei, F., Raston, C. L., *In situ* coating of diatom frustules with silver nanoparticles. *Green Chemistry*, 15(8), 2060, 2013.

[4.138] Vickers, N. J., Animal communication: when I'm calling you, will you answer too?. *Current Biology*, 27(14), R713-R715, 2017.

[4.139] Vinayak, V., Gordon, R., Gautam, S., Rai, A., Discovery of a diatom that oozes oil. *Advanced Science Letters*, 20(7-8), 1256, 2014.

[4.140] Vinayak, V., Joshi, K. B., Sarma, P. M., Diafuel™ (diatom biofuel) vs electric vehicles, a basic comparison: a high potential renewable energy source to make India energy independent. *Diatoms: Fundamentals and Applications*, 537, Wiley-Scrivener, 2019.

[4.141] Vinayak, V., Manoylov, K. M., Gateau, H., Blanckaert, V., Hérault, J., Pencréac'h, G., Schoefs, B., Diatom milking: a review and new approaches. *Marine Drugs*, 13(5), 2629, 2015.

[4.142] Vrieling, E. G., Sun, Q., Tian, M., Kooyman, P. J., Gieskes, W. W., van Santen, R. A., Sommerdijk, N. A., Salinity-dependent diatom biosilicification implies an important role of external ionic strength. *Proceedings of the National Academy of Sciences*, 104(25), 10441, 2007.

[4.143] Volkman, J. K., Johns, R. B., The geochemical significance of positional isomers of unsaturated acids from an intertidal zone sediment. *Nature*, 267(5613), 693, 1977.

[4.144] Wang, Y., Cai, J., Jiang, Y., Jiang, X., Zhang, D., Preparation of biosilica structures from frustules of diatoms and their applications: current state and perspectives. *Applied Microbiology and Biotechnology*, 97(2), 453, 2013.

[4.145] Wei, S., Li, Y., Zhan, J., Wang, S., Zhu, J., Tolerant mechanisms of *Rorippaglobosa* (Turcz.) Thell. hyperaccumulating Cd explored from root morphology. *Bioresource Technology*, 118, 455, 2012.

[4.146] Wishkerman, A., Arad, S., Production of silver nanoparticles by the diatom *Phaeodactylum tricornutum*. In *Nanotechnology VIII* (Vol. 10248, pp. 94-100). SPIE, 2017.

[4.147] Wu, S., Zhang, B., Huang, A., Huan, L., He, L., Lin, A., Wang, G., Detection of intracellular neutral lipid content in the marine microalgae *Prorocentrum micans* and *Phaeodactylum tricornutum* using Nile red and BODIPY 505/515. *Journal of Applied Phycology*, 26(4), 1659, 2014.

[4.148] Yi, D. K., Selvan, S. T., Lee, S. S., Papaefthymiou, G. C., Kundaliya, D., Ying, J. Y., Silica-coated nanocomposites of magnetic nanoparticles and quantum dots. *Journal of the American Chemical Society*, 127(14), 4990, 2005.

[4.149] Yi, Z., Xu, M., Di, X., Brynjolfsson, S., Fu, W., Exploring valuable lipids in diatoms. *Frontiers in Marine Science*, 4, 17, 2017.

[4.150] Yu, E. T., Zendejas, F. J., Lane, P. D., Gaucher, S., Simmons, B. A., Lane, T. W., Triacylglycerol accumulation and profiling in the model diatoms *Thalassiosira pseudonana* and *Phaeodactylum tricornutum* (Baccilariophyceae) during starvation. *Journal of Applied Phycology*, 21(6), 669, 2009.

[4.151] Yu, Y., Addai-Mensah, J., Losic, D., Synthesis of self-supporting gold microstructures with three-dimensional morphologies by direct replication of diatom templates. *Langmuir*, 26(17), 14068, 2010.

[4.152] Zachleder, V., van den Hoek, C., Mann, D. G., Jahns, HM (ed.), Algae. An introduction to phycology. *Photosynthetica*, 4(35), 506, 1998.

[4.153] Zeb, S., Ali, N., Ali, Z., Bilal, M., Adalat, B., Hussain, S., Iqbal, H. M., Silica-based nanomaterials as designer adsorbents to mitigate emerging organic contaminants from water matrices. *Journal of Water Process Engineering*, 38, 101675, 2020.

5

Diatom Cell Wall: Nature Engineered Nanostructures

Sakshi Phogat, Rashi Tyagi, Abhishek Saxena, Pankaj Kumar Singh and Archana Tiwari*

Diatom Research Laboratory, Amity Institute of Biotechnology, Amity University, Noida, Uttar Pradesh, India

Abstract

The global economy is moving toward green and sustainable development to tackle health concerns and climate change issues. Diatoms can survive in difficult environments because of their diverse pore architectures and distinct cell walls, making them an attractive alternative for industrial products. The diatom cell wall is one of the most unique structures and the pore size of diatom frustules' 3D nano-architecture spans from nanometer to micrometer. Diatom frustule structure varies in terms of homogeneity, space, and complexity depending on environmental conditions and silicon absorption efficiency. Though the biosilicification process is fairly complicated, it includes the role of silicic acid transporters, silica transport, and polymerization of silica monomers, among other processes that have been well characterized. This allows them to be used in the healthcare and nanotechnology sectors. Because of their amorphous nature and form, several diatom species, such as *Coscinodiscus concinnus* and *Thalassiosira weissflogii*, are prospective drug carrier possibilities. Furthermore, several studies have demonstrated that diatom microcapsules are good carriers for poorly soluble and water-soluble medicines, which may be used in both oral and implant applications. This chapter elaborates on the cell wall of diatoms and the diverse applications associated with diatom frustules.

***Keywords*:** Biosilica, diatoms, frustules, nanotechnology, mesoporous

**Corresponding author*: panarchana@gmail.com

Vandana Vinayak and Richard Gordon (eds.) Diatom Cultivation for Biofuel, Food and High-Value Products, (115–138) © 2025 Scrivener Publishing LLC

5.1 Introduction

Today, substantial changes are being made to shift the economy and society towards more sustainable and greener ways to tackle the climate and health crisis. Biotechnology and engineering research is investing large sums of resources, time and funds to investigate and exploit microorganisms for various purposes [5.1]. We are constantly surrounded by a vast range of microbial organisms that regulate prominent biochemical cycles and subsequently end up with the potential of generate great amounts of sustainable goods [5.2]. Much of the microbial research work has been focused on the industrial administration of microbes like yeast or bacteria and not enough attention has been offered to diatoms [5.1] [5.3]. Diatoms have been recently identified as a commercially important group of micro-algae and have left the scientific community in awe of their industrial potential [5.3].

Diatoms are microscopic, single-celled photosynthetic organisms ranging from 10–200 μm [5.3] [5.6]. They are present in two general shapes, i.e., pennate or centric [5.4]. Diatoms are the most diverse unicellular algae with an estimated 200,000 known species, each uniquely possessing silica in their cell walls [5.5]. These microalgal species are generally at their most abundant during the start of the autumn and spring seasons when there are unlimited nutrients and light needed for photosynthesis [5.4] [5.6]. They generally sink out of their photic zone when they aggregate into clumps due to nutrient exhaustion [5.7]. Although they fundamentally reproduce through mitosis, they seem to be interrupted by rare sexual occurrences [5.8]. Many factors have influenced diatom evolution, for instance, their potential to adapt and cultivate in numerous different kinds of water reservoirs. Their suitability for various multiple applications is due to their abundance and their ability to adapt to almost any weather and geographical conditions [5.7] [5.8].

Diatoms are unique and different from other algae in terms of production; they generate oil while other algae generate starch [5.9]. Diatoms are neither animals nor plants; various pieces of evidence suggest that they have an exclusive range of attributes that link them to a junction between photosynthetic algae and a heterotrophic host [5.10]. For instance, unlike plants, diatoms have a complete and intact urea cycle and, similar to animals, they have the capability to break down fats to produce chemical energy along with a plant-like ability to produce metabolic intermediates from material disintegration [5.10]. This combination of features allows diatoms to outlive long durations of darkness present at the poles and proceed with division and growth in the presence of light. They appear to possess intricate transmission and communication abilities; they use a nitric-oxide messenger system that regulates the signaling between the cells as well as inside the cells [5.11].

Diatoms generally grow in two different types; benthic diatoms grow on the soil, sediments or macroalgae whereas planktonic forms are found floating freely in open seas [5.12]. They exist in sizes from micrometers to millimeters and as either solitary (single cells) or colonial (connected cell chains). However, the presence of larger-sized diatoms (2–5 mm) signals low seasonal variability [5.13]. A large proportion of them is also associated with ice in the Arctic and the Antarctic. Furthermore, in warmer oligotrophic water masses they could also be found living in a symbiotic relationship with Cyanobacteria and nitrogen-fixing bacteria, contributing to the ecosystem with a considerate amount of nitrogen [5.14].

These microalgae contribute to about 20–30% of the oxygen produced on Earth each year and invisibly participate in the process of recycling gases enclosing our planet [5.2] [5.19]. Diatoms potentially exhale more oxygen than all the world's rainforests combined [5.15]. However, contrary to the carbon dioxide produced by trees, the carbon generated by the diatoms is rapidly exhausted in the marine food webs, from fisheries to deep-water organisms. In open seas, a large proportion of the diatoms sink from the surface and resolve on the sea floor, exposing themselves to petroleum reserves [5.16] [5.17]. They strongly influence the functions of the marine ecosystem, the concentrations of CO_2 and the global climate. Their response to the swiftly changing ocean conditions is very important to observe because it will reflect the health of the surrounding environment [5.17].

Diatoms produce the majority of the organic matter that provides food for sea life [5.15]. Due to the endosymbiotic incident between heterotrophic flagellate and red algae, they now inhabit almost every aquatic environment [5.16]. In comparison to other microalgae, diatoms dominate the marine phytoplankton because of their silicious cell wall, called a frustule [5.18]. This ornamented cell wall is the reason behind its varied sizes and structures. It has an intricate and self-assembled porous structure that satisfies various functionalities [5.19].

5.2 Nature of Diatom Cell Wall

Diatoms are diverse in many of their features, from frustule architecture to organelle grouping, ultrastructure, shape, number of chloroplasts, and type of sexual reproduction [5.21]. The frustule architecture is one of the most elegant, naturally occurring 3D structures. The diatom cell wall is surrounded by silicon dioxide, is systematically arranged, and has an intricate porous structure. The silica production happens on the cell surface, starting from its morphogenesis in the silica disposition vesicles [5.22]. These vesicles act as a template that manages the development of frustule, assists in forming holes and slits, and also hold multiple organic macromolecules. The frustule is flexible and can tolerate deformation to allow the loading of the material inside it [5.23]. According to a study conducted by Mark and Richard, the range of the Young's modulus of diatom frustules is around 22 GPa [5.20]. For a better understanding of the elasticity, it was compared to calcite, and the Young's modulus of calcite is 76 GPa, i.e., 54 GPa higher [5.20].

The frustule is responsible for multiple important roles, it acts as a mechanical barrier against grazers, supplies nutrients, regulates the sinking rate, and provides protection against photoinhibition [5.19]. The constructional architectonics of diatom's frustule enriched with spikes, pores, ridges, channels, and spines can be genetically influenced. However, for diatoms to be employed in the medicinal industry they must be biodegradable. A silicon reduction technique known as the magnesiothermic reduction method is generally used to transform diatoms into biodegradable silicon replicas [5.29]. Nowadays, the advanced medicinal technology in oral control release is expanding to greater levels because of enlarged polymeric systems. Despite that, there are still debates on targeted drug delivery without destroying the normal tissues in cancer therapies [5.30]. Surface modifications can be targeted to control the drug release through mesoporous systems. The material's pore size can be manipulated according to the requirements, i.e., for slow release, face tuning can be performed to reduce the pore size. Diatom's exceptional drug release profile makes them the material of precedence in the medical field [5.31]. The drug loading and release strategies can be very easily modulated and fit according to specific regulations through surface modifications.

5.2.1 Biosilicification

The mechanism by which microorganisms assimilate soluble, isomeric silica acid $Si(OH)_4$ in the form of polymerized, insoluble silica, SiO_2, is known as biosilicification [5.36]. On a worldwide scale, diatoms dominate the process of biosilicification, which converts soluble silicon into the solid phase of silica [5.17]. Diatoms use silica as a structural component in their frustules, or cell walls, and may create a vast variety of silica structures with exquisite nanoscale properties. Diatoms have become a model system for investigating biosilicification because their metabolism is naturally geared toward interaction with silicon [5.32]. *In vivo,* biosilicification occurs at or near pH 7.0 and under normal temperature and pressure circumstances. Synthetic polymers were used to demonstrate silica synthesis under mild circumstances *in vitro,* with control over the shape of the resultant silica. The findings reveal that silica synthesis *in vitro* is not limited to enzymes/polypeptides because of their chemical structure and activity, but that many other synthetic macromolecules can also facilitate silica creation at neutral pH.

Recent investigations have demonstrated that translational or post-translational regulation of SIT gene expression is congruent with the specific ability of silicic acid transporter proteins (SITs) to transport silicon into diatom cells. Based on the identification of conserved amino acids, metabolic models for the recognition and transport of silicic acid by SITs have been constructed. Three stages of structure creation—micro, meso, and nano—have been recognized based on the sizes of the silica particles formed during diatom biosilicification [5.33]. The internal compartment where silicification occurs, the silica deposition vesicle (SDV), interacts with components of the cytoskeleton to grow at the microscale. Long-chain polyamines and polypeptides, known as silaffins, act to start the initial silica polymerization process at the nanoscale (LCPAs). Silaffins and LCPAs work together to influence the rate of silica polymerization and the type of structures that form. Silaffins and LCPAs do not, however, self-assemble into higher-order structures as the diatom frustule does. This suggests that several mesoscale variables affect silicification [5.33]. Examples of mesoscale impacts include modifications to the shape of silica concerning the lowest free energy form and the emergence of certain substructures inside the confines of the SDV. The final silica structure is determined by the interaction of the several processes and scales, albeit the three scales make useful differences. Despite the genetic underpinnings of diatom silica structure, a variety of factors can influence the biochemical silicification process [5.33].

5.2.2 Applications

Investigating the mechanics underlying the production of diatom silica structures may be useful in addition to pinpointing the cell and molecular biology of the process. Nanotechnology, which relies on the fabrication and manipulation of materials or devices with qualities ranging from 1–100 nm, is the subject of a thorough investigation at the moment.

Finding ways to create complex three-dimensional structures precisely, in large quantities, and at a reasonable cost is crucial for developing materials with the necessary size properties. A desirable source of materials with all the desired properties is diatoms. As part of early research into the technological potential of diatoms, a functioning enzyme was first integrated and stabilized within the silica of a living diatom's cell wall. A functioning enzyme was first inserted and stabilized by genetic engineering into the silica cell wall of a living diatom. During early investigations into the technological potential of diatoms, girdle bands in *Coscinodiscus granii* were also demonstrated the ability to steer light.

Even while a range of natural diatom silica structures could be helpful in various nanotechnological applications, a more adaptable strategy would be to advance genetic modification techniques to the point where the structure could be particularly tailored for an application [5.34]. It may be possible to produce enormous numbers of readily replicable structures that can be transformed into usable chemistry for nanotechnological applications by combining the capacity to alter the structure of diatom silica with biological reproduction [5.35].

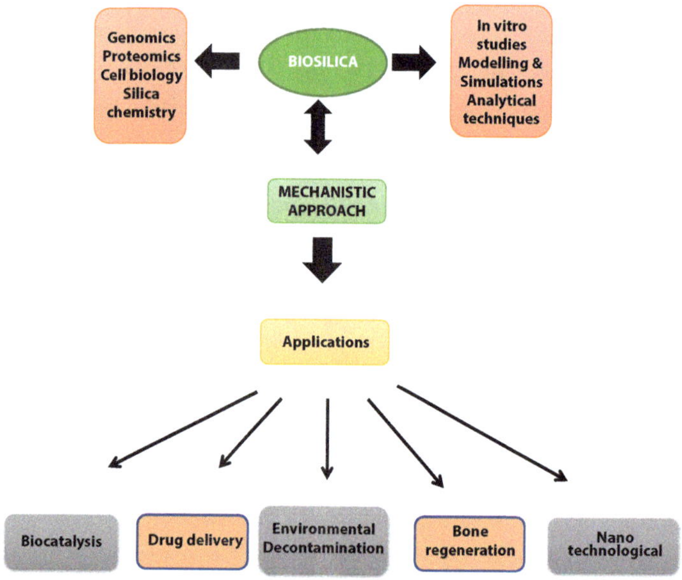

Figure 5.1 Application areas of biosilica.

Bone regenerative technologies are a particularly interesting part of biomedical applications. Even in its natural state, the frustule's silica has been demonstrated to be useful; consequently, adding healing elements like magnesium and strontium could represent a significant advancement in the field of bone treatment and remove existing barriers. Finally, diatoms are simple to grow and their frustules may be processed without the use of chemicals or a lot of energy, making them a more environmentally friendly material overall [5.53].

5.3 Purification of Diatoms

Growing diatoms demonstrate some application in several fields, particularly for micro or nanotechnologies. They have stability, unbroken cell walls, and pure culture availability, thus this is possible. Natural or organic matter, as well as inorganic pollutants, can be efficiently removed in this method. Furthermore, high-quality frustules were produced when the samples were cleaned in an excess of boiling 10–15% HCl and subsequently heated to 600 °C at a heating rate of 3 °C per minute for 6 hours. The majority of amorphous silica and other natural materials make up frustules. Frustules are a kind of composite material because of the combination of their main components, amorphous silica, and natural materials [5.39].

5.3.1 Principle

Since chloride particles are widely distributed in seawater, hydrochloric acid (HCl) was identified as crucial because it does not add other ions to the frustules. Additionally, the unpredictable nature of HCl helps to remove any remaining acid throughout the baking process. Thermal gravimetric analysis (TGA) and energy spectrum analysis (ESA) were used to identify the appropriate baking temperature and period. The method described in Subsection 5.3.2 produces a huge quantity of undamaged, extraordinarily pure, and highly SiO_2-containment diatom frustules [5.39].

5.3.2 Process

At the point when the biomass concentration (dry weight) reached more than 300 mg/L, the diatom cell was collected. To accomplish the best cleansing impact, a grouping of four purification stages was utilized to create pure diatom frustules. The specific steps were as follows: (i) Collection:

A funnel containing a piece of 10-micrometer filter cloth and the culture medium was used (cloth was adjusted to the size of the diatoms). The diatom cells were allowed to remain on the filter cloth before being rinsed with deionized water and placed in a tiny conical flask. A vacuum syphon was used in conjunction with the filter flask to hasten the filtration process. (ii) Acid cleaning (removal of organic matter and metal oxides): The materials in the conical flask were well mixed with 10–15% HCl (diatoms in deionized water were blended in with 37% HCl at a volume proportion of 2:1). Under constant stirring with boiling water, the mixture was boiled for 1 hour. Some released HCl fumes are directed into a flask containing NaOH during this boiling step to be neutralized. (iii) Filtration (acid removal): The gear in stage 1 was replaced with a rotator that rotated at 4500 rpm for 3 minutes to reduce the possibility of the experimenters being exposed to hot hydrochloric acid. The diatom frustules were rinsed in deionized water a few times (often 3 to 5 times) until the supernatant's pH was close to 7.0. (iv) Baking (removal of organic matter): Before being produced, diatom frustules were in this way dried at 120 °C in a drying broiler until the water had fully evaporated. To guarantee that the dry sample thickness was under 2 mm and allow for the heating of frustules, the samples were spread out over the level base surface of the crucible. Therefore, the completely dry samples were heated in an electric heater for 2, 4, and 6 hours, heating to 600 °C at a rate of 3 °C per minute. For the examination, hot frustules without acid washing were used.

5.3.3 Purification of Raw DE Silica

The DE mining enterprises and the mineral processing sector supply the raw DE in the form of rocks or powder, which is widely available around the world (500.000 per year). Among the impurities in these raw DE minerals alkaline earth, alkaline metals, iron, and organic compounds are included, which may clash with the characteristics of DE required for some particular applications. To remove these contaminants and purify the DE into material with the purity needed for various uses, it is crucial to process the mined raw DE mineral. Crushing, filtration, and size-based separation are typically included in the industrial DE processing of the raw DE [5.53].

5.4 Nutritive and High-End Product

Microalgae productivity is heavily influenced by environmental stress. Many environmental challenges, such as light stress and nutritional

Table 5.1 High-value products from microalgae.

Serial number	Microalgal species	Carbohydrate content (%)	Protein content (%)	Lipid content (%)	References
1.	*Anabaena* sp.	25–30	48	4–7	[5.39]
2.	*A. plentesis*	–	42.33	11	[5.47]
3.	*Botryococcus braunii*	18.5	39.9	34.4	[5.48]
4.	*Chlamydomonas reinhardtii*	17	48	21	[5.52]
5.	*Chaetoceros gracilis*	5	12	7	[5.51]
6.	*Chaetoceros calcitrans*	6	34	16	[5.51]
7.	*Chlorella vulgaris*	12–17	52–58	5–58	[5.52]
8.	*Cantella repens*	28.96	8.42	5.29	[5.40]
9.	*Dunaliella* sp.	4–32	49–57	6–8	[5.44]
10.	*Dunaliella salina*	32	57	6–25	[5.52]
11.	*Enteromorpha intestinalis*	30.58	6.15	7.13	[5.40]
12.	*Euglena gracilis*	14–18	39–61	14–20	[5.52]
13.	*Lola capillaris*	22.32	4.87	4.05	[5.40]
14.	*Nostoc* sp.	25	56	18	[5.52]
15.	*Nitzchia closterium*	10	26	13	[5.51]
16.	*Polysiphonia mollis*	25.81	16.59	5.79	[5.40]
17.	*Phaeodactylum tricornutum*	8	30	14	[5.51]
18.	*Rhizoclonium riparium*	15.34	21.09	3.37	[5.40]
19.	*Scenedesmus* sp.	10–52	50–56	12–14	[5.43]

(Continued)

Table 5.1 High-value products from microalgae. (*Continued*)

Serial number	Microalgal species	Carbohydrate content (%)	Protein content (%)	Lipid content (%)	References
20.	*Scenedesmus dimorphus*	21–52	8–18	16–40	[5.52]
21.	*Skeletonema costatum*	5	25	10	[5.51]
22.	*Ulva lactura*	35.27	8.44	4.36	[5.40]
23.	*Synechoccus* sp.	15	63	11	[5.49]
24.	*Prymnesium* sp.	25–33	28–45	22–38	[5.45]
25.	*Chlorella* sp.	51–58	12–17	14–22	[5.50]
26.	*Chlamydomonas* sp.	2.9–17	43–56	14–22	[5.50]
27.	*Porphyridium* sp.	50–57	28–39	–	[5.39]
28.	*Pavlova* sp.	6–9	24–29	9–14	[5.39]
29.	*Euglena* sp.	14–18	39–61	14–20	[5.41]
30.	*Spirulina maxima*	13–16	60–71	6–7	[5.39]
31.	*Spirogyra*	33–64	6–20	11–21	[5.39]
32.	*Tetraselmis* sp.	15	52	16–45	[5.51]
33.	*Tetraselmis chuii*	25	46.5	12.3	[5.48]
34.	*Thalassiosira pseudonana*	9	34	19	[5.51]

scarcity, are naturally abiotic. It includes nitrogen, phosphorus, iron deficiency, silicon depletion, and light quality and intensity. Deficiency of nutrients (N, P, Fe) also has a significant effect on diatom physiology [5.54].

Various microalgal species serve as a good source of proteins, carbohydrates, and lipids, as shown in Table 5.1. These microalgae species are useful in the production of other high-value products such as carotenoids, EPA, and DHA.

- **Nitrogen:** The lipid content of diatoms is higher than that of other algal groups. Under N and P nutritional deprivation, the production of polyunsaturated fatty acids (PUFAs) or neutral lipids was enhanced and optimized [5.55].

  In ocean ecosystems, nitrogen has a considerable impact on algal productivity. Nitrogen deprivation is a common method for causing algae to accumulate lipids to make biodiesels.
- **Phosphorus:** Triacylglycerols, which are the main building blocks of biodiesel, increased 1.65-fold in the model diatom *P. tricornutum* when P stress conditions were present. Oil bodies are cellular organelles found in algae that store neutral lipids as a source of energy. When P stress was applied during *P. tricornutum*'s stationary phase of growth, the number, and size of oil bodies increased more than they did in control cells [5.57].
- **Iron Deprivation:** At low light levels, iron insufficiency hindered diatom growth and productivity. This was attributed to a higher Fe requirement for the synthesis of proteins rich in iron. The expression and performance of silicon transporters in cell membranes, as well as diatom silicification, can be significantly influenced by availability [5.59]. It was discovered that *Thalassiosira pseudonana* (TpDSP1), a death-specific protein located in the plastid of the coastal diatom, improves growth performance by enhancing the efficiency of photosynthetic fixation of CO_2 during acute iron restriction at subsaturating lighting [5.58].
- **Silicon (Si):** Due to its important function in cell wall formation, silicon is a crucial limiting micronutrient/part for diatom growth. Diatom silica created by a manufacturing process can also be considered a marketable product [5.34].

Under Si deficiency, rapid accumulation of triacylglycerol is caused in most diatoms, altering their shape; however, photosynthesis-related gene expression and protein content are unaffected, as they are under N deficiency [5.60]. Diatoms are thought to have the ability to produce enough lipids to replace petrol-based oils in the future, as well as value-added products [5.64].

5.5 Biofuel Industry

Around 95% of the transportation industry is based on a non-renewable source of energy. As a result, producing sustainable and carbon-neutral fuels could help to reduce the current reliance on fossil fuels while also contributing to the restoration of natural balance without interrupting current economic progress [5.46].

To enhance the quality of biodiesel, the concentration of various fatty acids that influence factors like cetane number, level of emissions, cold flow, oxidative stability, viscosity, and lubricity can be regulated [5.46]. Modern genetic technologies can help in completing the gaps in the development of diatom cells as biofuel machinery. The use of microalgae-based biofuel is supported by statistical data. However, several technological constraints prevent this project from being implemented on a wide scale. As a result, using a blended type of biodiesel is one of the options. For large-scale operation, a mixture of petrodiesel and microalgae/diatom-based fuel would be more efficient. There are no significant differences in performance between blended (20% microalgae fuel + 80% petrodiesel) and 100% petrodiesel tests. Additionally, when compared to pure diesel, CO, unburned HC, and smoke emissions were reported to be lower in blended form [5.28].

5.6 Factors of Diatom for Producing Biofuel

Diatoms have several other qualities in addition to effective photosynthesis and nutrient utilization that are directly or indirectly related to desirable traits for a biofuel production system. Diatoms are extremely diversified and differ significantly from chlorophytes in terms of cellular structure and metabolic functions based on evolution [5.34]. Diatoms are extremely productive compared to other algae classes and have exceptional ecological success, which may be related to the cellular processes covered in this research [5.65]. Diatoms may rapidly produce triacylglycerol under Si constraint and accumulate lipid to an equivalent or greater amount than other algal classes, avoiding the negative effects of N limitation on photosynthesis, gene expression, and protein content. Diatoms have been produced in production amounts by aquaculture for many years since they provide products with added value and are accessible to omics and genetic manipulation techniques [5.34].

5.7 Biomedical Industry

Antibodies and hydrophobic drug molecules are attached to diatom biosilica without the use of crosslinking, resulting in a specific nanoporous biosilica delivery system for chemotherapeutic drugs (Organism: *T. pseudonana*). Innovative biomedical research techniques and LED technology have the potential to greatly enhance diatom growth performance and production. To support the current light-collection system in *P. tricornutum* cells, programmable modules may be created with heterologous expression of enhanced green fluorescence proteins (eGFPs) as light absorbers and transmitters. Then, eGFPs can be excited and efficient development can be facilitated by LED illumination with various light spectra. By improving the light-harvesting mechanism in diatom culture for all sizes of photobioreactors, LED-aided synthetic biology may hold promise for the effective generation of carotenoids [5.62]. Tyr-ZnII serves as a zinc carrier for controlled release to bacteria and inhibits bacterial growth in a modified frustule with self-assembled antibacterial aromatic amino acid conjugates (Organism: *N. palea*) [5.54]. Using a crosslinked amine-functionalized diatom photoluminescent biosensor, typhoid may be detected quickly and selectively (Organism: *Amphora sp.*). For non-enzymatic detection of methyl parathion, a nanocomposite of nanoporous diatoms and ZrO2 has been developed (Organism: *P. tricornutum*). For lithium-ion batteries, a biomaterial made of a 3D-structured diatom biosilica for the negative electrode showed enhanced charge capacity when compared to graphite (Organism: *P. trainori*) [5.61] [5.63]. FeOx *in situ* coating on live diatoms is a viable material for supercapacitor electrodes which improves capacitor performance (Organism: *P. tricornutum).*

5.8 DE Silica for Tissue Engineering

The function of osteoblast cells is improved by silicon, which also stimulates the production of new bone cells' mineralization. A silica/silicon deficit is frequently linked to long bones that are malformed and deformed. A naturally occurring diatom skeleton deposit known as diatomite, a low-cost and biogenic silica, is a plentiful resource with applications in regenerative medicine [5.53]. Due to their distinctive 3D silica structures, which were produced by genetically controlled identity methods under environmentally friendly conditions, diatoms provide an outline for the formation of a new generation of biomaterials with properties that are not possible with synthetic porous silica materials. Due to their distinctive hierarchical porous structure, high surface area, adaptable surface modification, high biocompatibility, chemical stability, and unique optical and photoluminescence properties, diatom frustules have been used as a novel, affordable scaffold for a variety of biomedical applications. The most recent uses and hypotheses for using diatoms as drug carriers and for biological applications were examined in an article [5.53]. Surface modifications, drug loading, and therapeutic molecular delivery abilities of the produced diatom carriers were all carried out in a sustained and controlled manner. Diatom carrier synthesis, surface modifications, drug loading, and therapeutic methods were also examined [5.53]. The ability to deliver molecules was highlighted in a controlled and continuous way. The most recent development, which transformed converted diatom silica into silicon-degradable micro- and nanocarriers, was demonstrated to be a promising and inspiring strategy to further explore diatom silica material as a valuable low-cost substitute for synthetic porous silicon for the development of a new generation of naturally produced drug delivery systems. It will take a lot of work to examine the biological response of diatom silica as a new pharmaceutical drug delivery carrier required for clinical applications.

5.9 Nanotechnologically Derived Smart Drug Delivery System

Nanoporous synthetic silica is arguably the most beneficial mesoporous material that continues to be applied in drug delivery applications. However, their synthetic route is known to be demanding, expensive, and time-consuming. In addition, its synthesis demands highly skilled labor and involves a high level of toxic and chemical waste [5.19]. There is a critical need to substitute the synthetic silica material with a natural surrogate. The never-ending benefits of diatom silica make it a promising applicant for overcoming conventional pharmaceutical limitations and a definitive replacement for synthetic silica. The diatom shells have the potential to conquer the challenges currently being faced by the medical industry. The abundantly existing diatom silica has numerous advantages over synthetic silica, it is amorphous, mechanically resistant, with high chemical inertness and thermal stability. This biocompatible material has low density and high permeability. It has great internal space that allows it to carry a higher quantity of the drug. Its surface chemistry can be modified according to the requirements and it also has an exceptional ability to imitate the nature of natural bone and therefore can be utilized for medical implants [5.19].

All these exceptional features make diatom frustules a great scaffold for the engineering and advancement of nanostructured devices for not only drug delivery but also numerous other applications, i.e., liquid filtration, photonics, DNA purification, and immunoprecipitations [5.32] [5.33]. Diatoms, because of their amorphous nature, are considered harmless and a potential alternative for the medical as well as food supplement industries [5.33]. The exemplary structural morphology and substantial features of *Thalassiosira weissflogii*, *Nitzschia* sp., *Coscinodiscus concinnus* and microporous *Aulacoseira* sp. have allowed them to stand out as potential nanocarriers for drug delivery applications. The large surface area of the diatom allows very easy modification of their nanopores for the controlled drug delivery and release process [5.34]. They can also be easily cast with an ultra-fine layer of polymer. Their nanopores can be very uniquely altered to fit the commitments of drugs, their perfusion rates, and durations. For sustained drug release, they can be inorganically regulated and transformed into a potential device (Figure 5.2). There have also been multiple studies concerning the chemical modifications of diatoms' surfaces [5.35].

In terms of controlled drug delivery applications, the use of diatoms has escalated greatly because of their unique modifiable features. The very first

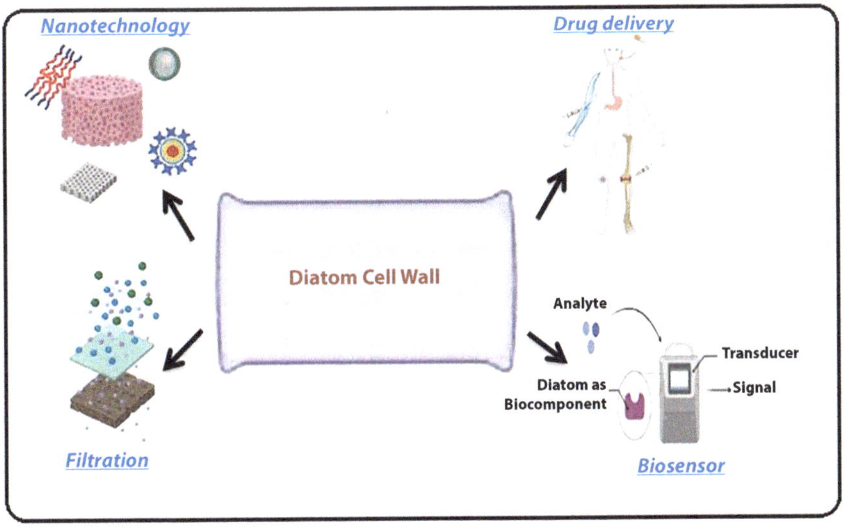

Figure 5.2 Diatoms: Nature's engineers for nanotech, drug delivery, filtration, and biosensing.

engagement of diatoms with drug delivery was communicated in 2010 by Uthappa *et al.* [5.38]. For magnetized drug delivery applications, the drug release profile of diatoms must be fine-tuned and the surface modifications must be put in place. The first magnetically guided diatom was synthesized by electrostatically altering the diatom's surface with dopamine-altered iron oxide nanoparticles [5.19]. They were employed to conduct tests with indomethacin. Indomethacin has a 0.28 g/g drug loading capacity [5.39]. In the tests conducted to test their drug release profile, the first-phase drug release was around 60% for six hours and the remaining drug release lasted for the remaining two weeks [5.39]. However, when the drug was delivered with no surface modification and indomethacin was directly supplied in the diatom pores, 65–70% of the drug's burst release was observed in the first two hours [5.39]. This difference is due to the existence of weak bonds connecting drug molecules to the silica surface. On the other hand, the slow drug release that occurred through the modified diatoms was because the drug was stored in empty internal micropores of the diatoms [5.19]. The drug-release operation can be best described by zero-order kinetics.

To study the desired functionality of the diatom's exterior alterations, organosilanes can be utilized. Organosilanes are silicon-based monomers that have a minimum of one direct bond between a carbon and a silicon atom [5.41]. They can govern the hydrophobic and hydrophilic features of the diatom surface [5.41]. A substantial observation was made in regards to the polarity of the diatom surface. Hydrophilic surfaces were noticed to act more accountable for the magnification in controlled drug loading and release [5.42]. Multiple antibiotic drugs like gentamicin and indomethacin have been used in multiple studies to investigate pharmacokinetics with surface modifications. It was not hard to see those hydrophilic surfaces like epoxy-rich-GPTMS 3-trimethoxysilane that intensified the loading capacity of hydrophobic drugs [5.43]. It also provided a higher level of regulation for drug release. Compared to this, less drug loading capacity can be seen in diatoms modified with hydrophobic surfaces. Due to the addition of PEG-Silane and OTS-like substances on the diatom's surface, the drug release rate can be intensified [5.41].

Along with the physical modifications, there have also been studies on chemical modifications of diatoms. In one of the studies, it was observed that the drug loading capacity can be doubled; the drug release can be extended up to 23 days using a nano-hybrid complex of diatom's surface with graphene oxide [5.38]. This was because of improved interaction between the layers of graphene oxide and the drug molecule. The presence of weaker bonds between the diatom's surface and the drug is the prominent reason for its lower loading capacity and faster release [5.44].

This study reassured the continuous drug release from graphene-dioxide modified diatoms, which is extremely beneficial in drug delivery applications. Biosilica is preferred for drug delivery processes due to its natural existence and nontoxic behavior with hydroxyl groups. For tissue-regeneration processes, amorphous silica is preferred over any other material due to its biocompatibility and its nontoxic prolonged drug release [5.19].

Diatoms are regarded highly in the medicinal and research industries because they are non-cytotoxic and have high permeability [5.45]. Multiple studies have been conducted to evaluate and corroborate their cytotoxicity and their impact on drug permeation. During safety and evaluation examinations, it was found that diatoms, modified or original, did not initiate any toxic reaction or death on cells, but rather improved the drug permeability and provided extended drug release [5.46]. The motivation behind applying diatoms for drug delivery is to tackle the general problems that arise in clinical therapies. Diatom's surface is extremely easy to modulate according to the drug delivery requirements and can be specifically designed to meet the needs of the patient. The drug loading and release rate can be monitored and controlled according to the therapy demands [5.5]. They are also heavily used in the process of magnetically guided smart drug delivery because of their unique features. Continuous development in nanotechnology has carved out a path for advancing strategies in disease prevention and treatments. The development of clinically potent and novel nanodevices that highly influence the physiological interactions at the systematic and molecular level is aimed at the advancement of biomedical applications [5.19]. Strict guidelines must be put in place, however, when characterization engineered nanodevices address their safety, quality, and toxicity. A thorough clinical investigation must be performed to fully diagnose the nano-platforms for their application in the biomedical industry.

Medicinal drugs to be taken orally are conventionally produced for instant release and absorption. These are the most extensively operated routes. However, they come with a range of disadvantages. These conventional drugs have a limited half-life and require continuous regulation, consequently increasing the possibility of missing the dose and causing poor patient compliance [5.47]. A sustained drug release system was designed to overcome these issues and transform the medical treatment process. Controlled drug delivery needs a structured drug release for a pre-established and definite rate and time [5.48]. The drug release can be prolonged to shorten the dose repetition, ultimately avoiding the side effects of the medications and limiting the drug plasma concentrations in a defined range [5.48]. Sustained drug delivery systems are superior in terms of controlling the drug dosage and release patterns, tolerance, comfort, and patient compliance [5.49]. Compared to conventional drug delivery systems, they have little to no control over the transport and distribution of the dose in the body, especially in the targeted area [5.19] [5.48]. Diatom frustules are promising nanodevices that can be applied to provide sustained drug release [5.7]. Their biocompatibility and high maneuverability serve the pharmaceutical industries adequately. They are presented as the next medical revolution for targeted drug delivery systems due to their extraordinary characteristics, i.e., large surface area, great mechanical resistance, biocompatibility, and biodegradability. Significant modification strategies, when applied, can create opportunities for targeted insoluble drug delivery without the risk of precipitation. To advance the bioavailability of medical treatments, nontoxic features of biosilica can be enhanced and applied.

5.10 Future Perspective

Diatom biosilica is an influential instrument that can revolutionize the drug delivery systems in the medical industry. They have been regarded as the best agent for delivering drugs because they dissolve into degradable compounds that are nontoxic and can easily be cleared out of the body. There is a need for comprehensive studies further exploring the cost-effective diatom scaffold for smart mediated drug delivery systems. Further investigations must be carried out to test its clinical efficacy, biocompatibility, and toxicity for multiple drugs. Now, after the approval of the FDA (Food and Drug Administration), EPA (Environmental Protection Agency), and USDA (United States Department of Agriculture), the next big step is to be taken by the pharmaceutical and medicinal industries to conduct and implement a regulatory approval for biosilica's application in medical therapies.

5.11 Conclusion

Starting with their contribution to the oceans and earth, diatoms have come a long way. Researchers have investigated and corroborated their competence in therapeutic applications, for instance, cancer therapy, biosensors, biophotonics, cartilage, and dermal renovations, and now individually designed drug carriers. Modified biosilica has proven to be the best biocompatible agent for gene transfer, ultimately improving the process of non-viral gene delivery and its imaging. Novel drug carriers are required and believed to progress in the upcoming future. However, exhaustive measures need to be put in place to make them available to the genomics and proteomics platforms. Progress must be accelerated to understand them throughout and to apply them extensively in therapeutics.

References

[5.1] Sharma, N., Simon, D. P., Diaz-Garza, A. M., Fantino, E., Messaabi, A., Meddeb-Mouelhi, F., Germain, H., & Desgagné-Penix, I. (2021). Diatoms Biotechnology: Various Industrial Applications for a Greener Tomorrow. *Frontiers in Marine Science*, 8. https://doi.org/10.3389/fmars.2021.636613

[5.2] Berg, G., Rybakova, D., Fischer, D., Cernava, T., Vergès, M.-C. C., Charles, T., Chen, X., Cocolin, L., Eversole, K., Corral, G. H., Kazou, M., Kinkel, L., Lange, L., Lima, N., Loy, A., Macklin, J. A., Maguin, E., Mauchline, T., McClure, R., … Schloter, M. (2020). Microbiome definition re-visited: old concepts and new challenges. *Microbiome*, 8(1), 103. https://doi.org/10.1186/s40168-020-00875-0

[5.3] Sachin, K., & Karn, S. K. (2021). Microbial Fabricated Nanosystems: Applications in Drug Delivery and Targeting. *Frontiers in Chemistry*, 9, 617353. https://doi.org/10.3389/fchem.2021.617353

[5.4] Liu, H., Chen, M., Zhu, F., & Harrison, P. J. (2016). Effect of Diatom Silica Content on Copepod Grazing, Growth and Reproduction. *Frontiers in Marine Science*, 3. https://doi.org/10.3389/fmars.2016.00089

[5.5] Serôdio, J., & Lavaud, J. (2020). *Diatoms and Their Ecological Importance* (pp. 1–9). https://doi.org/10.1007/978-3-319-71064-8_12-1

[5.6] Malviya, S., Scalco, E., Audic, S., Vincent, F., Veluchamy, A., Poulain, J., Wincker, P., Iudicone, D., de Vargas, C., Bittner, L., Zingone, A., & Bowler, C. (2016). Insights into global diatom distribution and diversity in the world's ocean. *Proceedings of the National Academy of Sciences*, 113(11). https://doi.org/10.1073/pnas.1509523113

[5.7] Saxena, A., Tiwari, A., Kaushik, R., Iqbal, H. M. N., & Parra-Saldívar, R. (2021). Diatoms recovery from wastewater: Overview from an ecological and economic perspective. *Journal of Water Process Engineering*, 39, 101705. https://doi.org/10.1016/j.jwpe.2020.101705

[5.8] Sánchez, C., Cristóbal, G., & Bueno, G. (2019). Diatom identification including life cycle stages through morphological and texture descriptors. *PeerJ*, 7, e6770. https://doi.org/10.7717/peerj.6770

[5.9] Maeda, Y., Nojima, D., Yoshino, T., & Tanaka, T. (2017). Structure and properties of oil bodies in diatoms. *Philosophical Transactions of the Royal Society of London. Series B, Biological Sciences*, 372(1728). https://doi.org/10.1098/rstb.2016.0408

[5.10] Falciatore, A., Jaubert, M., Bouly, J.-P., Bailleul, B., & Mock, T. (2020). Diatom Molecular Research Comes of Age: Model Species for Studying Phytoplankton Biology and Diversity. *The Plant Cell*, 32(3), 547–572. https://doi.org/10.1105/tpc.19.00158

[5.11] Vardi, A. (2008). Cell signaling in marine diatoms. *Communicative & Integrative Biology*, 1(2), 134–136. https://doi.org/10.4161/cib.1.2.6867

[5.12] Scholz, B. (2014). Purification and culture characteristics of 36 benthic marine diatoms isolated from the Solthörn tidal flat (Southern North Sea). *Journal of Phycology*, 50(4), 685–697. https://doi.org/10.1111/jpy.12193

[5.13] Leblanc, K., Quéguiner, B., Diaz, F., Cornet, V., Michel-Rodriguez, M., Durrieu de Madron, X., Bowler, C., Malviya, S., Thyssen, M., Grégori, G., Rembauville, M., Grosso, O., Poulain, J., de Vargas, C., Pujo-Pay, M., & Conan, P. (2018). Nanoplanktonic diatoms are globally overlooked but play a role in spring blooms and carbon export. *Nature Communications*, 9(1), 953. https://doi.org/10.1038/s41467-018-03376-9

[5.14] Caputo, A., Nylander, J. A. A., & Foster, R. A. (2019). The genetic diversity and evolution of diatom-diazotroph associations highlights traits favoring symbiont integration. *FEMS Microbiology Letters*, 366(2). https://doi.org/10.1093/femsle/fny297

[5.15] Benoiston, A.-S., Ibarbalz, F. M., Bittner, L., Guidi, L., Jahn, O., Dutkiewicz, S., & Bowler, C. (2017). The evolution of diatoms and their biogeochemical functions. *Philosophical Transactions of the Royal Society of London. Series B, Biological Sciences*, 372(1728). https://doi.org/10.1098/rstb.2016.0397

[5.16] Shukla, S.K. and Mohan, R. (2012) The contribution of diatoms to worldwide crude oil deposits. In: *The Science of Algal Fuels: Phycology, Geology, Biophotonics, Genomics and Nanotechnology*. R. Gordon and J. Seckbach, (eds.) Springer, Dordrecht: 355–382.

[5.17] Armbrust, E. V. (2009). The life of diatoms in the world's oceans. *Nature*, 459(7244), 185–192. https://doi.org/10.1038/nature08057

[5.18] Pančić, M., Torres, R. R., Almeda, R., & Kiørboe, T. (2019). Silicified cell walls as a defensive trait in diatoms. *Proceedings. Biological Sciences*, 286(1901), 20190184. https://doi.org/10.1098/rspb.2019.0184

[5.19] Phogat, S., Saxena, A., Kapoor, N., Aggarwal, C., & Tiwari, A. (2021). Diatom mediated smart drug delivery system. *Journal of Drug Delivery Science and Technology*, 63, 102433. https://doi.org/10.1016/j.jddst.2021.102433

[5.20] Mark, H., & Richard, W. (2003). *Components and Control of Silicification in Diatoms* (pp. 11–57). https://doi.org/10.1007/978-3-642-55486-5_2

[5.21] Tiffany, M. A. (2005). Diatom Auxospore Scales and Early Stages in Diatom Frustule Morphogenesis: Their Potential for Use in Nanotechnology. *Journal of Nanoscience and Nanotechnology*, 5(1), 131–139. https://doi.org/10.1166/jnn.2005.011

[5.22] Heintze, C., Formanek, P., Pohl, D., Hauptstein, J., Rellinghaus, B., & Kröger, N. (2020). An intimate view into the silica deposition vesicles of diatoms. *BMC Materials*, 2(1), 11. https://doi.org/10.1186/s42833-020-00017-8
[5.23] Topal, E., Rajendran, H., Zgłobicka, I., Gluch, J., Liao, Z., Clausner, A., Kurzydłowski, K., & Zschech, E. (2020). Numerical and Experimental Study of the Mechanical Response of Diatom Frustules. *Nanomaterials*, 10(5), 959. https://doi.org/10.3390/nano10050959
[5.24] Von Dassow, P., Petersen, T. W., Chepurnov, V. A., & Virginia Armbrust, E. (2008). Inter- and intraspecific relationships between nuclear DNA content and cell size in selected members of the centric diatom genus *Thalassiosira* (Bacillariophyceae). *Journal of Phycology*, 44(2), 335–349. https://doi.org/10.1111/j.1529-8817.2008.00476.x
[5.25] Fuhrmann, T., Landwehr, S., el Rharbi-Kucki, M., & Sumper, M. (2004). Diatoms as living photonic crystals. *Applied Physics B*, 78(3–4), 257–260. https://doi.org/10.1007/s00340-004-1419-4
[5.26] Lang, Y., Monte, F. del, Rodriguez, B. J., Dockery, P., Finn, D. P., & Pandit, A. (2013). Integration of TiO2 into the diatom *Thalassiosira weissflogii* during frustule synthesis. *Scientific Reports*, 3(1), 3205. https://doi.org/10.1038/srep03205
[5.27] Gale, D. K., Gutu, T., Jiao, J., Chang, C.-H., & Rorrer, G. L. (2009). Photoluminescence Detection of Biomolecules by Antibody-Functionalized Diatom Biosilica. *Advanced Functional Materials*, 19(6), 926–933. https://doi.org/10.1002/adfm.200801137
[5.28] Mishra, M., Arukha, A. P., Bashir, T., Yadav, D., & Prasad, G. B. K. S. (2017). All New Faces of Diatoms: Potential Source of Nanomaterials and Beyond. *Frontiers in Microbiology*, 8, 1239. https://doi.org/10.3389/fmicb.2017.01239
[5.29] Bao, Z., Weatherspoon, M. R., Shian, S., Cai, Y., Graham, P. D., Allan, S. M., Ahmad, G., Dickerson, M. B., Church, B. C., Kang, Z., Abernathy III, H. W., Summers, C. J., Liu, M., & Sandhage, K. H. (2007). Chemical reduction of three-dimensional silica micro-assemblies into microporous silicon replicas. *Nature*, 446(7132), 172–175. https://doi.org/10.1038/nature05570
[5.30] Vasir, J. K., & Labhasetwar, V. (2005). Targeted Drug Delivery in Cancer Therapy. *Technology in Cancer Research & Treatment*, 4(4), 363–374. https://doi.org/10.1177/153303460500400405
[5.31] Delasoie, J., & Zobi, F. (2019). Natural Diatom Biosilica as Microshuttles in Drug Delivery Systems. *Pharmaceutics*, 11(10). https://doi.org/10.3390/pharmaceutics11100537
[5.32] Townley, H. E., Parker, A. R., & White-Cooper, H. (2008). Exploitation of Diatom Frustules for Nanotechnology: Tethering Active Biomolecules. *Advanced Functional Materials*, 18(2), 369–374. https://doi.org/10.1002/adfm.200700609

[5.33] Wenkai Jiang *et al.* (2013). Purification of biosilica from living diatoms by a two-step acid cleaning and baking method. *Journal of applied phycology* 26(3). 1511-1518.
[5.34] Hildebrand M, Davis AK, Smith SR, Traller JC, Abbriano R. (2012). The place of diatoms in the biofuels industry. *Biofuels*;3:221–40.
[5.35] M.A. Borowitzka *et al.* (eds.), The Physiology of Microalgae, *Developments in Applied Phycology* 6.
[5.36] Davis, A. K., Hildebrand, M. (2010). Molecular Processes of Biosilicification in Diatoms. *Biomineralization*, 255–294.
[5.37] Siddharth V., Stephan J (2002). Silicification and Biosilicification. *Journal of Inorganic and Organometallic Polymers*, Vol. 12.
[5.38] Uthappa UT, Bhat MP, Jung HY, Kurkuri MD. (2022). Surface-functionalized diatoms for drug delivery and tissue engineering applications. *In Bioprospecting Algae for Nanosized Materials*. (pp. 275-289). Cham: Springer International Publishing.
[5.39] Becker, E.W. (1994). Microalgae: Biotechnology and microbiology. Cambridge: Cambridge University Press.
[5.40] Chakraborty, S., and Santra, S.C. (2008). Biochemical composition of eight benthic algae collected from Sunderban. *Indian J. Mar. Sci.* 37(3):329–332
[5.41] Collyer, D.M., and G.E. Fogg. (1955). Studies on fat accumulation by algae. *J. Exp. Bot.* 6:256.
[5.42] Eddy, B.P. (1956). The suitability of some algae for mass cultivation for food with special reference to *Dunaliella bioculata. J. Experimental Bot.* 7: 372.
[5.43] Hindak, F., and Probil, S. (1968). Chemical composition, protein digestibility and heat of combustion of filamentous green algae. *Plant Biol.* 10:234.
[5.44] Parson, T. R., Stephens, K. and Strickland, J. D. H., (1961). On the chemical composition of eleven species of marine phytoplankters. *Journal of the Fisheries Research Board Canada,* 18: 1001–1016.
[5.45] Ricketts, T.R. (1966). On the chemical composition of some unicellular algae. *Phytochemistry* 5: 67.
[5.46] Rodrigue, J.-P., and Notteboom, T. (2013). The geography of cruises: itineraries, not destinations. *Appl. Geography* 38, 31–42.
[5.47] Sydney, E. B., Sturm, W., de Carvalho, J. C., *et al.* (2010). Potential carbon dioxide fixation by industrially important microalgae. *Bioresour. Technol.* 101(15):5892–5896.
[5.48] Tibbetts, S. M., Milley, J. E., and Lall, S. P. (2015). Chemical composition and nutritional properties of freshwater and marine microalgal biomass cultured in photobioreactors. *J. Appl. Phycol.* 27(3):1109–1119.
[5.49] Trubachev, N.I., Gitel'zon, I.I., Kalacheva, G.S., Barashkov, V.A. Belyanin, V.N. and Andreeva, R.I. (1976). Biochemical composition of several

blue green algae and *Chlorella*. *Prikladnaia Biohimiiai Mikrobiologiia* 12:196–202.

[5.50] Renaud S. M., Parry D. L., Thinh L.-V. (1994). Microalgae for use in tropical aquaculture I: gross chemical and fatty acid composition of twelve species of microalgae from the Northern Territory, Australia. *J. Appl. Phycol.* 6, 337–345.

[5.51] Brown M. R. (1991). The amino acid and sugar composition of sixteen species of microalgae used in mariculture. *J. Exp. Mar. Biol. Ecol.* 145, 79–99.

[5.52] Mata T. M., Martins A. A., Caetano N. S. (2010). Microalgae for biodiesel production and other applications: A review. 14(1), 217–232.

[5.53] Maher, S., Kumeria, T., Aw, M. S., & Losic, D. (2018). Diatom Silica for Biomedical Applications: Recent Progress and Advances. *Advanced Healthcare Materials*, 1800552.

[5.54] Fu, W., Wichuk, K., & Brynjólfsson, S. (2015). Developing diatoms for value-added products: challenges and opportunities. *New Biotechnology*, 32(6), 547–551.

[5.55] Rezanka T., Lukavsky J., Nedbalova L., Kolouchova I. (2012). Sigler. Effect of starvation on the distribution of positional isomers and enantiomers of triacylglycerol in the diatom *Phaeodactylum tricornutum*. *Phytochemistry*, 80, pp. 17.

[5.56] United States Patent No. 7892580. Padmawar A. Process for producing a stable concentrated dietary supplement and supplement produced thereby. May 12, 2011.

[5.57] Chen Z, Yang M, Li C, Wang Y, Zhang J, Wang D, *et al.* (2014). Phosphoproteomic analysis provides novel insights into stress responses in *Phaeodactylum tricornutum*, a model diatom. *J Proteome Res* 13:2511–23.

[5.58] Thamatrakoln K., Bailleul B., Brown C.M., Gorbunov M.Y., Kustka A.B., Frada M., Joliot P.A., Falkowski P.G., Bidle K.D. (2013). Death-specific protein in a marine diatom regulates photosynthetic responses to iron and light availability. *Proc Natl Acad Sci*; 110(50):20123-20128.

[5.59] De La Rocha CL, Hutchins DA, Brzezinski MA, Zhang Y. Effects of iron and zinc deficiency on elemental composition and silica production by diatoms. *Mar Ecol Prog Ser* 2000;195:71–9.

[5.60] Martin-Jezequel V, Hildebrand M, Brzezinski M. (2000). Silicon metabolism in diatoms: implications for growth. *J Phycol* 36:821–40.

[5.61] Schulze P.S., Barreira L.A., Pereira H.G., Perales J.A., Verala J.C. (2014). Light emitting diodes (LEDs) applied to microalgal production. *Trends Biotechnol*, 32, pp. 422-430.

[5.62] Yam FK, Hassan Z. (2005). Innovative advances in LED technology. *Microelectron J*;36:129–37.

[5.63] Yeh N, Chung JP. (2009). High-brightness LEDs-energy efficientlighting sources and their potential in indoor plant cultivation. *Renew Sust Energ Rev* 13:2175–80.
[5.64] Jin P, Agustí S. (2018). Fast adaptation of tropical diatoms to increased warming with trade-offs. *Scientific Reports.* 11;8(1):17771.
[5.65] Brennan L, Owende P. (2010). Biofuels from microalgae—a review of technologies for production, processing, and extractions of biofuels and co-products. *Renewable and sustainable energy reviews.* 1;14(2):557-77.

6

Biochemical Engineering of Diatoms for Health Benefits

Rishabh Rathore[‡], Pragati Verma[‡], Sonali Raghdale[‡], Avishek Kumar, Mohd Jahir Khan* and Vandana Vinayak[†]

Diatom Nanoengineering and Metabolism Laboratory (DNM), School of Applied Science, Dr. Harisingh Gour Central University, Sagar, Madhya Pradesh, India

Abstract

Diatom frustules are the best source of natural silica. The frustules are nanoporous having nano arrayed architecture which varies from species to species. They are incredibly promising for use in medical applications due to their unique characteristics such as high specific surface area, thermal stability, and biocompatibility. Diatoms are among the largest primary producers in our environment since they can synthesize a wide range of amino acids for building proteins and other high-value compounds, such as lipids, fatty acids, pigments, carbohydrates, polyhydroxyalkanoates, and silica. The potential health benefits associated with these compounds make them ideal for use in the medical field. The purpose of this chapter is to examine the chemical composition of diatoms and their antioxidant, anti-inflammatory, and therapeutic properties for use in the treatment of many diseases. It also aims to determine the potential future direction of this diverse field.

Keywords: Diatoms, silica, value added products, fatty acids, fucoxanthin, health benefits

**Corresponding author*: khanmohdjahir@gmail.com

[†]*Corresponding author*: kapilvinayak@gmail.com

[‡]These authors contributed equally to this work.

Vandana Vinayak and Richard Gordon (eds.) Diatom Cultivation for Biofuel, Food and High-Value Products, (139–166) © 2025 Scrivener Publishing LLC

6.1 Introduction

Diatoms are unicellular microalgae belonging to the class Bacillariophyceae [6.12]. They can be found in moist soil, fresh and marine environments. Benthic diatoms predominate mostly in estuarine and shallow systems [6.67]. They also play an important part in the biogeochemical cycle, feeding a variety of higher trophic species and acting as primary producers in the food web. According to Kale and Karthick [6.48], they make up the majority of phytoplankton in aquatic habitats and are responsible for 20–25% of global carbon fixation. They also have the ability to extract biological oxides of nitrogen from combustion gases [6.36]. The cell wall of diatoms is characterized by their siliceous cell walls. Silica is a key aspect of diatoms' distinctiveness. Silica causes the cell walls to be hard, limits certain elements of reproduction, and helps to preserve diatom frustules. The nexus, chloroplasts, mitochondria, etc., in diatoms are similar to those found in other eukaryotic algae as are other cell components, for example Chlorophyll *a*, *c1*, and *c2* and fucoxanthin are found in the chloroplasts of diatoms.

6.1.1 Diatom Pigments

Fucoxanthin, a golden-brown pigment in diatoms, gives them their golden color [6.15]. According to Xia *et al.* [6.103], diatoms and brown algae are the main sources of this pigment, which also has some promising physiological qualities such as anti-inflammatory, anti-tumor, anti-obesity, anti-diabetes, anti-malarial, and others. They can be used as a renewable energy source and as "drop-in-fuel" to meet fuel needs in the near future [6.59].

Brown algae and the majority of other diatoms exhibit carotenoids mainly fucoxanthin, diadinoxanthin, and diatoxanthin. Fucoxanthin is one of the primary cell colorants in diatoms, and makes up between 1.5% and 2.5% of the dry cell weight, which is significantly more than in other fucoxanthin producing algae [6.43]. These pigments have a variety of possible health advantages and are engaged in both photosynthetic and photoprotective responses. According to Tokarek *et al.* [6.97], they can be extracted via sonication, extraction with a chloroform-methanol solvent mixture, and preparative thin layer chromatography.

6.1.2 Diatoms' Nutritional Value

Over the past few decades, the long-chain polyunsaturated omega-3 fatty acids (ω-3 PUFAs) have been recognized as compounds of great physiological importance for humans and other consumers [6.96]. The most nutritionally relevant biomolecules produced by diatoms are unsaturated fatty acids like eicosapentaenoic acid (EPA), 3.9–5% of dry weight in *Phaeodactylum tricornutum* [6.58] [6.70], arachidonic acid [6.58], docosahexaenoic acid [6.52], and other omega-3 fatty acids. The major carbon storage compound in diatoms is lipids, among which triacyl glycerides (TAGs) and fatty acids usually make up 15–25% of dry biomass [6.65]. They reduce cardiovascular disease [6.58], and are precursors for important tissue hormones, which are present in 20% of the brain's grey matter [6.52]. EPA and DHA (docosahexaenoic acid) play important roles in protecting the cardiovascular system [6.88], vision [6.37], and treating psychiatric disorders [6.56].

Notably, lipid fractions has been reported to be as high as 70–85% in some diatoms, but 15–25% is more common [6.86]. So, lipids are the major constituents of diatom cells and the average lipid content could be up to 25% of dry weight in diatoms [6.59]. Many valuable lipids and their derivatives, such as essential fatty acids, steroids, and oxylipins, can be produced in diatoms in natural growth environments [6.107]. Nitrogen is vital to diatom's proper physiological functions and is a key component in metabolism of amino acids, nucleic acids, and photosynthetic pigments, and its uptake is at the highest level among all nutrients [6.106]. Exposing diatoms to abiotic stresses, like nitrogen (N) deficiency or cold stress, not only boosts lipid synthesis and yield but also results in a modified composition of long-chain PUFAs [6.106].

6.1.3 Diatoms as Bio-Indicators

Diatoms are appropriate indicators that can provide eco-friendly tools in detecting contamination problems quality of water in aquatic environments because they are very sensitive to environmental stress [6.47]. They act as important bio-indicators to monitor the metal concentrations in diverse habitats [6.21]. Their tolerances and preferences for pH, conductivity, salinity, humidity, organic matter, saprobity, trophic state, oxygen requirements, nutrients, and current velocity in freshwater streams, rivers, lakes, wetlands, and estuaries have been defined, and diatoms have also been used in paleolimnological studies [6.62]. A significant negative correlation between pennate diatoms density and heavy metal concentration (available Co) is observed, while significant negative correlations were found between diatom density and Co, Cu, and Zn [6.101].

6.1.4 Metal Toxicity

Diatoms use complex defense mechanisms, which include the production of metal exclusion, regulation of metal ligand production, cell surface ligand modification, and antioxidant molecules production, in order to counter metal toxicity. Photosynthetic pigments chlorophyll *a*, *b* and *c* are affected by heavy metal toxicity, which is countered by diatoms' in terms of increasing pigment production [6.67]. Detoxified heavy metals are transported into membranes; heavy metals are utilized by cellular organelles for different metabolic activities. Diatoms have the ability to store up to 10% of their biomass in trace metals due to the presence of high-affinity and metal-binding functional groups in their silica cell wall [6.67]. They regulate their enzymatic activity, which enables them to detoxify heavy metals by dehalogenation and denitrification [6.105]. The quality of water not only decides the type of diatom but also alters the metabolism as well changes the shape of frustules. There is large biodiversity of diatoms in terms of their shape and size in different qualities of water. There are at least 30,000, and probably triple that, different diatom species in existence, each characterized by different forms and structures of their outer silica shell [6.66].

6.2 Chemical Composition of Diatom Biomass

Diatoms are single-celled microalgae with a silica, polypeptide, and protein-based cell wall [6.51]. Due to its high lipid content and distinct fatty acid composition, the marine diatom *Chaetoceros muelleri* is one of the species that is most frequently suggested for feeding crustaceans and mollusk larvae around the world [6.99]. Further diatoms, cell wall is rich in proteins, phospholipids, pigment and cholesterol [6.79]. The frustule, or cell wall, of diatoms, which is composed of silica and organic materials, is one of its distinctive characteristics. Each of the cell walls has two sections, the hypotheca and epitheca, which are made up of valve structures that are connected by girdle bands that normally ring the cell. Due to their short life cycles, rapid development, and basic nutritional needs, diatoms are well known as a reserve source for biofuel, marine pharmaceuticals, and various high and low value compounds [6.61]. Diatoms also play a role in the nitrogen, iron, and silica cycles of oceans [6.34]. Due to their use in the biofuel, food, and pharmaceutical industries, diatoms are nature's undiscovered source of a number of biosynthetic compounds [6.51]. Callose is an insoluble polymer of β-1,3-glucose that is connected to the frustule [6.94]. Diatoms have wide applications in photonics, sensing, optoelectronics, biomedicine, energy conversion and storage due to their high surface area, mechanical resistance, special optical properties, and biocompatibility [6.69]. Diatoms can be used for biomineralization, synthesis of biomaterials, and the degradation of wastes, and serve as bio-indicators of metal toxicity [6.47] [6.94].

6.2.1 Carbohydrate

Most of the diatom cell wall is composed of silica but as mentioned above, it is rich in other metabolites also, of which carbohydrate is major constituent. The frustule of a diatom is made up of three successive layers: the innermost organic layer, known as the diatotepum, which is in contact with the plasmalemma; the second layer is a mineralized, silicified shell that also contains organic matter; and the third layer is an external organic coat that is encased in secreted mucilage and is referred to as "cell wall-bound exopolysaccharides (EPSs)" [6.34]. Exopolysaccharides from diatoms are bioactive substances dispersed in the environment. Polysaccharide-rich EPS contain uronic acids, sulfated sugars, and protein. Pennate diatoms and other microbes respond to inorganic nutrient constraint, salinity, and low temperature stress by altering the production rates and chemical composition of their EPS [6.8]. Chrysolaminarin, a nonlinear β-glucan with a linear 1,3-chain and 1,6-branches that is stored in the cytoplasmic vacuole, is the primary storage carbohydrate of diatoms [6.44]. They secrete uncharacterized carbohydrate-rich extracellular polymeric compounds that enable them to adhere to surfaces underwater (EPS) [6.55]. A study has shown how embellishment of diatoms via EPS on different metal, glass and plastics surfaces can help to reduce degradation of metal surfaces into smaller microscopic particles harmful for the aquatic life [6.91]. The study showed that *Nitzschia* sp-had strong adherence properties with hydrophobic surfaces like plastics and PDMS. Although X-ray photoelectron spectroscopy measurements of cell surface components in *T. pseudonana* seem to indicate that polysaccharides are not predominant, experimental studies on the chemical composition of the organic matter in diatom cell walls have demonstrated that polysaccharide content dominates that of proteins and lipids [6.34]. A number of biological processes, including the transport and modification of organic matter, the complexation of dissolved metals, and the biogeochemical cycling of elements, are carried out by EPS

[6.104]. Additionally carbohydrate in microalgal biomass makes it as good substrate for the production of biodiesel [6.1]. Further energy for microalgae metabolic operations is provided by carbohydrates, which also serve as storage components. If necessary, they also enable microalgae to temporarily survive in low light [6.68]. The raphe, a unique cleft in the silica cell wall of several pennate species, secretes adhesive EPS strands that provide their substrate-dependent gliding motility and anchorage on surfaces. Diatoms cells can optimize their location in the environment in terms of light intensity and nutrition availability, as well as prevent desiccation, thanks to its gliding motility aided by EPS to provide the traction it needs [6.55].

6.2.2 Polyunsaturated Fatty Acid (PUFA)

Polyunsaturated fatty acids are straight-chain fatty acids with an average carbon chain length of 18 to 22 and at least two double bonds [6.61]. Since microalgae are capable of synthesizing a number of substances necessary for human nutrition, they have been identified as a source of functional components for nutraceuticals with favorable health effects [6.81]. Polyunsaturated fatty acid has been explored to produce EPA on a large scale, which accumulates eicosapentaenoic acid (EPA, 20:5n-3) as its primary fatty acid component [6.82]. The two primary PUFA families are n-6 and n-3 families, which are biosynthesized from linoleic acids (18:2n-2) and linolenic acids (18:3n-3), respectively. These two essential fatty acids (EFA) are important for maintaining human health [6.61]. For the synthesis of value-added bioactive compounds like polyunsaturated fatty acids, microalgae may offer a number of benefits. Some key PUFAs, such as DHA (docosahexaenoic acid), EPA (eicosapentaenoic acid), and AA (arachidonic acid), perform vital physiological tasks like lowering blood pressure, lowering cholesterol, and avoiding myocardial infarction [6.79]. Eicosanoids, such as resolvins and protectins, which protect our cardiovascular system and treat conditions, including depression, inflammation, and tumor activity, are made in humans from EPA and DHA [6.81]. DHA, an essential fatty acid for human nutrition, is crucial in preventing age-related cognitive losses, including Alzheimer's, multiple sclerosis, and Parkinson's disease. The role of DHAs in the development of the nervous system makes it an active component in the treatment of mental illnesses as well [6.61]. The quantity and varieties of lipids found in diatoms have an effect on the growth and survival rates of the marine animal species that consume them. Due to their high lipid concentration, diatoms play a significant part in the global carbon cycle and contribute to a rich diet for marine organisms [6.19]. Long-chain polyunsaturated omega-3 fatty acids are now understood to be substances that are very important for humans and other consumers in terms of physiology [6.81]. The marine diatom *Phaeodactylum tricornutum* is regarded as a good source for the industrial synthesis of EPA since it may collect up to 30% of the omega-3 long-chain polyunsaturated fatty acid (LC-PUFA) and EPA [6.38]. Diatom *Odontella aurita* is also cultured to produce omega-3 fatty acid at industrial scale [6.80]. Currently, fish oil is the main supplement source for omega-3s; additionally, the endocrinological system of consumers may be adversely affected by hazardous metals and persistent organic pollutants that bioaccumulate in fish oil, which makes oil made from diatoms an alternative to meet the requirement for omega-3 in the human diet [6.93].

6.2.3 Pigments

Diatoms are a major group of phytoplankton in freshwater and marine habitats [6.77] that need light for growth because the metabolism of diatoms is based on photosynthesis as a primary energy source [6.25]. In addition to using solar energy to perform photosynthesis, they also participate in photoprotective processes and exhibit antioxidant activity, both of which are essential for producing oxygen and biomass effectively [6.54]. Chlorophylls a and c, beta-carotene, fucoxanthin, diatoxanthin, and diadinoxanthin are some of the major pigments found in diatoms [6.41].

- ***Fucoxanthin:*** A chlorophyll (Chl) *a*/*c*-binding protein (FCP) serves as the peripheral antenna in diatoms, which harvest light [6.54]. In the deep water layers, FCP shows distinctive absorptions in the blue-green spectral band (450–580 nm) [6.77]. FCP is a huge antennae having light-harvesting pigment protein which is linked to 70% of the total chlorophyll in diatoms [6.45]. When proteins attach to fucoxanthin, which has a carbonyl moiety conjugated to the polyene backbone, the distinct fucoxanthins suffer dramatic bathochromic shifts, dividing them into more red, green, and blue absorbing ones [6.35]. The species *Phaeodactylum tricornutum* has a variety of carotenoid species, including fucoxanthin, diadinoxanthin, diatoxanthin, and beta-carotene [6.28]. All diatoms have FCP existing as trimers. The polypeptide compositions differ significantly from the trimerizations, which were relatively comparable [6.76]. Three categories of fucoxanthin-chlorophyll proteins (FCPs) are found in diatoms: the primary light-harvesting antennas (Lhcf), the photosystem I-specific Lhcr, and the photoprotective Lhcx [6.35]. Trimeric (FCPa) and hexameric or nonameric complexes (FCPb) have been discovered in the centric diatom *Cyclotella meneghiniana*. On the other hand, stable FCP complexes exist mostly as trimers in the pennate diatom *Phaeodactylum tricornutum*, despite variations in polypeptide and pigment composition [6.32]. Humans may benefit from fucoxanthin's possible health-promoting properties, which include antimalarial, anticancer, and anti-obesity effects [6.28].

- ***Carotenoid:*** The most common class of naturally occurring pigments, carotenoids are made up of more than 700 structurally distinct molecules [6.53]. Carotenoids, which alter the motional freedom of lipid alkyl chains and thus alter the physical characteristics of the membranes, are the pigments that play the most significant function [6.14]. They are crucial for the food, cosmetic, and pharmaceutical industries due to their therapeutic properties, such as antioxidant, anticancer, anti-inflammatory, anti-obesity, and neuroprotective benefits [6.53]. Beta-carotene is a significant primary source of vitamin A, which is essential for the retina's functioning and has an impact on a variety of tissue types by controlling gene expression [6.28]. Due to its antioxidant properties, beta-carotene also assists in preventing skin from photoaging [6.28].

6.3 Microalgae as Hidden Treasure of Novel Drugs for Good Health

Proteins containing essential amino acids, lipids containing PUFAs, carbohydrates, pigments, and many additional bioactive substances which have probiotic, anti-aging, antimicrobial, anti-inflammatory, antioxidant, immunomodulatory, as well as anticancer attributes, are some of the specialized substances present in algae that have been shown to have a constructive influence on human wellbeing and risk reduction of illnesses [6.6] [6.83]. By lowering the incidence of illness or improving wellness, such microorganisms are beneficial. The need for healthy ingredients including nutraceuticals that are naturally derived is rising as a consequence of rapidly changing views on human nutrition with a focus on wellness and endurance [6.9]. Additionally, several ranges of essential nutrients can be found in abundance in microalgae; K (0.9%–1.6%), Mg (0.4%–0.8%), S (0.3%–0.4%), Fe (0.6%), P (1%–2%), and other (0.05%) are within this range.

Microalgae can be encouraged to synthesize and retain a desired compound by altering the culturing parameters. They can also bloom quickly, utilize both environmental and synthetic light proficiently, and quickly [6.9].

6.3.1 Drugs from Microalgae

A field called blue biotechnology, sometimes known as marine biotechnology, is centered on the utilization of marine resources in biotechnology applications, both as the source or the target. The speedy discovery of novel and more potent medicines is one of the main tasks of marine biotechnology [6.22]. Marine species are distributed differently depending on the kind of open sea ecosystem-pelagic habitat, epipelagic habitat, mesopelagic zone, bathypelagic zone, abyssopelagic zone, and benthic environment. These diverse subsets of the aquatic environment are home to a wide variety of marine microflora and microalgae populations [6.74]. Microalgae have been used in the production of high-value products with a variety of medical and pharmacological uses. The creation of modified proteins is carried out by microalgae. The silica-based frustule structure has noteworthy characteristics for drug delivery, including highly organized three-dimensional (3D) pores, chemical stability, optimized geometry, and uniform nanopore architecture. Additionally, it was simple to functionalize, preserve, and modify frustules for subsequent drug loading and distribution [6.50].

Algal compounds have demonstrated anti-inflammatory properties, particularly by inhibiting the synthesis of pro-inflammatory cytokines and also by lowering the expression of pro-inflammatory genes. These algal metabolites work in a variety of ways, by modulating the activity of enzymes, including phospholipase A2, cyclooxygenase-2 (COX-2), and nitric oxide synthase (NOS), regulating cellular processes, and blocking two important signaling pathways. These include the routes for nuclear factor B (NF-B) and mitogen-activated protein kinase, which are crucial for the creation of a number of pro-inflammatory mediators. In addition to their antioxidant capabilities, algal chemicals in extracts might potentially have anti-inflammatory activities [6.10]. Additionally, microalgae pigments have health advantages, including the ability to fend off diseases of the eyes, heart, brain, and blood vessels, as well as cancer. Microalgae are a perfect host for the creation of recombinant proteins due to their special characteristics, such as their quick pace of growth, easy-use, and inexpensive medium. Additionally, their post-translational adaptations are more like those of mammalian cells than bacterial cells [6.50].

Few marine medicinal compounds have received commercial licences, and few components are undergoing preclinical and Phase I and III clinical trials. It is noteworthy that four marine chemicals, including cytarabine (Cytosar), trabectedin (Yondelis), eribulin mesylate (Halaven), and the coupled antibody brentuximab vedotin, are being utilized to cure cancer (Acentris) [6.74].

Since ancient times, people have used marine microalgae as food and medicine as they are crucial to health. Marine phytoplankton are eukaryotic flora that aid in the discovery of new drugs because of their metabolic versatility, which can cause the synthesis of different chemicals that could be used to treat a variety of ailments, including cancer [6.74]. Microalgae, including the diatoms *Phaeodactylum tricornutum, Skeletonema marinoi* and *Chaetoceros calcitrans* and the dinoflagellates *Ostreopsis ovata* and *Amphidinium operculatum*, have demonstrated antiproliferative and anti-cancer effects against human cancer cell lines. Driven by the continuous rise of antibiotic-resistant bacteria as well as illnesses, microalgae have also lately been investigated to find new anti-infective chemicals [6.57].

A few microalgal lipids, such as polyunsaturated fatty acids, have been shown to have antitumor effects on breast and cervical cancer. A phycobiliprotein called phytocyanin, which is derived from microalgae, including *Arthronema africanum, Spirulina platensis,* and *Porphyra haitanensis,* prevents the proliferation of leukaemia, lung cancer, and hepatocellular carcinoma in humans [6.50]. Loliolide and epiloliolide, two cytotoxic and selective chemicals that are most likely the breakdown products of fucoxanthin and/or zeaxanthin recovered from either the crude extracts or biomass of *T. lutea*, can be a source of compounds for the control as well as therapy of human hepatocarcinoma [6.30]. Sulfated polysaccharide known as fucoidan, which is extracted from a variety of microalgae, including *Fucus vesiculosus, Sargassum henslowianum, Cladosiphon fucoidan*, and

Coccophora longsdorfii, inhibits angiogenesis and metastasis by upregulating caspase-3/7 and downregulating kinase action in human lymphoma cell line [6.24]. Proteins called lectins connect to particular mono- and oligosaccharides. A lectin called cyanovirin-N, which has been shown to have antiviral properties against the Ebola, HIV, and influenza viruses, was isolated from the cyanobacterium *Nostoc ellipsosporum*. A sulfated polymer of microalgal origin called carrageenan can prevent viruses from attaching to, reproducing, and being transcribed in host cells. In a research study, Koenighofer *et al.* created a nasal spray containing carrageenan together with zanamivir, in which carrageenan was found to have synergistic effects against the influenza virus [6.109]. Different marine diatoms have previously been reported to have anti-inflammatory effects, including *Pseudo-nitzschia, Attheya longicornis, Odontella mobiliensis, Porosira glacialis, Cylindrotheca closterium, Amphidinium carterae, Pseudo-nitzschia pseudodelicatissima* and *Phaeodactylum tricornutum*, respectively, whereas the extracts and metabolites from *Dunaliella* displayed antioxidant, antimicrobial, cytotoxic, cardioactive, antidiabetic and hepatoprotection properties [6.2].

6.3.2 As a Feed for Aquaculture

In aquaculture feed, microalgae can be employed as a vital and sustainable feed component. Fish feed has recently been studied for economic expansion, metabolic activity, as well as for nutrient content using live algal biomass, complete algae, and lipid-extracted algae (LEA) [6.6]. Algal biomass is produced on a global scale on a yearly basis in amounts of 5 million kilograms, with a market price per kilogram of roughly 330 US dollars [6.42]. Microalgae are becoming increasingly important as a feed/feed additive because of their high nutritious makeup, possibly displacing food meal and various traditional ingredients in aquaculture and livestock fodder [6.6]. The ability of microalgae to supply protein, lipids, vitamins, pigments, and energy in feed is very promising. In late logarithmic development period, microalgae typically include 30–40% protein, 10–20% lipid, and 5–15% carbohydrates [6.90].

In aquaculture hatcheries, one fifth of this biomass is fed to the fish and shellfish which are raised there [6.42]. Protein and polyunsaturated fatty acids, such as eicosapentaenoic acid (EPA), arachidonic acid (ARA), and docosahexaenoic acid (DHA), play a major role in determining the nutritive potential of microalgae [6.90]. Microalgae provide a source of food for the larvae of mollusks, echinoderms, crustaceans and even certain fish larvae's live prey [6.75]. Due to their outstanding nutritious content and adaptability, algae, including spirulina, *Chlorella* sp., *Nannochloropsis* sp., *Dunaliella* sp., and *Scenedesmus* sp., have been frequently utilized as aquaculture feed [6.6].

The requirement for expensive amino acid supplements was eliminated by certain microalgae because they have excellent amino acid content. Methionine levels are significant in algae such as *Nannochloropsis, Chlorella, Chlamydomonas, Porphyridium, and Isochrysis* [6.78]. Astaxanthin and other pigments found in microalgae have antioxidant characteristics, and certain microalgae create large amounts of vitamins and immunostimulants in their cells, which can benefit aquatic life [6.78]. A marine haptophyte called *Tisochrysis lutea* is abundant in carotenoids and omega-3 polyunsaturated fatty acids, such as DHA and fucoxanthin. This microalga is used in aquaculture to feed oyster and shrimp larvae due to the nutraceutical applications of these chemicals [6.30].

Due to its higher dietary value, favorable impact on the pace of development of aquatic lifeforms, enhanced triglyceride and protein accumulation in muscle, enhanced tolerance to disease, reduced nitrogen released into surroundings, omega-3 fatty acid composition, metabolic function, and carcass quality, the utilization of microalgae might have a considerable positive impact and might effectively substitute or minimize common feed stuff. Aquafeed could become less reliant on traditional raw materials if microalgae are used [6.90]. To complement aquaculture feed, one might potentially utilize entire dried algae, fats, and biomass with pigments removed. Animal feed is made from 30% of the algal biomass harvested. This is a result of the various benefits that algae have, including their year-round growth and ability to grow in wastewater and seize 1.83 kilos CO_2 for every kilogram of algal biomass produced. Algae also have a quick proliferation rate and don't need agricultural soil to thrive, which is also a benefit [6.6].

The fundamental justification for the microalgae claim is that it contains the proper ratio of protein, fat, and carbohydrate, which is necessary to safeguard the health of the fish. In comparison to other substitute components like yeast and bacteria, microalgae specifically offer a high protein source and fat [6.78]. High-oxygen-producing microalgae can function as a bio-pump, enabling oxygenation in aquaculture and regulating the microbial ecology in an aquatic environment. In order to prevent an algal bloom or oxygen depletion, it is possible to correctly maintain the water condition in aquaculture practises [6.39].

6.3.3 Diatoms in Drug Delivery

Biotemplating is a time and money-efficient strategy for creating the next era of biomaterials-based drug delivery mechanisms since biological matters have a wide range of chemical functionalization techniques and a great deal of structural diversity. For the creation of delivery systems vectors, biosensing supports, and photonic crystals, diatoms have proven to be a useful substitute for synthetic silica that meets several pharmacological requirements. The biological matters are harvested from living algae or from fossil sediments of diatomaceous earth. These biocompatible and nontoxic matters provide an exceptional potential in the field of micro or nano-devices, drug delivery, therapy, diagnostics, and other medical applications. Although cultivating algae in a laboratory could be a challenging process, diatomaceous earth is the most abundant resource of silicon dioxide in the world; hence, it is a very inexpensive and eco-friendly starting material for the fabrication of theranostic devices and drug delivery systems [6.100]. The U.S. Food and Drug Administration has approved diatom biosilica for use in the production of foods and pharmaceuticals, and, most importantly, it satisfies the physical, chemical, and optical requirements for technological and scientific applications. Diatom biosilica can be cultivated in massive amounts from diatomite in an environmentally friendly manner. Diatom biosilica has a variety of 3D layouts that are presently unrivaled by human engineering; it is mechanically and thermally inert, the chemical modification of its surface is relatively simple, it is biocompatible, stable, has a high specific surface area, and has regular porous structures. These properties all contribute to the growing research interest in its use.

Materials composed of micrometer size particles capable of drug delivery are known as microshuttles. With this particle, microshuttles, both natural or synthetic porous silica-based materials have gained rapid interest in the field over the last decades [6.63]. Diatoms are ideal micro- or nanocarrier materials, as they permit retention of the nature and bioactivity of the loaded drug, while being capable of carrying it to the site of action. Thus, the chemical characteristics, various available morphologies, specific size and porosity, customizable surface functionalization, biocompatibility, and inertness of diatoms make frustules harvested from microalgae unique in the field of drug delivery applications [6.102]. Furthermore, the particular morphologies of diatoms and properties, such as large surface area, wide porosity, and highly organized hierarchical structure, present valuable features in the field of drug delivery. Their biomineralization process has also been taken as an inspirational model for the elaboration of synthetic materials and efficient drug carriers.

Indeed, silica micro/nanoparticles present several benefits compared to other materials due to the ease of particle size control during their synthesis, their chemical inertness, high surface porosity, flexible surface modification, thermal stability, and biocompatibility [6.18]. However, the fabrication of synthetic mesoporous silica requires advanced skills, often involves the use of toxic chemicals, and results in the formation of non-reusable polluting byproducts, leading to a poor cost-effective process. To overcome this problem, 20 years ago, Morse proposed an alternative to the synthetic production of mesoporous silica by using naturally occurring diatoms [6.18]. The most frequent reagent utilized in silica bonding technology is 3-aminopropyltriethoxysilane (APTES), which enables additional chemical modulation of the diatom surface using conventional synthetic processes by making use of APTES's free, surface-exposed amino functionalities. These groups are utilized in the creation of biosensors, receptor-targeting particles specifically aimed for site specific drug delivery, stimuli-responsive deposition that enable, among other things, restricted degradation, cellular uptake, and drug loading/release, as well as chemical immobilization of molecules like antibodies, enzymes, and DNA aptamers.

Oral administration is a common and convenient way to deliver therapeutics in the body, and it is generally limited to small molecules as pharmaceuticals that are more complex could be easily degraded through digestion in the stomach. Parenteral administration regroups the routes involving injections as intravenous (IV) or intramuscular (IM) administrations, offering the advantage of targeted effect and reduced toxicity when compared to other routes of delivery [6.3] [6.95]. The ideal size of particles in a suspension to be administrated by injection should range between 100 to 300 nm, a scale enhancing their permeability and retention effect in solid tumors. Diatoms are used as a food additive and could be safely administrated through the oral route; however, their size should be decreased in a controllable manner to be eligible for parenteral administration. Fine grinding and ball milling techniques allow reaching the specific nanometric scale required for injectable nanosuspension [6.16]. The poorly water-soluble chemotherapeutic drugs camptothecin and its stronger derivative 7-ethyl-10-hydroxycamptothecin were explicitly delivered to neuroblastoma and Burkitt's lymphoma HR1K cells by

Kröger and Voelcker using genetically modified biosilica derived from the diatom *Thalassiosira pseudonana* [6.110]. The immunoglobulin G (IgG)-binding domain was inserted into the diatoms' DNA to create the species through genetic engineering (GB1-bearing fusion protein). The GB1 fusion protein was discovered in the final build to be persistently linked to the biosilica interface and available for IgG affinity, allowing the adhesion of cell-targeting antibodies. In a xenograft mouse model of neuroblastoma, medication of animals with intraperitoneal injection of the identical drug-loaded biosilica resulted in malignant progression or regression in mice.

Diatomaceous earth (diatomite; DE) could undergo modifications that change and enhance its characteristics. For example, the oligo (ethylene glycol) methacrylate copolymers on diatom microcapsules make stimulus-responsive carriers for the delivery of levofloxacin, an antibiotic being used to treat *S. aureus* and *Pseudomonas aeruginosa* [6.98]. Additionally, diatom frustules treated with DOPA/Fe_3O_4 (also known as dopamine-terminated Fe_3O_4 NPs) were able to regulate an outside magnetic field thanks to their magnetic characteristics. In this regard, Khavari *et al.* [6.50] functionalized *Amphora subtropica* with chitosan for doxorubicin (DOX) loading and delivery [6.50].

The drug encapsulation of ophiobolin A, an anticancer chemical generated from fungi, was shown to be boosted by enhancing silanol at the exterior of diatoms, based on a study by Khavari *et al.* [6.50]. In the aforementioned work, ophiobolin A production was seen to be prolonged and surface technique for the preparation was revealed to be an efficient method for drug loading on diatoms. In a study by Cicco *et al.*, antibiotic (ciprofloxacin) was loaded onto nanoporous silica-based particles that were functionalized from *Thalassiosira weissflogii* with cyclic nitric oxide 2,6,6-tetramethyl piperidine-N-oxyl to have antioxidant and antibacterial effects on fibroblast and osteoblast-like cell development [6.20] [6.23].

6.4 Microalgal Drugs in Preventing Viral Pandemics

The coronavirus disease 2019 (COVID-19) spread all around the world in 190 countries and was declared a pandemic on March 11, 2020 (WHO report, 2020). The whole world's healthcare systems are still struggling to overcome this pandemic. Due to mandatory isolation or quarantines, millions of lives have been affected [6.33] [6.7]. The world's medical systems are worked hard and fast to control this pandemic by every means. Since there are no specific treatments for COVID infection, preventive vaccines are still being explored. Both marine and fresh water algae diatoms as well as other microalgae have certain crucial compounds that can be effective or act as an inhibitory against many viruses [6.73]. In this scenario, which needs prompt and practical solutions, microalgae could emerge as a promising model to partially fight the effects of COVID-19.

It is necessary to integrate new strategies for medical therapies and economic strengthening [6.11]. The bioeconomy that revolves around microalgae is widely known. Microalgae can contribute to this economy by producing valuable biomass for human-related applications (pharmaceuticals, cosmetics, food, and feed). However, it can contribute much more to the bioeconomy by increasing the current production capacity and developing new applications [6.27], including employment generation in this path [6.11].

To combat COVID-19, research communities continuously develop and implement rapid diagnoses methods, vaccinations, and other alternative therapeutic procedures. Algal metabolites can be immune boosters against the COVID-19 pandemic. This is obvious because side effects and high costs of synthetic drugs have piqued scientists' interest in natural product-based therapies and medicines. In this regard, antiviral substances derived from natural resources have seen a boom in popularity. For instance, algae are a rich source of compounds such as lectins and sulfated polysaccharides [6.87], which have potent antiviral and immunity-boosting properties. Moreover, algae-derived compounds or metabolites can be used as antibodies and vaccine raw materials against COVID-19. Furthermore, some algal species can boost immunity, reduce viral activity in humans and be recommended for usage as a COVID-19 preventative measure. To overcome the effects of COVID-19, the functionalities of microalgae could be exploited to boost the areas most affected by the pandemic. Among the most promising benefits are the various biomolecules derived from microalgae that could be adjuvant therapies or preventive agents, which have the potential to become biofactories of antibodies or vaccines through genetic engineering [6.11]. On the offset, microalgae (diatoms) are the most diverse and abundant group which, in addition to containing most of the therapeutic components [6.64], have a distinct siliceous cell wall that is luring scientists into jockeying these organisms into position to also be potential drug delivery systems. Diatoms like *Phaeodactylum tricornutum* are a good example of such algae which are a rich source of bioactive metabolites such as carotenoids like xanthophylls [6.4], β-carotene, fucoxanthin, zeaxanthin, phycocyanin, flavonoid, and fatty acids; and *Nitzschia* sp. are rich in omega-3 polyunsaturated fatty acids (PUFAs) [6.89].

Diatoms contain traces of unconventional lipids such as phosphatidyl sulfocholine [6.60] and lipids such as halogenated fatty acids [6.49]. Fatty acids play a major role in maintaining the stability of membranes along with proper organization of cells in the human body. Unfavorable conditions manifest the production and release of fatty acids from algal cells in order to counteract the action of pathogenic microbes that cause loss in cell integrity. Reports suggests that carbon unsaturation in the fatty acid structure directly correlates with the efficient antibacterial property of the algae [6.17]. The intricate organization of fatty acids and lipids as a whole contributes to its diversified efficiency in the therapeutics hall of fame [6.108]. High concentration of Vitamin C was obtained from marine diatom species *Skeletonema marinoi* and *Chaetoceros* sp. A high concentration of Vitamin A was also reported from *Chaetoceros* sp. [6.13]. Vitamin E was reported in the algal species diatom *Skeletonema marinoi*, and the extracts are also rich in vitamin supplements which are necessary for maintaining stimulation of various physiological body functions [6.71].

Additionally, sulfated polysaccharide, naviculan extracted from the diatom *Navicula directa* displayed a potential antiviral response to enveloped viruses such as influenza virus and herpes simplex virus type 1 and type 2 by inhibiting the viral entry into and proliferation in the host cells [6.5].

The sulfated polysaccharides are the most viable source of antiviral agents from marine algae. The activity against the sulfated polysaccharides on the respiratory disease-causing viruses shows the potentiality of sulfated polysaccharides. Hence, these sulfated polysaccharides from marine algal sources can be explored to produce the vaccine against the new COVID-19 [6.40]. From the previous studies, it is proved that the sulfated polysaccharides from marine algae could be used against the enveloped RNA viruses and it could be well exploited against the novel coronavirus which also belongs to the same group. The enveloped RNA viruses like HIV RSV, influenza-type viruses had been inhibited by the algal polysaccharides in several studies [6.26]. Hence, it is possible to produce antiviral drugs from marine algae for treating novel coronavirus COVID-19 [6.31].

As per the previous literature, a variety of algal polysaccharides, including carrageenan, galactan, alginate, fucan and fucoidan, laminaran, naviculan, calcium spirulan, and nostaflan, etc., have been shown to exert antiviral activities. The majority of these polysaccharides exhibiting virucidal effects on influenza, herpes simplex, HIV, hepatitis C, corona virus, etc., are the sulfated polysaccharides [6.87]. Thus, the diversified constitution and complicacy of algal polysaccharide and its derivatives account for their antiviral activity at various stages of viral infection. The past few years witnessed outstanding results of the virucidal activity of algal polysaccharides and have endorsed them as a valuable pharmaceutical or biomedical element [6.85]. Naviculan extracted from the diatom species *Navicula directa* has potential antiviral activity against HSV-1 and HSV-2 [6.4]. Furthermore, *P. tricornutum* is a scalable system with minimal biocontainment requirements for the inducible production of SARS-CoV-2 or other coronavirus antigens for pandemic diagnostics. The results provide proof-of-principle that *P. tricornutum* expression plasmids based on the HASP1 promoter represent an orthogonal, simple, scalable and regulated system for the production of SARS-CoV-2 antigens for pandemic diagnostics [6.92]. The algal polysaccharides and other components can be utilized for the development of antiviral agents against the novel coronavirus, COVID-19. The sulfated polysaccharides have high potential to act against viral infections. At the same time, the development of vaccines and commercializing such vaccine would take a certain amount of time for completing the clinical trials and other international standard procedures [6.84].

6.5 Conclusions

Microalgae have long been seen as potential platforms because they can be easily grown in existing natural aquatic or synthetic environments to generate bioactive molecules using only carbon dioxide, water, and sunlight. Microalgae have yielded high-value compounds with a variety of potential pharmacological and medicinal benefits. Fucoxanthin is an active component generated from diatoms that has antioxidant, antiviral, antifungal, anticancer, and antimalarial effects. The liver, blood arteries in the brain, bones, skin, and eyes are all protected by this photopigment found in diatoms. Fucoxanthin is further noticeably more effective than β-carotene and astaxanthin at preventing obesity, inhibiting the growth of adult T-cell leukemia cells [6.46], and reducing the viability of HL-60 cells [6.29]. Fucoxanthin has a special ability to inhibit adipocyte development due to its structural characteristics, while carotenoids without an allenic link do not display this activity [6.72]. This impact is related to its structural qualities. Additionally, it is thought that marine diatoms represent a significant source of cutting-edge anticancer drugs with enormous potential to be turned into diatoms-based anticancer drug delivery systems. Monoacylglycerides, oxylipins, chrysolaminaran polysaccharide, adenosine, fucoxanthin, fatty alcohol ester, marennine, stigmasterol, and haslenes are some of the chemicals generated from diatoms that have shown strong anticancer properties.

References

[6.1] Abate, R., Song, S., Patil, V., Chen, C., Liang, J., Sun, L., Li, X., Huang, B. and Gao, Y. (2020) Enhancing the production of a marine diatom (Skeletonema costatum) with low-frequency ultrasonic irradiation. *Journal of Applied Phycology* 32(6), 3711-3722.

[6.2] Lauritano, C., Helland, K., Riccio, G., Andersen, J.H., Ianora, A., Hansen, E.H., Lysophosphatidylcholines and chlorophyll-derived molecules from the diatom cylindrotheca closterium with anti-inflammatory activity. *Marine Drugs*, 18, 166, 2020.

[6.3] Ahire, E., Thakkar, S., Darshanwad, M. and Misra, M. (2018) Parenteral nanosuspensions: a brief review from solubility enhancement to more novel and specific applications. *Acta Pharmaceutica Sinica B* 8(5), 733-755.

[6.4] Ahirwar, A., Kesharwani, K., Deka, R., Muthukumar, S., Khan, M.J., Rai, A., Vinayak, V., Varjani, S., Joshi, K.B. and Morjaria, S. (2022) Microalgal drugs: A promising therapeutic reserve for the future. *Journal of Biotechnology* 349, 32-46.

[6.5] Ahmadi, A., Zorofchian Moghadamtousi, S., Abubakar, S. and Zandi, K. (2015) Antiviral potential of algae polysaccharides isolated from marine sources: a review. *BioMed Research International* 2015, 825203.

[6.6] Ansari, F.A., Guldhe, A., Gupta, S.K., Rawat, I. and Bux, F. (2021) Improving the feasibility of aquaculture feed by using microalgae. *Environmental Science and Pollution Research* 28(32), 43234-43257.

[6.7] Anthony, S.J., Johnson, C.K., Greig, D.J., Kramer, S., Che, X., Wells, H., Hicks, A.L., Joly, D.O., Wolfe, N.D., Daszak, P., Karesh, W., Lipkin, W.I., Morse, S.S., Mazet, J.A.K. and Goldstein, T. (2017) Global patterns in coronavirus diversity. *Virus Evol* 3(1), vex012.

[6.8] Aslam, S.N., Cresswell-Maynard, T., Thomas, D.N. and Underwood, G.J. (2012) Production and characterization of the intra-and extracellular carbohydrates and polymeric substances (EPS) of three sea-ice diatom species, and evidence for a cryoprotective role for EPS. *Journal of Phycology* 48(6), 1494-1509.

[6.9] Assunção, M.F., Varejão, J.M. and Santos, L.M. (2017) Nutritional characterization of the microalga *Ruttnera lamellosa* compared to *Porphyridium purpureum*. *Algal Research* 26, 8-14.

[6.10] Seth, K. Kumar, A. Rastogi, R.P. Meena, M. Vinayak, V. Harish. (2021) Bioprospecting of fucoxanthin from diatoms-Challenges and perspectives. *Algal Research*, 60, 102475.

[6.11] Barrantes, M.G., Espinoza, O.G. and Montero, K.N. (2022) Microalgae-based approaches to overcome the effects of the COVID-19 pandemic. *Tecnología en Marcha* 35(1), 84-93.

[6.12] Battarbee, R.W., Jones, V.J., Flower, R.J., Cameron, N.G., Bennion, H., Carvalho, L. and Juggins, S. (2002) Diatoms. In: *Tracking Environmental Change Using Lake Sediments*. Springer: 155-202.

[6.13] Bhattacharya, D., Price, D.C. (2020). The algal tree of life from a genomics perspective. In: Photosynthesis in algae: Biochemical and physiological mechanisms. Advances in Photosynthesis and Respiration, vol 45. Springer, 11-24.

[6.14] Bojko, M., Olchawa-Pajor, M., Goss, R., Schaller-Laudel, S., Strzałka, K. and Latowski, D. (2019) Diadinoxanthin de-epoxidation as important factor in the short-term stabilization of diatom photosynthetic membranes exposed to different temperatures. *Plant, Cell & Environment* 42(4), 1270-1286.

[6.15] Ahirwar, A., Kesharwani, K., Deka,R., Muthukumar, S., Khan, M.J., Rai, A., Vinayak, V., Varjani, V., Koshi, K.B., Morjaria, S. (2022) Microalgal drugs: A promising therapeutic reserve for the future. *Journal of Biotechnology* 349, 32–46.

[6.16] Cauda, V., Schlossbauer, A. and Bein, T. (2010) Microporous and Mesoporous materials bio-degradation study of colloidal mesoporous silica nanoparticles: effect of surface functionalization with organo-silanes and poly (ethylene glycol). *Microporous Mesoporous Mater.* 132, 60-71.

[6.17] Cepas, V., Gutiérrez-Del-Río, I., López, Y., Redondo-Blanco, S., Gabasa, Y., Iglesias, M.J., Soengas, R., Fernández-Lorenzo, A., López-Ibáñez, S. and Villar, C.J. (2021) Microalgae and cyanobacteria strains as producers of lipids with antibacterial and antibiofilm activity. *Marine Drugs* 19(12), 675.

[6.18] Chao J. T., Biggs M. J.P., Pandit A. S. (2014) Diatoms: a biotemplating approach to fabricating drug delivery reservoirs. *Expert Opinion on Drug Delivery* 11(11), 1687-1695.

[6.19] Chen, Y.-C. (2012) The biomass and total lipid content and composition of twelve species of marine diatoms cultured under various environments. *Food Chemistry* 131(1), 211-219.

[6.20] Cicco, S.R., Vona, D., De Giglio, E., Cometa, S., Mattioli-Belmonte, M., Palumbo, F., Ragni, R. and Farinola, G.M. (2015) Chemically modified diatoms biosilica for bone cell growth with combined drug-delivery and antioxidant properties. *ChemPlusChem* 80(7), 1104-1112.

[6.21] da Silva, E.F., Almeida, S.F., Nunes, M.L., Luís, A.T., Borg, F., Hedlund, M., de Sá, C.M., Patinha, C. and Teixeira, P. (2009) Heavy metal pollution downstream the abandoned Coval da Mó mine (Portugal) and associated effects on epilithic diatom communities. *Science of the Total Environment* 407(21), 5620-5636.

[6.22] De Vera, C.R., Díaz Crespín, G., Hernández Daranas, A., Montalvão Looga, S., Lillsunde, K.-E., Tammela, P., Perälä, M., Hongisto, V., Virtanen, J. and Rischer, H. (2018) Marine microalgae: promising source for new bioactive compounds. *Marine Drugs* 16(9), 317.

[6.23] Delasoie, J. and Zobi, F. (2019) Natural diatom biosilica as microshuttles in drug delivery systems. *Pharmaceutics* 11(10), 537.

[6.24] Deniz, I., García-Vaquero, M. and Imamoglu, E. (2017) Trends in red biotechnology: Microalgae for pharmaceutical applications. In: *Microalgae-Based Biofuels and Bioproducts*. Elsevier: 429-460.

[6.25] Duarte, B., Feijão, E., Goessling, J.W., Caçador, I. and Matos, A.R. (2021) Pigment and fatty acid production under different light qualities in the diatom Phaeodactylum tricornutum. *Applied Sciences* 11(6), 2550.

[6.26] Elaya Perumal, U. and Sundararaj, R. (2020) Algae: A potential source to prevent and cure the novel coronavirus–A review. *Int. J. Emerg. Technol* 11(2), 479-483.

[6.27] Fernández, F.G.A., Reis, A., Wijffels, R.H., Barbosa, M., Verdelho, V. and Llamas, B. (2021) The role of microalgae in the bioeconomy. *New Biotechnology* 61, 99-107.

[6.28] Fu, W., Wichuk, K. and Brynjólfsson, S. (2015) Developing diatoms for value-added products: Challenges and opportunities. *New Biotechnology* 32(6), 547-551.

[6.29] Ganesan, P., Noda, K., Manabe, Y., Ohkubo, T., Tanaka, Y., Maoka, T., Sugawara, T. and Hirata, T. (2011) Siphonaxanthin, a marine carotenoid from green algae, effectively induces apoptosis in human leukemia (HL-60) cells. *Biochimica et Biophysica Acta (BBA)-General Subjects* 1810(5), 497-503.

[6.30] Gangadhar, K.N., Rodrigues, M.J., Pereira, H., Gaspar, H., Malcata, F.X., Barreira, L. and Varela, J. (2020) Anti-hepatocellular carcinoma (HepG2) activities of monoterpene hydroxy lactones isolated from the marine microalga Tisochrysis lutea. *Marine Drugs* 18(11), 567.

[6.31] Geahchan, S., Ehrlich, H. and Rahman, M.A. (2021) The anti-viral applications of marine resources for COVID-19 treatment: an overview. *Marine Drugs* 19(8), 409.

[6.32] Giovagnetti, V. and Ruban, A.V. (2017) Detachment of the fucoxanthin chlorophyll a/c binding protein (FCP) antenna is not involved in the acclimative regulation of photoprotection in the pennate diatom *Phaeodactylum tricornutum*. *Biochimica et Biophysica Acta (BBA)-Bioenergetics* 1858(3), 218-230.

[6.33] Gorbalenya, A.E., Baker, S.C., Baric, R.S., de Groot, R.J., Drosten, C., Gulyaeva, A.A., Haagmans, B.L., Lauber, C., Leontovich, A.M., Neuman, B.W., Penzar, D., Perlman, S., Poon, L.L.M., Samborskiy, D.V.,

Sidorov, I.A., Sola, I., Ziebuhr, J. and Coronaviridae Study Group of the International Committee on Taxonomy of, V. (2020) The species Severe acute respiratory syndrome-related coronavirus: classifying 2019-nCoV and naming it SARS-CoV-2. *Nature Microbiology* 5(4), 536-544.

[6.34] Gügi, B., Le Costaouec, T., Burel, C., Lerouge, P., Helbert, W. and Bardor, M. (2015) Diatom-specific oligosaccharide and polysaccharide structures help to unravel biosynthetic capabilities in diatoms. *Marine Drugs* 13(9), 5993-6018.

[6.35] Gundermann, K. and Büchel, C. (2014) Structure and functional heterogeneity of fucoxanthin-chlorophyll proteins in diatoms. In: *The Structural Basis of Biological Energy Generation*. Springer: 21-37.

[6.36] Haese, R.R., Murray, E.J., Smith, C.S., Smith, J., Clementson, L. and Heggie, D.T. (2007) Diatoms control nutrient cycles in a temperate, wave-dominated estuary (southeast Australia). *Limnology and Oceanography* 52(6), 2686-2700.

[6.37] Hallahan, B. and Garland, M.R. (2005) Essential fatty acids and mental health. *The British Journal of Psychiatry* 186(4), 275-277.

[6.38] Hamilton, M.L., Warwick, J., Terry, A., Allen, M.J., Napier, J.A. and Sayanova, O. (2015) Towards the industrial production of omega-3 long chain polyunsaturated fatty acids from a genetically modified diatom *Phaeodactylum tricornutum*. *PloS One* 10(12), e0144054.

[6.39] Han, P., Lu, Q., Fan, L. and Zhou, W. (2019) A review on the use of microalgae for sustainable aquaculture. *Applied Sciences* 9(11), 2377.

[6.40] Hans, N., Malik, A. and Naik, S. (2021) Antiviral activity of sulfated polysaccharides from marine algae and its application in combating COVID-19: Mini review. *Bioresource Technology Reports* 13, 100623.

[6.41] Hasle, G.R., Syvertsen, E.E., Steidinger, K.A., Tangen, K. and Tomas, C.R. (1996) *Identifying Marine Diatoms and Dinoflagellates*. Elsevier.

[6.42] Hemaiswarya, S., Raja, R., Ravi Kumar, R., Ganesan, V. and Anbazhagan, C. (2011) Microalgae: a sustainable feed source for aquaculture. *World Journal of Microbiology and Biotechnology* 27(8), 1737-1746.

[6.43] Hu, X., Li, Y., Li, C., Fu, Y., Cai, F., Chen, Q. and Li, D. (2012) Combination of fucoxanthin and conjugated linoleic acid attenuates body weight gain and improves lipid metabolism in high-fat diet-induced obese rats. *Archives of Biochemistry and Biophysics* 519(1), 59-65.

[6.44] Huang, W., Río Bártulos, C. and Kroth, P.G. (2016) Diatom vacuolar 1, 6-β-transglycosylases can functionally complement the respective yeast mutants. *Journal of Eukaryotic Microbiology* 63(4), 536-546.

[6.45] Ishihara, T., Ifuku, K., Yamashita, E., Fukunaga, Y., Nishino, Y., Miyazawa, A., Kashino, Y. and Inoue-Kashino, N. (2015) Utilization of light by fucoxanthin–chlorophyll-binding protein in a marine centric diatom, Chaetoceros gracilis. *Photosynthesis Research* 126(2), 437-447.

[6.46] Ishikawa, C., Tafuku, S., Kadekaru, T., Sawada, S., Tomita, M., Okudaira, T., Nakazato, T., Toda, T., Uchihara, J.N. and Taira, N. (2008) Antiadult

T-cell leukemia effects of brown algae fucoxanthin and its deacetylated product, fucoxanthinol. *International Journal of Cancer* 123(11), 2702-2712.
[6.47] Khan, M.J. Rai, A. Ahirwar, A. Sirotiya, V. Mourya, M. Mishra, S. Schoefs B. Marchand, J. Bhatia S.K. Varjani S. Vinayak V. (2021) Diatom microalgae as smart nanocontainers for biosensing wastewater pollutants: recent trends and innovations. *Bioengineered*, 12, 9531-9549.
[6.48] Kale, A. and Karthick, B. (2015) The diatoms. *Resonance* 20(10), 919-930.
[6.49] Kartik, A., Akhil, D., Lakshmi, D., Gopinath, K.P., Arun, J., Sivaramakrishnan, R. and Pugazhendhi, A. (2021) A critical review on production of biopolymers from algae biomass and their applications. *Bioresource Technology* 329, 124868.
[6.50] Khavari, F., Saidijam, M., Taheri, M. and Nouri, F. (2021) Microalgae: therapeutic potentials and applications. *Molecular Biology Reports* 48(5), 4757-4765.
[6.51] Koedooder, C., Stock, W., Willems, A., Mangelinckx, S., De Troch, M., Vyverman, W. and Sabbe, K. (2019) Diatom-bacteria interactions modulate the composition and productivity of benthic diatom biofilms. *Frontiers in Microbiology* 10, 1255.
[6.52] Kroth, P. (2007) Molecular biology and the biotechnological potential of diatoms. In: *Transgenic Microalgae as Green Cell Factories*, Springer: 23-33.
[6.53] Kuczynska, P. and Jemiola-Rzeminska, M. (2017) Isolation and purification of all-trans diadinoxanthin and all-trans diatoxanthin from diatom *Phaeodactylum tricornutum*. *Journal of Applied Phycology* 29(1), 79-87.
[6.54] Kuczynska, P., Jemiola-Rzeminska, M. and Strzalka, K. (2015) Photosynthetic pigments in diatoms. *Marine Drugs* 13(9), 5847-5881.
[6.55] Lachnit, M., Buhmann, M.T., Klemm, J., Kröger, N. and Poulsen, N. (2019) Identification of proteins in the adhesive trails of the diatom *Amphora coffeaeformis*. *Philosophical Transactions of the Royal Society B* 374(1784), 20190196.
[6.56] Lafourcade, M., Larrieu, T., Mato, S., Duffaud, A., Sepers, M., Matias, I., Smedt-Peyrusse, D., Labrousse, V.F., Bretillon, L. and Matute, C. (2011) Nutritional omega-3 deficiency abolishes endocannabinoid-mediated neuronal functions. *Nature Neuroscience* 14(3), 345-350.
[6.57] Lauritano, C., Ferrante, M.I. and Rogato, A. (2019) Marine natural products from microalgae: An -omics overview. *Marine Drugs* 17(5), 269.
[6.58] Lebeau, T. and Robert, J.-M. (2003) Diatom cultivation and biotechnologically relevant products. Part I: Cultivation at various scales. *Applied Microbiology and Biotechnology* 60(6), 612-623.
[6.59] Nair, A., Ahirwar, A., Singh, S., Lodhi, R., Lodhi, A., Rai, A., Jadhav D.A., Harish., Varjani, S., Singh, G., Marchand J., Schoefs, B., Vinayak, V. (2023) Astaxanthin as a King of Ketocarotenoids: Structure, Synthesis, Accumulation, Bioavailability and Antioxidant Properties. *Mar. Drugs*, 21, 176.

[6.60] Li-Beisson, Y., Thelen, J.J., Fedosejevs, E. and Harwood, J.L. (2019) The lipid biochemistry of eukaryotic algae. *Progress in Lipid Research* 74, 31-68.

[6.61] Li, H.-Y., Lu, Y., Zheng, J.-W., Yang, W.-D. and Liu, J.-S. (2014) Biochemical and genetic engineering of diatoms for polyunsaturated fatty acid biosynthesis. *Marine Drugs* 12(1), 153-166.

[6.62] Lobo, E.A., Heinrich, C.G., Schuch, M., Wetzel, C.E. and Ector, L. (2016) Diatoms as bioindicators in rivers. In: River algae. Springer: 245-271.

[6.63] Maher, S., Kumeria, T., Aw, M.S. and Losic, D. (2018) Diatom silica for biomedical applications: Recent progress and advances. *Advanced Healthcare Materials* 7(19), 1800552.

[6.64] Malviya, S., Scalco, E., Audic, S., Vincent, F., Veluchamy, A., Poulain, J., Wincker, P., Iudicone, D., De Vargas, C. and Bittner, L. (2016) Insights into global diatom distribution and diversity in the world's ocean. *Proceedings of the National Academy of Sciences* 113(11), E1516-E1525.

[6.65] Mangas-Sánchez, J. and Adlercreutz, P. (2015) Highly efficient enzymatic biodiesel production promoted by particle-induced emulsification. *Biotechnology for Biofuels* 8(1), 1-8.

[6.66] Mann, D.G. and Vanormelingen, P. (2013) An inordinate fondness? The number, distributions, and origins of diatom species. *Journal of Eukaryotic Microbiology* 60(4), 414-420.

[6.67] Marella, T.K., Saxena, A. and Tiwari, A. (2020) Diatom mediated heavy metal remediation: A review. *Bioresource Technology* 305, 123068.

[6.68] Markou, G., Angelidaki, I. and Georgakakis, D. (2012) Microalgal carbohydrates: an overview of the factors influencing carbohydrates production, and of main bioconversion technologies for production of biofuels. *Applied Microbiology and Biotechnology* 96(3), 631-645.

[6.69] McNair, H.M., Brzezinski, M.A. and Krause, J.W. (2018) Diatom populations in an upwelling environment decrease silica content to avoid growth limitation. *Environmental Microbiology* 20(11), 4184-4193.

[6.70] Meiser, A., Schmid-Staiger, U. and Trösch, W. (2004) Optimization of eicosapentaenoic acid production by *Phaeodactylum tricornutumin* the flat panel airlift (FPA) reactor. *Journal of Applied Phycology* 16(3), 215-225.

[6.71] Milito, A., Orefice, I., Smerilli, A., Castellano, I., Napolitano, A., Brunet, C. and Palumbo, A. (2020) Insights into the light response of *Skeletonema marinoi*: involvement of ovothiol. *Marine Drugs* 18(9), 477.

[6.72] Miyashita, K., Nishikawa, S., Beppu, F., Tsukui, T., Abe, M. and Hosokawa, M. (2011) The allenic carotenoid fucoxanthin, a novel marine nutraceutical from brown seaweeds. *Journal of the Science of Food and Agriculture* 91(7), 1166-1174.

[6.73] Vinayak, V. Bhaskar, P.V., Pandey, L.K. and Khan, M.J. (2022) Diatoms: the Living Jewels and their Biodiversity, Phycosphere and Associated Phenotypic Plasticity: A Lesson to Learn from the Current Pandemic of Coronavirus. In: Biodiversity in India: Status, Issues and Challenges. Springer Nature: Singapore: 385–429.

[6.74] Mondal, A., Bose, S., Banerjee, S., Patra, J.K., Malik, J., Mandal, S.K., Kilpatrick, K.L., Das, G., Kerry, R.G. and Fimognari, C. (2020) Marine cyanobacteria and microalgae metabolites—A rich source of potential anticancer drugs. *Marine Drugs* 18(9), 476.

[6.75] Muller-Feuga, A. (2000) The role of microalgae in aquaculture: situation and trends. *Journal of Applied Phycology* 12(3), 527-534.

[6.76] Nagao, R., Takahashi, S., Suzuki, T., Dohmae, N., Nakazato, K. and Tomo, T. (2013) Comparison of oligomeric states and polypeptide compositions of fucoxanthin chlorophyll a/c-binding protein complexes among various diatom species. *Photosynthesis Research* 117(1), 281-288.

[6.77] Nagao, R., Ueno, Y., Yokono, M., Shen, J.-R. and Akimoto, S. (2018) Alterations of pigment composition and their interactions in response to different light conditions in the diatom *Chaetoceros gracilis* probed by time-resolved fluorescence spectroscopy. *Biochimica et Biophysica Acta (BBA)-Bioenergetics* 1859(7), 524-530.

[6.78] Nagappan, S., Das, P., AbdulQuadir, M., Thaher, M., Khan, S., Mahata, C., Al-Jabri, H., Vatland, A.K. and Kumar, G. (2021) Potential of microalgae as a sustainable feed ingredient for aquaculture. *Journal of Biotechnology* 341, 1-20.

[6.79] Niu, Y.-F., Zhang, M.-H., Li, D.-W., Yang, W.-D., Liu, J.-S., Bai, W.-B. and Li, H.-Y. (2013) Improvement of neutral lipid and polyunsaturated fatty acid biosynthesis by overexpressing a type 2 diacylglycerol acyltransferase in marine diatom *Phaeodactylum tricornutum*. *Marine Drugs* 11(11), 4558-4569.

[6.80] Pasquet, V., Ulmann, L., Mimouni, V., Guihéneuf, F., Jacquette, B., Morant-Manceau, A. and Tremblin, G. (2014) Fatty acids profile and temperature in the cultured marine diatom *Odontella aurita*. *Journal of Applied Phycology* 26(6), 2265-2271.

[6.81] Peltomaa, E., Hällfors, H. and Taipale, S.J. (2019) Comparison of diatoms and dinoflagellates from different habitats as sources of PUFAs. *Marine Drugs* 17(4), 233.

[6.82] Peng, K.-T., Zheng, C.-N., Xue, J., Chen, X.-Y., Yang, W.-D., Liu, J.-S., Bai, W. and Li, H.-Y. (2014) Delta 5 fatty acid desaturase upregulates the synthesis of polyunsaturated fatty acids in the marine diatom *Phaeodactylum tricornutum*. *Journal of Agricultural and Food Chemistry* 62(35), 8773-8776.

[6.83] Pina-Pérez, M.C., Brück, W., Brück, T. and Beyrer, M. (2019) Microalgae as healthy ingredients for functional foods. In: *The Role of Alternative and Innovative Food Ingredients and Products in Consumer Wellness*. Elsevier: 103-137.

[6.84] Plotkin, S., Robinson, J.M., Cunningham, G., Iqbal, R. and Larsen, S. (2017) The complexity and cost of vaccine manufacturing–an overview. *Vaccine* 35(33), 4064-4071.

[6.85] Ptak, S.H. and Fretté, X.C. (2021) Algae Polysaccharides as Nutraceuticals: Isolation, Characterization and Bioactivities of Fucoidans. In: *Algae for Food*. CRC Press: 211-238.

[6.86] Rodolfi, L., Chini Zittelli, G., Bassi, N., Padovani, G., Biondi, N., Bonini, G. and Tredici, M.R. (2009) Microalgae for oil: Strain selection, induction of lipid synthesis and outdoor mass cultivation in a low-cost photobioreactor. *Biotechnology and Bioengineering* 102(1), 100-112.

[6.87] Sangtani, R., Ghosh, A., Jha, H.C., Parmar, H.S. and Bala, K. (2021) Potential of algal metabolites for the development of broad-spectrum antiviral therapeutics: Possible implications in COVID-19 therapy. *Phytotherapy Research* 35(5), 2296-2316.

[6.88] Sapieha, P., Stahl, A., Chen, J., Seaward, M.R., Willett, K.L., Krah, N.M., Dennison, R.J., Connor, K.M., Aderman, C.M. and Liclican, E. (2011) 5-Lipoxygenase metabolite 4-HDHA is a mediator of the antiangiogenic effect of ω-3 polyunsaturated fatty acids. *Science Translational Medicine* 3(69), 69ra12-69ra12.

[6.89] Saranya, G. and Ramachandra, T. (2021) Scope for biodiesel and bioactive compounds production in the diatom *Nitzschia punctata*. *Fuel* 300, 120985.

[6.90] Shah, M.R., Lutzu, G.A., Alam, A., Sarker, P., Chowdhury, K., Parsaeimehr, A., Liang, Y. and Daroch, M. (2018) Microalgae in aquafeeds for a sustainable aquaculture industry. *Journal of Applied Phycology* 30(1), 197-213.

[6.91] Khan, M.J. Singh, R. Shewani, K. Shukla, P. Bhaskar, P.V. Joshi, K.B. Vinayak, V. (2020) Exopolysaccharides directed embellishment of diatoms triggered on plastics and other marine litter. *Scientific Reports* 10, 18448.

[6.92] Slattery, S.S., Giguere, D.J., Stuckless, E.E., Shrestha, A., Briere, L.-A.K., Galbraith, A., Reaume, S., Boyko, X., Say, H.H. and Browne, T.S. (2022) Phosphate-regulated expression of the SARS-CoV-2 receptor-binding domain in the diatom *Phaeodactylum tricornutum* for pandemic diagnostics. *Scientific Reports* 12(1), 7010.

[6.93] Sprynskyy, M., Monedeiro, F., Monedeiro-Milanowski, M., Nowak, Z., Krakowska-Sieprawska, A., Pomastowski, P., Gadzała-Kopciuch, R. and Buszewski, B. (2022) Isolation of omega-3 polyunsaturated fatty acids (eicosapentaenoic acid-EPA and docosahexaenoic acid-DHA) from diatom biomass using different extraction methods. *Algal Research* 62, 102615.

[6.94] Rai. A., Sirotiya. V., Mourya M, Khan, M.J., Ahirwar. A., Sharma, A.K., Kawatra, R., Marchand, J., Schoefs, B., Varjani, S., Vinayak V. (2022) Sustainable treatment of dye wastewater by recycling microalgal and diatom biogenic materials: Biorefinery perspectives. *Chemosphere*, 305, 135371.

[6.95] Tian, X., Li, H., Zhang, D., Liu, G., Jia, L., Zheng, D., Shen, J., Shen, Y. and Zhang, Q. (2013) Nanosuspension for parenteral delivery of a p-terphenyl derivative: preparation, characteristics and pharmacokinetic studies. *Colloids and Surfaces B: Biointerfaces* 108, 29-33.

[6.96] Tocher, D.R., Betancor, M.B., Sprague, M., Olsen, R.E. and Napier, J.A. (2019) Omega-3 long-chain polyunsaturated fatty acids, EPA and DHA: bridging the gap between supply and demand. *Nutrients* 11(1), 89.

[6.97] Tokarek, W., Listwan, S., Pagacz, J., Leśniak, P. and Latowski, D. (2016) Column chromatography as a useful step in purification of diatom pigments. *Acta Biochimica Polonica* 63(3), 443-447.

[6.98] Vasani, R., Losic, D., Cavallaro, A. and Voelcker, N. (2015) Fabrication of stimulus-responsive diatom biosilica microcapsules for antibiotic drug delivery. *Journal of Materials Chemistry B* 3(21), 4325-4329.

[6.99] Wang, X.-W., Liang, J.-R., Luo, C.-S., Chen, C.-P. and Gao, Y.-H. (2014) Biomass, total lipid production, and fatty acid composition of the marine diatom *Chaetoceros muelleri* in response to different CO_2 levels. *Bioresource Technology* 161, 124-130.

[6.100] Vinayak, V. Khan, M.J., and Joshi, K.B. (2023) Nanoengineering diatoms in microfluidic lab on chip devices. In: *Applications of Multifunctional Nanomaterials*. Elsevier: 95-124.

[6.101] Wanner, M., Birkhofer, K., Fischer, T., Shimizu, M., Shimano, S. and Puppe, D. (2020) Soil testate amoebae and diatoms as bioindicators of an old heavy metal contaminated floodplain in Japan. *Microbial Ecology* 79(1), 123-133.

[6.102] Wee, K.M., Rogers, T.N., Altan, B.S., Hackney, S.A. and Hamm, C. (2005) Engineering and medical applications of diatoms. *Journal of Nanoscience and Nanotechnology* 5(1), 88-91.

[6.103] Xia, S., Wang, K., Wan, L., Li, A., Hu, Q. and Zhang, C. (2013) Production, characterization, and antioxidant activity of fucoxanthin from the marine diatom *Odontella aurita*. *Marine Drugs* 11(7), 2667-2681.

[6.104] Xiao, R. and Zheng, Y. (2016) Overview of microalgal extracellular polymeric substances (EPS) and their applications. *Biotechnology Advances* 34(7), 1225-1244.

[6.105] Xiong, J.-Q., Kurade, M.B. and Jeon, B.-H. (2018) Can microalgae remove pharmaceutical contaminants from water? *Trends in Biotechnology* 36(1), 30-44.

[6.106] Yang, Z.-K., Niu, Y.-F., Ma, Y.-H., Xue, J., Zhang, M.-H., Yang, W.-D., Liu, J.-S., Lu, S.-H., Guan, Y. and Li, H.-Y. (2013) Molecular and cellular mechanisms of neutral lipid accumulation in diatom following nitrogen deprivation. *Biotechnology for Biofuels* 6(1), 1-14.

[6.107] Yi, Z., Xu, M., Di, X., Brynjolfsson, S. and Fu, W. (2017) Exploring valuable lipids in diatoms. *Frontiers in Marine Science* 4, 17.

[6.108] Zhang, C. and Liu, P. (2017) The lipid droplet: A conserved cellular organelle. *Protein & Cell* 8(11), 796-800.

[6.109] Koenighofer, M., Lion, T., Bodenteich, A., Prieschl-Grassauer, E., Grassauer, A., Unger, H., Mueller, CA., Fazekas, T., Carrageenan nasal spray in virus confirmed common cold: individual patient data analysis of two randomized controlled trials. *Multidisciplinary Respiratory Medicine*, 9, 1, 1–12, 2014.

[6.110] Delalat, B., Sheppard, V.C., Rasi Ghaemi, S., Rao, S., Prestidge, C.A., McPhee, G., Rogers, M.L., Donoghue, J.F., Pillay, V., Johns, T.G. and Kröger, N., Targeted drug delivery using genetically engineered diatom biosilica. *Nature Communications*, 6, 1, 8791, 2015.

7

Metabolism and Transcriptome Stress in Diatom *Phaeodactylum tricornutum* for Value-Added Products

Urvashi Soni†, Sonali Rahangdale†, Megha Mourya and Vandana Vinayak*

Diatom Nanoengineering and Metabolism Laboratory (DNM), School of Applied Sciences, Dr. Harisingh Gour Central University, Sagar, Madhya Pradesh, India

Abstract

Phaeodactylum tricornutum is one of the most ecologically successful diatoms dwelling in the oceans. Unlike other diatoms, it has the capability to exist in different morphotypes and cytostructural adaptations that may be stimulated by environmental conditions. Lipid and pigment mainly fucoxanthin aggregation in algae diatoms is often stimulated due to environmental stress manifested at the genomic level. Environmental stress includes nitrogen starvation, phosphate depletion, high salinity conditions, light stress, and even pollutants like microplastics. 3-Ketoacyl-CoA synthase and fatty acid elongase 2 are the two genes concerned in fatty acid elongation that confirmed better transcription stages in high-salt situations. The lipidome profile is characterized by the increase of lipid levels; for example, triacylglycerol (TAG) is observed in phosphorus starved cells of *P. tricornutum*. The most prevalent finding from transcriptome analysis of algal cells undergoing nitrogen starvation was lipid accumulation as a consequence of remodeling central metabolism. Genes of *P. tricornutum* are involved in chlorophyll biosynthesis pathway, and certain genes are expressed to produce light-harvesting proteins, which are more sensitive to light stress. There are several treatments to upregulate the mRNA of encoding antenna proteins involved in light protection. Due to the existence of complete transcriptomics, proteomics and genomics databases, this diatom species is highly suitable for studying effects of oxidation, mild stress, nickel and naphthenic environment stress and surrounding physiochemical and light conditions for enhancement of value added products.

**Corresponding author*: kapilvinayak@gmail.com
†These two authors contributed equally to this work.

Vandana Vinayak and Richard Gordon (eds.) Diatom Cultivation for Biofuel, Food and High-Value Products, (167–202)

***Keywords*:** Diatoms, transcriptomics, environment stress, pigments, *Phaeodactylum tricornutum*, metabolism, triacylglycerol, genomics

7.1 Introduction

Phaeodactylum tricornutum is a marine pennate diatom, single-celled brown microalgae [7.31]. *P. tricornutum* diatom differs from other diatoms in its pleomorphism, which is oval, cruciform, spindle-shaped, and three-pointed. Of the other three morphologies, the fusiform has a high growth rate and is therefore of great importance for commercialization [7.65]. Interestingly *P. tricornutum* is the first pennate diatom which has gone through complete genomic sequencing [7.16]. There is high demand for commercial value-added products generated by *P. tricornutum* like eicosapentaenoic acid (EPA), carotenoids such as fucoxanthin, lipids mostly triacylglycerols (TAGs), and β-1,3-glucan (crysolaminarin), proteins, terpenes such as botulin and lupeol, and initiators of bioplastics [7.42] [7.49] [7.120]. The significant pigments produced by diatoms are fucoxanthin, diadinoxanthin, beta-carotene, and chlorophyll with high metabolic activities, Carotenoids are divided into two broad classes, which are primary carotenoids and secondary carotenoids [7.40]. Primary carotenoids include xanthophylls and β-carotene and are present in the chloroplast; and secondary carotenoids are involved in photosynthesis [7.26]. It has great potential for the production of biofuels and biodiesel [7.42] [7.49] [7.120]. The EPA (C20:5) from *P. tricornutum* is a crucial source of commerce for the pharmaceutical and nutraceutical sectors [7.39] [7.84]. These long-chain polyunsaturated fatty acids have recently been used in prospective adjuvant therapy for cardiovascular issues linked to COVID-19 [7.93]. These long-chain PUFAs have a bioactive role against a range of illnesses, including coronary heart disease and thrombosis. Up to 3–5% of the dry weight *P. tricornutum* is reported to contain EPA, with an of average productivity of 56 mg $L^{-1}D^{-1}$ [7.98] [7.107]. *P. tricornutum* is also rich in PUFA's (polyunsaturated fatty acids), EPA, palmitoleic acid, and hexadecatrienoic acid, which have antibacterial properties [7.20] [7.21]. Under nitrogen deficiency the cell organelles start to degrade and produce about 40% of lipid, which can further be used in biodiesel production. Biofuel from diatoms mainly produced by fatty acid methyl ester (FAME), a clean, sustainable fuel that can be created from a variety of bioresources. Biodiesel is produced by the conversion of bioalgal lipid into fatty acid through FAME [7.7] [7.8] [7.44]. In contrast to conventional fuels, biofuel from diatoms also known as Diafuel™, has zero carbon footprint [7.103]. Not only that, under nitrogen starvation, branched-chain amino acids begin to lyze and enter the TCA cycle to produce triacylglycerols [7.72]. The diatom contains approximately 36% protein, 26% available carbohydrates, 18.0% biofat, 0.25% mild biodetergent fiber, 5.19% biosilica and 0.86% fucoxanthin by dry weight [7.83]. Figure 7.1 shows the evidence of several value added products obtained from *P. tricornutum*. This diatom has great potential to grow and divide under conditions of silicon deficiency [7.65]. Having also a mixed trophic mode, *P. tricornutum* synthesizes both glucose (autotrophic) and inorganic carbon (heterotrophic). At the exponential phase, when cell growth is maximal, 90% of the carbon in the biomass is solely from the aldose-hexose monosaccharide glucose ($C_6H_{12}O_6$) [7.121]. This diatom has inserted several genes from bacteria via horizontal gene transfer and its genomic pattern contains small intergenic regions but no noncoding sequences (introns) [7.19] [7.88]. It contains about 127–130 genes, of which 27 are tRNA genes, 3 are rRNA genes, and 2 are sRNA genes [7.88]. The mitochondrial genome of *P. tricornutum* is approximately 77,356 bp in size [7.70]. It also synthesizes a good amount of omega-3 fatty acid content [7.19]. Transcriptome studies of *P. tricornutum* are of great importance in discovering the biological functions of multiple genes. This is because these studies focus only on transcripts or only regions of genes that are fully expressed. This chapter provides an understanding of the commercial use of diatoms, fucoxanthin metabolism and transcriptomics studies.

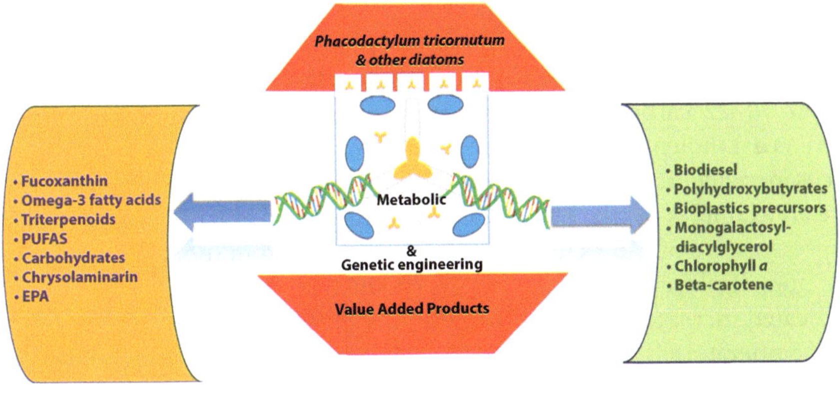

Figure 7.1 Value-added products produced diatoms.

7.2 Commercial Market Value

Diatoms are viewed as a promising candidates for unsaturated fats and the mass creation of triacylglycerol (TAG) for the biofuel business [7.46]. As of now, microalgae are financially applied for the heterotrophic creation of unsaturated fats for the food and feed industry [7.5] [7.74]. They are photoautotrophic in nature, directly using the sun's energy and CO_2 to produce biomass and valuable products like TAG [7.104]. Diatoms generally produce lipids up to 60% of dry load to the detriment of development [7.81]. The diminished photosynthetic productivity under sub-par development condition is one of the critical impediments in commercializing photosynthetic algal mass for lipid production. According to geoscientists, among the several species of microalgae now in existence, diatoms have a huge potential to provide 30% of the world's crude oil [7.102]. Diatoms besides being used as biodiesel feedstock, have been the main focus for the manufacture of a variety of secondary metabolites for the personal care, health, and food industries [7.48]. They can be economically cultured, and researchers are working to capture their valuable metabolites at the same time [7.78].

7.2.1 Industrial Applications

Diatoms have three ways to aid in the production of ethanol, methane and biodiesel.

1. Ethanol production via carbohydrates,
2. Methanogenesis via protein,
3. Biodiesel production via unsaturated fatty acids [7.91].

Polyunsaturated fatty acids (PUFAs) are straight-chain fatty acids of about 18–22 carbons containing two or more double bonds. Among PUFAs α-Linolenic acid (ALA, 18:3n-3), stearidonic acid (SDA, 18:4n-3), eicosapentaenoic acid (EPA, 20:5n-3), and docosahexaenoic acid (DHA) are all examples of n-3 PUFA (DHA, 22:6n-3) and arachidonic acid is a representative of n-6 PUFAs. *P. tricornutum* grown in tubular photobioreactors displayed declining biomass productivity, while the EPA content indicated increased output. *P. tricornutum* grows at its fastest pace when the temperature is 20 °C [7.87]. However, temperature had no impact on *P. tricornutum* growth response in short-term temperature treatments, but in long-term temperature treatments, growth rate decreased gradually at lower temperatures and abruptly at higher temperatures [7.97]. The concentrations of EPA and PUFAs were found to be up to 120% higher than the control when the temperature was lowered from 25 °C to 10 °C for 12 hours. The yields of EPA and PUFAs were found to be as high as 4.9% and 2.6%, respectively [7.11]. *P. tricornutum* is thought to be a promising candidate for biodiesel production and is characterized by high fatty acid content [7.35]. In Shenzhen, China, a clear Plexiglas cylinder with a diameter of 0.5 m and a height of 1 m with an open top was used as the growth tank unit. A test system was set up to analyze long-term yield production. However, industrial convenience and cost dictated the use of much larger open ponds. In the reactor, each unit was filled with 1 L of fine air bubbles per minute and the medium was stirred. According to the data, a dry diatom production per acre per day is an average of 0.44 MT dry weight [7.105].

Assuming a 30% oil content, production costs are currently estimated at $2.80 per liter of algal biodiesel. This is almost 2.8 times the price per liter of oil and refinery announced by the US Energy Information Administration in the summer of 2008 [7.6]. However, diatoms thrive in aqueous suspensions and have easy access to water, carbon dioxide and other nutrients [7.80]. They can produce 30 times more oil than other biofuels and can convert up to 85% of their body weight into lipid reserves, which can yield 30 times more oil per acre than terrestrial oilseed crops. Even without gene modification or photobioreactor optimization, it is estimated that a 200 kHa open diatom/green algae pond can generate 1015 BTU (1018 kJ) of liquid fuel energy. The rise in demand for crude oil has thus made biofuel made from algae one of the primary sources of liquid fuel alternatives.

Diatoms can be used as a supplementary source for the production of carotenoids, proteins, biofuels, and other value-added-based products to meet the population's desire for large-scale cultivation. The main commercial uses of these biomolecules are in the food, cosmetics and pharmaceutical industries [7.3]. They may be used as biotechnological tools in various industrial sectors, such as biofuel production as a sustainable energy source, fertilizer production, secondary metabolite development, and animal feed production [7.3].

The differences in properties indicate the different structural types present in biodiesel from diatoms and petroleum diesel. Biodiesel consists of open-chain hydrocarbon molecules that are esters with one oxygen atom per molecule. Petroleum diesel, on the other hand, is composed of linear molecules and cyclic hydrocarbon chains that are not oxygenated in structure. One of the most promising other options and renewable feedstock sources for the production of third-generation biodiesel is microalgae. When comparing microalgae to other oilseeds such as soybeans and palm oil content, it was discovered that microalgae produce a surprising amount of oil content [7.82].

7.2.2 Types of Foods with Health Benefits from Diatoms

Omega-3 fatty acids found in diatoms have been associated with health benefits in many epidemiological and clinical studies [7.4]. EPA and DHA are beneficial for the cardiovascular tract, vision, and the treatment of psychiatric diseases. Essential fatty acid DHA is crucial in preventing age-related cognitive deficits, including those caused by Parkinson's disease, Alzheimer's disease, and multiple sclerosis. DHA's role in the development of the nervous system makes it an active component in the treatment of mental illnesses as well [7.99]. Additionally, commercial manufacturing of products having neonatal formulas, infant formulas, fortified snack bars, and other dairy products with added omega fatty acids has resulted in limited emphasis on the cardioprotective and cognitive benefits of omega fatty acids [7.90]. Figure 7.2 shows the evidence of commercial value of *P. tricornutum* in various sectors. The most nutritionally relevant biomolecules,

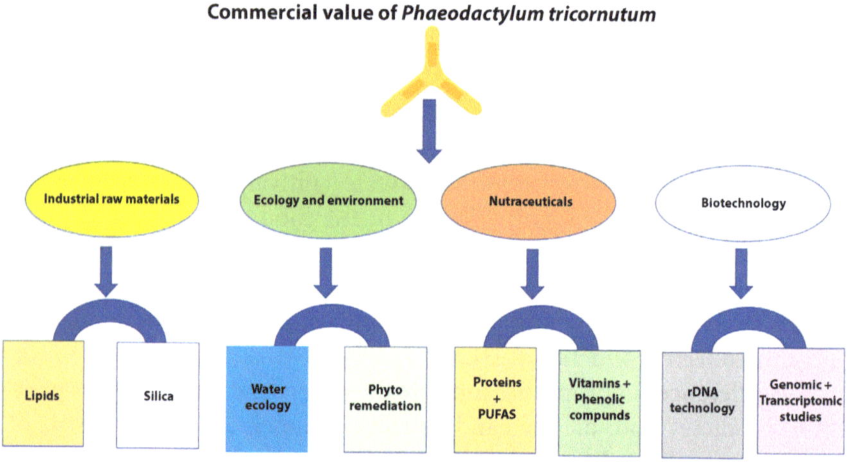

Figure 7.2 Flow chart showing commercial value of *P. tricornutum* in various sectors.

such as the omega-3 fatty acid, arachidonic acid omega-3 and other omega-3 fatty acids, are produced by diatoms and are unsaturated fatty acids [7.47].

7.3 Metabolic Pathways and Mechanisms for Synthesis of High Value Added Products in Diatoms

Diatoms produce the economically significant pigment molecule fucoxanthin, which has several properties, including antioxidant, anti-inflammatory, anticancer, anti-angiogenic, anti-obesity, and antimalarial activities [7.89]. Fucoxanthin is a keto-carotenoid with the following chemical compositions: (C=C=C=) bond, a unsaturated carbonyl group and a 5,6-monoepoxide (Figure 7.3) [7.71] [7.89].

Fucoxanthin is present in the light-capturing apparatus in photosystems as a part of the electron transport chain in the thylakoid membrane of diatoms [7.71]. Its bioconversion revealed that the ingested fucoxanthin was deacetylated in the intestinal lumen and transported through the blood in some hens [7.75]. Fucoxanthin was hydrolyzed to fucoxanthinol during the uptake by differentiated Caco-2 epithelial cells isolated from colon tissue derived from a 72-year-old, White, male with colorectal adenocarcinoma, a tissue culture model focusing on the absorption of dietary compounds by human intestinal epithelium [7.58]. Use of valsartan antihypertensive long-term use evaluation (VALUE) trial showed that the fucoxanthin was deacetylated to fucoxanthinol in the digestive system by lipase and pancreatic esterase or in intestine cells, and transmitted to the mammalian blood circulation system from the gastrointestinal tract [7.75]. The conversion of fucoxanthinol into amarouciaxanthin A was mostly observed in liver microsomes; fucoxanthin was added, and HepG2 cells also underwent

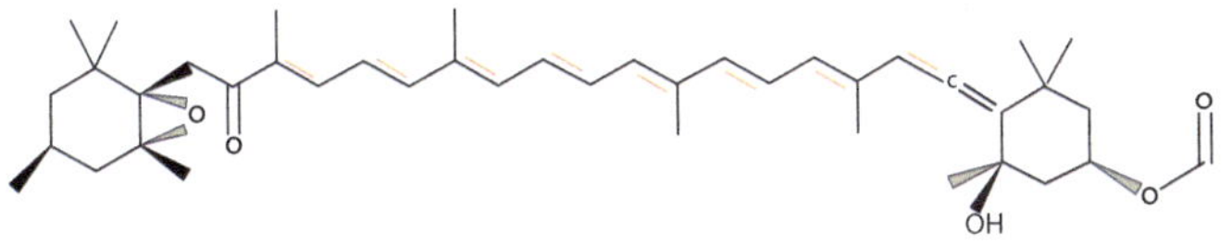

Figure 7.3 Chemical structure of the fucoxanthin molecule.

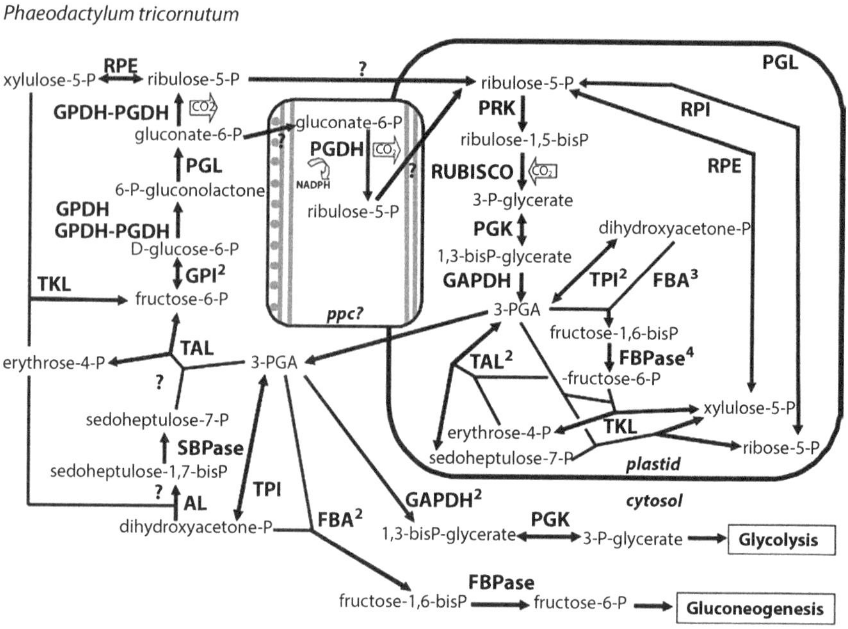

Figure 7.4 Model of the oxidative and reductive pentose phosphate pathways and related reactions in *P. tricornutum*. Reproduced with permission from [7.34].

amarouciaxanthin A formation [7.75]. Most of the studies have already been published related to different metabolic pathways and regulation processes in diatoms [7.12]. These involve the phosphogluconate pathway, the Entner-Doudoroff (ED) pathway, carbon dioxide metabolism [7.66], the presence of the urea cycle, enzyme functions in the second phase of glycolysis in cytoplasm, chloroplast and power house of the cell [7.33] and C4 pathways, etc. [7.12]. Figure 7.4 shows the evidence of oxidative and reductive pentose phosphate pathway in *P. tricornutum*.

7.3.1 Carbon Dioxide Fixation

Several biochemical studies were done on *P. tricornutum* and *Thalassiosira pseudonana* diatoms which suggested that there could be a C4 cycle through which a 4-carbon-containing compound could exist; for example, $C_4H_6O_5$ or oxaloacetate, which are present in chloroplast producing

Rubisco (Ribulose 1,5-bisphosphate carboxylase/oxygenase) that fixes atmospheric carbon dioxide [7.112]. The genetic material of *P. tricornutum* appears to synthesize the enzymes pyruvate orthophosphate dikinase (PPDK), phosphoenolpyruvate carboxykinase (PEPCK), and phosphoenolpyruvate carboxylase (PEPC) [7.117]. All these enzymes are essential for C4 metabolism, and some are used to play a role in C3 metabolism. Ribulose 1,5-biphosphate carboxylase/oxygenase (Rubisco) is the principal enzyme for CO_2 fixation in the vast majority of oxygenic photosynthetic organisms, including photosynthetic algae. Photosynthesis is a critical mechanism for primary productivity [7.111]. The primary result of two molecules of 3-phosphoglycerate, are formed when ribulose 1,5-bisphosphate (RuBP) is carboxylated by Rubisco [7.14]. The Michaelis-Menten constant (Km = 20-60 M) for CO_2 concentration in the vicinity of the diatom Rubisco enzyme is substantially higher than the value of the dissolved inorganic carbon (DIC) in seawater, which corresponds to a CO_2 concentration of 10-15 M at pH 8.2. Diatoms use CO_2 concentrating mechanisms (CCMs), which involve the active uptake and transport of inorganic carbon, to overcome CO_2 limitation. This increases the CO_2 concentration near Rubisco and boosts photosynthetic efficiency, especially in a carbon-limited environment. Biophysical and biochemical CCM can be distinguished from one another. Bicarbonate (HCO_3) is actively transported across biological membranes by biophysical CCMs, and the conversion of CO_2 to bicarbonate is promoted [7.112]. Aquaporins play a role in the carbon-concentrating processes that actively absorb HCO_3, leading to high levels of inorganic carbon (Ci) in microalgal cells and increased CO_2 levels for the photosynthetic mechanism with carbonic anhydrase. The cytoplasm, the chloroplast, and the outside of the plasma membrane all contain carbonic anhydrase [7.86]. In order to enrich CO_2 for photosynthesis, carbonic anhydrase accelerates the equilibrium between CO_2 and bicarbonate ion. Additionally, extracellular carbonic anhydrase (CA) facilitates the hydration of CO_2 that escapes from the cell, reducing CO_2 outflow into the environment. Rubisco's carboxylase activity is preferred above that of CA because all of the Ci that enters the cell is hydrated into bicarbonate ion in the cytoplasm before diffusing into the carboxysomes where CA converts it to CO_2 [7.61]. Phosphoenolpyruvate carboxylases (PEPCs) in biochemical CCMs carboxylate C3 molecules, fixes CO_2 to C4 acids [7.116].

7.3.2 Photorespiration and Glyoxylate Metabolism

Ribulose 1,5-bisphosphate carboxylase/oxygenase (RuBisCO), a dual-purpose enzyme, mediates photorespiration, a significant route of plant primary metabolism. One molecule of 2-P-glycolate and one molecule of 3-P-glycerate are created when O_2 outcompetes CO_2 for RuBP. 2-P-glycolate inhibit the Calvin cycle enzyme triosephosphate isomerase. The C2 cycle is used to break down 2-P-glycolate. Glycolate is exported from the chloroplast once the phosphate group is broken, where it enters the peroxisome where it is oxidized to glyoxylate and then transaminated to create glycine, which enters the mitochondrion. The condensation of two glycines into one serine, which results in the release of ammonium ion and CO_2, is catalyzed by the glycine decarboxylase complex and serine hydroxymethyltransferase. The serine undergoes additional metabolism, returning to P-glycerate, which can subsequently join the Calvin cycle in the chloroplast. As a result of the four carbon atoms that enter the C2 cycle, three are changed back into PGA and one is liberated as CO_2 [7.54].

7.3.3 Reductive/Oxidative Pentose Phosphate Pathway

In diatoms, the Calvin cycle, sometimes referred to as the reductive/oxidative pentose phosphate pathway, is responsible for photosynthetic carbon fixation [7.55]. Light-dependent redox regulation at the transcriptional or enzymatic level, as well as substrates and ions like Mg^{2+}, are all important regulators of carbon fixation in microalgae plastids. Microalgae have an oxidative pentose phosphate pathway [7.55]. This common procedure generates NADPH and pentose-phosphates for the in-dark production of nucleotides, amino acids, and fatty acids. By decarboxylating glucose-6-phosphate, it achieves, because the two processes in plastids are interrelated, attempting to use both at once would result in an inefficient cycle that would consume ATP without actually fixing any CO_2 [7.100]. In contrast, the key enzyme of the OPP, glucose-6-phosphate dehydrogenase (G6PD), is active in the dark but inhibited after reduction in the plastids of land plants and green algae. These enzymes include phosphoribulokinase (PRK), glyceraldehyde-3-phosphate dehydrogenase (GAP-DH), and fructose 1,6-bisphosphatase (FBP) [7.54]. In the plastids of certain diatoms, there does not appear to be a fully functional oxidative pentose phosphate pathway, indicating that diatom plastids generally lack this mechanism. The enzymes, 6-phosphogluconate dehydrogenase (6PGDH, 26934, and the 6PGDH component of a G6PDH/6PGDH fusion protein, 54663) and glucose-6-phosphate dehydrogenase (G6PDH, 30040), were the cytosolic enzymes, demonstrating that the entire OPP is only functional in the cytosol [7.111]. The only Calvin cycle enzymes encoded on the plastid genome are the small and the large subunit of the Rubisco (GenBank AY819643). The nuclear genomes of *P. tricornutum* and *T. pseudonana* contain all additional genes encoding primary Calvin cycle enzymes. The gene for sedoheptulose bisphosphatase (SBP) is the sole exception [7.54]. The SBP gene (56467) encodes a protein that doesn't have a plastid targeting sequence and thus seems to be located in the cytosol. But because this SBP is absent from the extensive collection of ESTs from *P. tricornutum*, it is clear that under the current circumstances, it is not being actively transcribed. In the Calvin cycle, SBP catalyzes the conversion of sedoheptulose-1,7-bisphosphate to sedoheptulose-7-phosphate. However, it is not yet known whether the SBP reaction takes place in the plastids of diatoms [7.60] or, more likely, whether one of the plastidic FBPs performs this conversion, as has been demonstrated for FBP I in cyanobacteria. A bipartite plastid targeting sequence is present in a gene that encodes a plastidic FBP (FBPC2, 42456) that interestingly is situated 500 bases upstream of the SBP gene in the same direction (similar as in *T. pseudonana*) [7.53].

7.3.4 Glycolysis

Glycolytic pathways occur in all living organisms that basically involve the breakdown of hexoses inside the cytosol resulting in pyruvate, which is transported to mitochondria in eukaryotes under anaerobic conditions, or might be utilized in some other anaerobic pathways [7.101].

7.3.5 Storage Products Synthesis and Degradation

Chrysolaminarin is the soluble carbohydrate with molecular mass of 7.75 kDa from marine diatoms [7.17]. Since chrysolaminarin is high in marine diatoms, it plays a significant role in the ocean carbon cycle. About 10–20% of total carbon in the diatoms is contributed by chrysolaminarin at exponential phase and nitrogen is a limiting factor for carbon [7.17].

7.3.6 Inositol and Propionate Pathway

Among the variety of cyclitoles, the most common is myoinositol [7.79]. Many vital substances are found in cells, including phosphatidylinositol-4, 5-bisphosphate, phosphatidylinositol, and cell signaling; hormone regulators such as indole acetic acid conjugate (IAA); tethering layer (glycerophosphoinositide anchors); and stress relievers such as ononitol and pinitol; which produce galactinol and store inositol. In order to create different noncellular pectic chemicals for the cell wall, D-glucuronic acid is employed [7.13]. Diatoms produce a number of carotenoids, such as fucoxanthin, diadinoxanthin, β-carotene, and diatoxanthin, among others, having a biosynthetic pathway, such as the non-mevalonate pathway used by *P. tricornutum* to produce isoprenoids [7.40] [7.43].

7.3.7 Biosynthesis Production of Carotenoid in *P. tricornutum* Diatoms

P. tricornutum grows quickly, doubling its biomass in a matter of hours, making it a particularly promising diatom for use in industrial operations [7.63]. Its development is affordable, sustainable, and easily manageable both indoors and outdoors. *P. tricornutum* is the most well-known diatom molecular model system, and it has state-of-the-art tools for genome editing, gene silencing, and overexpression of genes to change how transformed cells express their genes [7.1]. Carotenoids are produced by diatoms using methylerythritol phosphates (MEP) (Figure 7.5). Diatoms have fucoxanthin as one of the vital pigment formed after diadinoxanthin (Dd), diatoxanthin (Dt) and β carotene [7.63]. Fucoxanthin has qualities that fight cancer, diabetes, and obesity because of its antioxidant properties [7.2]. Their radical scavenging capacities are 13.5 and 1.7 times greater than those of α-tocopherol, respectively, and their anticancer effects are greater than those of beta-carotene, with a moderate increase in fucoxanthin amount. Seven genes, four for the de-epoxidase reactions (violaxanthin de-epoxidase VDE, VDE-like VDL1 and 2, and the VDE-related, VDR), which transform violaxanthin into zeaxanthin or Dd into Dt, and three for the reverse reactions, are putatively involved in the violaxanthin-zeaxanthin and Dd-Dt cycles in the *P. tricornutum* diatom (zeaxanthin epoxidases, ZEP1, 2 and 3) [7.63]. The catalytic activity of these enzymes varied; for

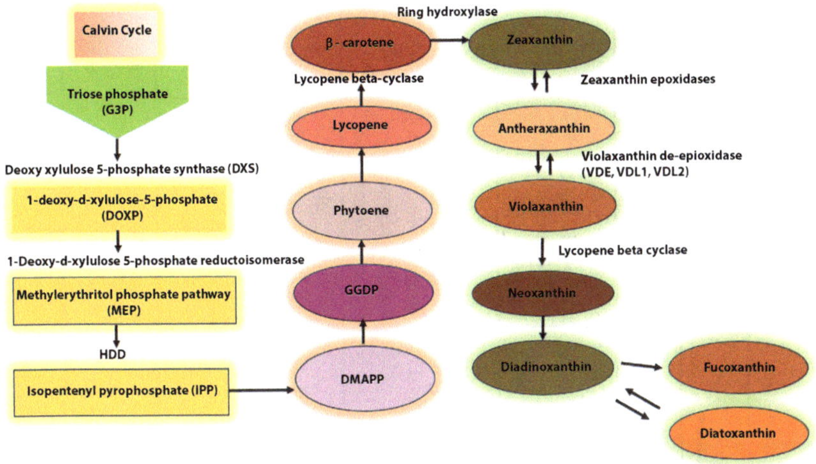

Figure 7.5 Flowchart showing the biosynthesis pathway of carotenoids *P. tricornutum* diatom.

instance, ZEP2 showed a broader substrate specificity than ZEP3, which raised the possibility that this enzyme might be engaged in the Dd-Dt cycle [7.63].

7.4 Light Stress in Diatoms and Fucoxanthin Biosynthesis

The higher fucoxanthin content of *P. tricornutum* compared to brown seaweeds and its ease of cultivation and extraction make it a promising producer for commercial fucoxanthin production [7.77]. A variety of marine diatom species are exposed to light stress for increased fucoxanthin production. Table 7.1 shows a list of several experiments performed on several light stress conditions on various diatoms and microalgae for fucoxanthin production.

In recent work, mixotrophic cultures of *P. tricornutum* CCMP 1327, were optimised at different light stress for enhancing fucoxanthin synthesis in *P. tricornutum* [7.25].

A 14-day culture is performed for the evaluation of a range of starting cell densities. After that, cells were cultivated at several light intensities for 12 days.

Table 7.1 Fucoxanthin production achieved in some of the diatoms and microalgal strains using light stress.

Microalgal strain/ diatoms	Media used	Temperature	Light intensity/ photoperiods	Cultures cultivated	Biomass	Fucoxanthin yield	References
Phaeodactylum tricornutum	Conway media	20 °C	60-W fluorescent bulbs with 2500 lx intensity	Plastic cylinders are used	–	15.7 mg g^{-1} DW.	[7.50]
Phaeodactylum tricornutum strain	ASP media	18 °C	40 μmol photons/16:8 h day-night cycle	–	–	genetically modify dxs reached 24.2 mg g^{-1} DW and PSY reached up to 18.4 mg g^{-1} DW	[7.23]
Phaeodactylum tricornutum	ASP Media	18 °C	40 μmol photons/16:8 h day-night cycle	–	–	3 mg g^{-1} DW in non-transformant/wild type	[7.23]
Odontella aurita	Modified L1 media	–	White fluorescent lamps with light intensity of 100 photons	Bubble column photobioreactor	6.36 g^{-1}.	Fucoxanthin content 18.47 mg g^{-1}, Fucoxanthin productivity 7.96 mg L^{-1} day^{-1}	[7.113]
Odontella aurita	Modified L1 media	–	White fluorescent lamps light intensity of 300 μmol photons	Flat plate photobioreactor	3.53 g^{-1}	21.67 mg g^{-1} DW	[7.143]
Isochrysis* aff. *galbana	f/2-Si media	20 °C	60 W fluorescent lamps at intensity of 2500 lx	30-L plastic cylinders	–	18.23 mg g^{-1} DW	[7.51]
Odontella aurita	F/2 media	–	30 μmol light intensity $Light^{-1}$ $dark^{-1}$ cycle of 12h:12h	Phototrophic culture	–	1.5% of DW	[7.37]

(*Continued*)

Table 7.1 Fucoxanthin production achieved in some of the diatoms and microalgal strains using light stress. (*Continued*)

Microalgal strain/ diatoms	Media used	Temperature	Light intensity/ photoperiods	Cultures cultivated	Biomass	Fucoxanthin yield	References
***P. tricornutum* and microalgal strain**	Nitrogen supplementation	22-24°C	R:B light (6:1) at 20 µmol with the addition of tryptone	Novel two-phase culture	5.53 g L^{-1} Biomass concentration in red-blue light.	16.28 mg g^{-1} DW under blue light	[7.114]
Navicula	Different nitrogen media (ammonium, urea, and nitrate medium)	22-24°C	Indoors and outdoors (sunlight)	–	Biomass productivity was 153 ± 3 for indoor culture and 93 ± 3 for outdoor culture	Fucoxanthin content decreased from 5.40 ± 0.05 (indoor) and 2.61 ± 0.06 (outdoor) mg g^{-1} dried biomass	[7.96]
Nitzschia laevis	LDM media	22-30°C	LEDs (white,blue,mix) were used for Mixotrophic cultures, Dark condition used for heterotrophic cultures remained.	Fed-batch culture	17.25 g^{-1}	16.5 mg^{-1} fucoxanthin concentration	[7.62]
Sellaphora minima	FD Media with CO_2 as carbon source	25 °C	100 µmol photons light intensity	–	7.5 mg g^{-1}	11.1 mg L^{-1}	[7.30]

(*Continued*)

Table 7.1 Fucoxanthin production achieved in some of the diatoms and microalgal strains using light stress. (*Continued*)

Microalgal strain/ diatoms	Media used	Temperature	Light intensity/ photoperiods	Cultures cultivated	Biomass	Fucoxanthin yield	References
Thalassiosira weissflogii	Modified f/2 Si media	–	100 μmol light intensity, blue LED	Mass culture	538 mg L^{-1} day^{-1} of biomass	30.6 mg L^{-1}	[7.64]
Nitzschia laevis	Liquid LDM media	22 °C	15 μmol light intensity	Mixotrophic cultures	2.34 g^{-1} cell density	1.56% of DW	[7.62]
Isochrysis zhangjiangensis	Modified f/2 medium	22-24°C	Low light intensity of 40 μmol and 300 μmol high intensity.	-	2.4 g^{-1} biomass in high light intensity	23.29 mg g^{-1} DW in low light intensity.	[7.108]
***Isochrysis* strain**	Modified f/2 media	22 °C, with 1.5% CO_2 enriched air	60 μmol light intensity	Column photobioreactors	2.72 g^{-1}	Fucoxanthin content of 14.5 mg g^{-1} dried weight and Fucoxanthin productivity of 5.85 mg L^{-1} day^{-1}	[7.92]
***Mallomonas* sp.**	f/2 media	24 °C with 5%	Light intensity of 140 μmol CO_2, 16:8 h light:dark photoperiod	–	3.75 g^{-1}	26.6 mg g^{-1} DW	[7.50]

(*Continued*)

Table 7.1 Fucoxanthin production achieved in some of the diatoms and microalgal strains using light stress. (*Continued*)

Microalgal strain/ diatoms	Media used	Temperature	Light intensity/ photoperiods	Cultures cultivated	Biomass	Fucoxanthin yield	References
Cylindrotheca closterium	f/2 media	22-24°C	100 µmol light intensity, blue light-emitting diode	Bottle PBRs, Bag PBRs	Under constant light 76 mg L^{-1}/ day of biomass is obtained in bottle PBRs whereas 57 mg L^{-1}/ day in bag PBRs	In bag PBRs: Fucoxanthin content is 25.5 mg g^{-1} DW Fucoxanthin productivity is 1.4 mg L^{-1} day^{-1} In bottle PBRs: Fucoxanthin content 23.6 mg g^{-1} DW Fucoxanthin productivity 1.8 mg L^{-1} day^{-1}	[7.106]
Cyclotella cryptica	Modified SK media and glucose is used as organic carbon source	22-24°C	–	Heterotrophic condition	–	0.77% of DW	[7.37]
Chaetoceros gracilis	Conway media	30 ± 1°C with 2% CO_2 gas	Light intensity of 300 µmol, LED lights	–	–	17.1 mg g-dw^{-1} of fucoxanthin and 41.5 percent EPA	[7.94]
Phaeodactylum tricornutum	f/2 media	22-24°C	Lowest 100 µmol photons and highest at 210 µmol photons intensity with CO_2		–	At low light intensities fucoxanthin concentration (42.8 ± 19.5 mg g^{-1}) and at a higher intensity (9.9 ± 4.2 mg gL^{-1})	[7.94]

(*Continued*)

Table 7.1 Fucoxanthin production achieved in some of the diatoms and microalgal strains using light stress. (*Continued*)

Microalgal strain/ diatoms	Media used	Temperature	Light intensity/ photoperiods	Cultures cultivated	Biomass	Fucoxanthin yield	References
Diatom *Stauroneis* sp.	f/2 media under nutritional limitation silica and nitrate	–	Light intensity 100 µmol photon, blue LEDs	Tubes	–	5.9 mg gL^{-1}	[7.73]
Fragilariopsis cylindrus	f/2	7 °C	Monochromatic blue light (445 nm) at a PUR of 11.7 µmol photons 12 h light:12 h dark photoperiod	–	–	Fucoxanthin productivity of 43.80 µg L^{-1} day^{-1} and the Fucoxanthin yield of 7.53 µg.	[7.36]
Chaetoceros gracilis	Modified Conway media	30 ± 1 °C and 2% CO_2 gas	Light intensity 300 µmol ,12 L:12 D, LED lights	Column reactor	–	17.1 mg g-dw^{-1} fucoxan-thin and 41.5% EPA	[7.94]
Conticribra weissfogii	f/2 medium	21 °C	Light intensity of 100 µmol photons and a 12-h light period	200-mL Erlenmeyer flask	–	1.00±0.24% AFDW	[7.18]
Amphora capitellata	f/2 medium	24±2 °C	20 µmol photons light intensity and 300 µmol photons	300-mL flask ,2-L bubble photo bioreactor	–	Fucoxanthin content was 41.83±0.92 mg g^{-1}.	[7.24]
Phaeodactylum tricornutum	nutrients (urea and glycerol	–	-	Phototrophic and mixotrophic condition	Biomass yield is 0.77 g L^{-1}	EPA and fucoxanthin productivity are 27 mg L^{-1} and 11 mg L^{-1}	[7.85]

(*Continued*)

Table 7.1 Fucoxanthin production achieved in some of the diatoms and microalgal strains using light stress. (*Continued*)

Microalgal strain/ diatoms	Media used	Temperature	Light intensity/ photoperiods	Cultures cultivated	Biomass	Fucoxanthin yield	References
Isochrysis zhangjiangensis	–	–	Constant light at 100 μmol		–	22.6 mg g^{-1} fucoxanthin content fucoxanthin productivity is 3.06 mg	[7.59]
Tisochrysis lutea	NutriBloom Plus	25 °C	Light intensity 140 μmol and 2% CO_2,18:6 day: night photoperiod	250 mL flasks,	–	fucoxanthin productivity (9.81 mg L^{-1} d^{-1}) fucoxanthin content (5.24 mg g^{-1})	[7.28]
T. lutea	Nutri Bloom Plus	25 °C	Light intensity 140 μmol, 18/6h day/night cycle, 2.0% CO_2, blue, red, led lights	250 mL Erlenmeyer flasks	Blue light increase productivity and red light decrease productivity	16.30 mg g^{-1} fucoxanthin content, 2.77 mg L^{-1} d^{-1} fucoxanthin productivity.	[7.29]

The initial cell density (1×10^7 cells mL^{-1}) and lower light intensity (20 mol m^{-2} s^{-1} m^{-2} s^{-1}) were advantageous for higher biomass and fucoxanthin accumulation.

This study developed two classic approaches to enhance biomass production in Red:Blue light (6:1) regime in ratio of 6:1 at 20 μmol m^{-2} s^{-1}, in 6 days culture [7.52].

Fucoxanthin content is maximal under blue light (BL) at 16.28 mg g^{-1} dry weight, but biomass concentration (5.53 g L^{-1}) is highest under red light (RL) for 6 h and BL for 6 h culture in batch.

It was reported that diatom *Phaeodactylum tricornutum* UTEX 646 genetic variant cultured at 18°C in ASP medium temperature, with 16:8 h L:D of photoperiodism, produced 10.3 mg g^{-1} DW of fucoxanthin [7.22].

7.5 Transcriptomics in Diatoms

The major difference between genomic studies and transcriptomic studies is that in transcriptomic studies it is easy to understand the expression of gene and which condition is favorable for the formation of mRNA.

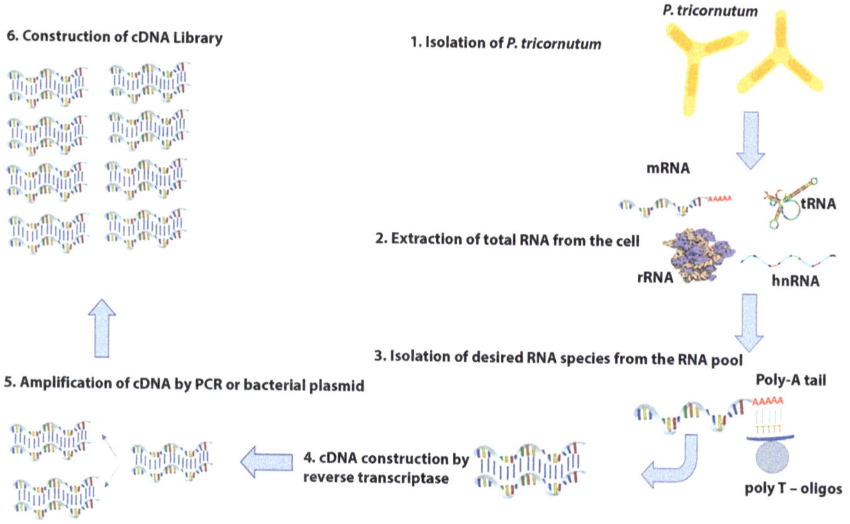

Figure 7.6 Schematic representation of transcriptomics in diatoms.

On the other hand, genomic studies will help us to understand the presence of genes in the genome. Both of the studies are equally important in genetics. Diatoms have multiple origins. The marine centric diatom *Thalassiosira pseudonana* and coastal pennate diatom *P. tricornutum* had undergone complete sequence annotations [7.10]. Figure 7.6 shows the evidence of basic process of transcriptomics and cDNA libraries in diatoms.

7.5.1 Steps in Transcriptomics Sequencing

1. *Separation of mRNA from an organism such as diatoms and microalgae:* Construction of a sequencing library requires a good amount of RNA with good quality. rRNA contributes about 95% of total RNA in the RNA pool. It is essential to remove rRNA before library construction as it will create problems in the detection of other types of RNA present in the cell in less quantity. Removal of rRNA is difficult, therefore many protocols are formed to increase the mRNA before the construction of library. One of the most appropriate protocols for the selection of mRNA is ligation of poly-T-oligos with the 3' poly A- tail of mRNA; further poly-T-oligos are covalently attached to a substrate like magnetic beads [7.56].
2. *Construction/formation of library for RNA sequencing:* Library construction for RNA sequencing is the cascade of steps which includes a) Separation of RNA of interest, b) Conversion of RNA into cDNA with the help of reverse transcriptase, c) Breakdown of cDNA into smaller fragments, d) Amplification of cDNA through PCR or bacterial/yeast machinery, e) Attaching sequencing adopters [7.56].

7.5.2 Transcriptomic Studies in *Phaeodactylum tricornutum* Under Various Influential Factors

Light Stress: Intertidal benthic diatom *P. tricornutum* goes through highly variable light environments. This makes it difficult for these diatoms to interact with high light energy at low tide. Nonphotochemical chlorophyll fluorescence quenching (NPQ) is a speedy process that diatoms have for quenching excess energy. Its capabilities have been elucidated by the xanthophyll cycle and LHCX proteins. The transcriptional regulation of light-harvesting complex X has been observed in the diatom *P. tricornutum*. The *P. tricornutum* genome has tetra *LHCX* genes (*LHCX*1, 2, 3 and 4). Table 7.2 shows the list of several genes in *P. tricornutum* and their functions. Of these four genes, *LHCX*1 mostly takes place under non-stress light conditions [7.32] [7.69]. As a result, the corresponding protein is present in low-light conditions, endowing the diatom with its NPQ basal ability. The NPQ is mainly localized near the PSII nucleus in *P. tricornutum* and interacts with abrupt changes in photoclimate [7.95]. Under strong light, RNA synthesis of both *LHCX*2 and *LHCX*3 is heavily increased. As both proteins co-aggregate in the de novo synthesis of Ddx + Dtx, they can provide extra Ddx/Dtx binding sites to activate antenna NPQ and increase the basal NPQ provided by *LHCX*1 [7.95]. The *LHCX4* transcript aggregates in constant darkness but its expression is heavily influenced by light [7.57].

Metal Stress: Metals like nickel are natural ingredient in the environment [7.41]. The concentration of nickel in an open ocean is not necessarily very high. In addition, anthropogenetic activities such as electroplating, nickel mining, cement production, steel production, and metalworking operations are rapidly increasing nickel concentrations in the environment [7.117]. The toxic effects of nickel in *P. tricornutum* affect biological functions and transcriptomic responses. The cell density and photosynthetic capacity of *P. tricornutum* decreased at

Table 7.2 Genes in *P. tricornutum* and their function.

Gene	Function	Reference
SPX	Negative controller of P uptake and P-stress reactions	[7.119]
bZIP14	Controller of the TCA cycle	[7.67]
PSY	The transformants increased levels of carotenoids	[7.45]
UDPGP	Inactivation of the UDP glucose results in a gene increase in lipid synthesis and decrease in the amount of chrysolaminarin	[7.122]

1 mgL^{-1} nickel. Photosystem II (PSII) oxygen evolution complex (OEC) efficiency (Fv/F0), photosystem II maximum photosynthetic efficiency (Fv/Fm), electron transport rate (ETR), and actual photosynthetic efficiency, which consists of PSII [Y(II)], non-photochemical quenching (NPQ), and photochemical quenching (qP), indicated that the OEC of PSII could be significantly affected by nickel. Transcriptomic data also reveal that the OEC apparatus gene encoding PSII oxygen generation enhancer protein 2 (PsbP) is regulated by nickel [7.15]. *P. tricornutum* cultures were grown in near white light with a photon flux density of 65 μmol photons/m^2s^{-1} with a 12:12 h light:dark period in f/2 medium at approximately 20 °C. Exponential cultures start at 5.2 x 10^5 cells/mL^{-1}. A series of predicted final concentrations (1.0, 2.5, 5.0, 10.0, 20.0, and 50.0 mg/L^{-1}) of nickel (nickel as $NiCl_2$) were added to cultures of *P. tricornutum*; and the actual nickel concentrations determined by ICP-MS analysis gave concentrations of (0.92, 2.38, 4.7, 9.36, 18.61, and 48.53 mg/L^{-1}). *P. tricornutum* cultures were collected after 24, 48, 72, and 96 hours to measure cell density, chlorophyll *a* content, and photosynthetic capacity [7.38].

Naphthenic acid (NA) stress: NA is a cyclic alkyl straight-chain carboxylic acid commonly derived from water of lipid tailing [7.109]. At the time of mining, transportation and processing, intensively acidic crude oil is formed and can release large amounts of naphthenic acid contaminants into some ecosystems such as Earth's crust, underground water, ponds and settlements [7.27]. NA influence photosynthesis in marine diatoms. Naphthenic acid-induced transcriptome modified marine *P. tricornutum* using RNA sequencing has been studied. An experiment was conducted to investigate the effect of naphthenic acid on diatoms. Seawater was filtered and dispersed with NA in the ratio of 9:1. Samples were loaded for transcriptome sequencing after 96 hours. There were 818 upregulated and 436 downregulated genes in LCG as compared to CG. The most upregulated gene was PHATRDRAFT_50740 and the most downregulated gene was PHATRDRAFT_4829 [7.118].

7.6 Conclusions

Transcriptomics studies are important for understanding the function of the expressed genome being transcribed. Transcriptomic research is a popular branch of molecular biology that reveals various metabolic processes that occur *in vivo*. This will help us to understand how different non-biochemicals and biochemicals spread their effects on transcription and ultimately affect metabolic processes. In this chapter, metabolics and transcriptomics stress factors on *P. tricornutum* have been studied. Various genes on in *P. tricornutum*, some of which are *PSY* gene, UDP-glucose pyrophosphorylase gene, and bZIP14, have an adverse influence on metabolism. This area needs further research to scale up to the industrial level to meet worldwide needs. There is a vast treasure trove of so many different types of microalgae in the depths of our oceans, many of which are yet to be tested for the production of value-added products. Although transcription is highly efficient, it is not used by all researchers due to its cost. The transcriptome relies on RNA sequencing, which requires RNA sequencers, expensive and sophisticated tools. Transcriptomics data give new insights on the annotation of diatom genes, enzymatic pathways involved in the generation of bioactive molecules and potential uses of *P. tricornutum*. This is important for both economic and commercial reasons.

References

[7.1] Abida, H., Ruchaud, S., Rios, L., Humeau, A., Probert, I., De Vargas, C., Bach, S., Bowler, C. 2013. Bioprospecting marine plankton. *Marine Drugs*, 11(11), 4594-4611.

[7.2] Ahirwar, A., Kesharwani, K., Deka, R., Muthukumar, S., Khan, M.J., Rai, A., Vinayak, V., Varjani, S., Joshi, K.B. and Morjaria, S., 2022. Microalgal drugs: A promising therapeutic reserve for the future. *Journal of Biotechnology*, 349, pp.32-46

[7.3] Ahirwar, A., Meignen, G., Khan, M.J., Khan, N., Rai, A., Schoefs, B., Marchand, J., Varjani, S. and Vinayak, V., 2021. Nanotechnological approaches to disrupt the rigid cell walled microalgae grown in wastewater for value-added biocompounds: commercial applications, challenges, and breakthrough. *Biomass Conversion and Biorefinery*, pp.1-26, DOI:10.1007/s13399-021-01965-1

[7.4] Ahirwar, A., Rai, A., Sirotiya, V., Khandelwal, P., Singh, G., Jhadav, D., Harish, Marchand, J., Schoefs, B., Varjani, S.J.E.T.R. 2024. Fermentation of algal biomass for its nutritional value: perspectives and revolutions in the food industry. 13(1), 1-28.

[7.5] Alam, M.A., Wang, Z. 2019. *Microalgae Biotechnology for Development of Biofuel and Wastewater Treatment*. Springer.

[7.6] Alam, S. 2022. Algae: An emerging feedstock for biofuels production. in: *Algal Biotechnology*, Elsevier, pp. 165-185.

[7.7] Ali, M.A., Al-Hydary, I.A., Al-Hattab, T.A. 2017a. Nano-magnetic catalyst CaO-Fe3O4 for biodiesel production from date palm seed oil. *Bulletin of Chemical Reaction Engineering & Catalysis*, 12(3), 460-468.

[7.8] Ali, S., Hassan, F., Abdul-Jabar, R. 2017b. Evaluation of water quality by trophic diatom index (TDI) in Tigris River within Wasit province. *Indian Journal of Ecology*, 44(4), 711-716.

[7.9] Angerer, P., Simon, L., Tritschler, S., Wolf, F.A., Fischer, D., Theis, F.J. 2017. Single cells make big data: New challenges and opportunities in transcriptomics. *Current Opinion in Systems Biology*, 4, 85-91.

[7.10] Armbrust, E.V., Berges, J.A., Bowler, C., Green, B.R., Martinez, D., Putnam, N.H., Zhou, S., Allen, A.E., Apt, K.E., Bechner, M., Brzezinski, M.A., Chaal, B.K., Chiovitti, A., Davis, A.K., Demarest, M.S., Detter, J.C., Glavina, T., Goodstein, D., Hadi, M.Z., Hellsten, U., Hildebrand, M., Jenkins, B.D., Jurka, J., Kapitonov, V.V., Kröger, N., Lau, W.W.Y., Lane, T.W., Larimer, F.W., Lippmeier, J.C., Lucas, S., Medina, M., Montsant, A., Obornik, M., Parker, M.S., Palenik, B., Pazour, G.J., Richardson, P.M., Rynearson, T.A., Saito, M.A., Schwartz, D.C., Thamatrakoln, K., Valentin, K., Vardi, A., Wilkerson, F.P., Rokhsar, D.S. 2004. The Genome of the Diatom *Thalassiosira Pseudonana*: Ecology, Evolution, and Metabolism. *Science*, 306(5693), 79-86.

[7.11] Arora, S., Mishra, G. 2021. Effect of gibberellin, methyl jasmonate and myoinositol on biomass and eicosapentaenoic acid productivities in the eustigmatophyte *Monodopsis subterranea* CCALA 830. *Journal of Applied Phycology*, 33(1), 287-299.

[7.12] Bauer, C., Schmitz, C., Corrêa, R., Herrera, C., Ramlov, F., Oliveira, E., Pizzato, A., Varela, L., Cabral, D., Yunes, R. 2019. *In vitro* fucoxanthin production by the *Phaeodactylum tricornutum* diatom. *Studies in Natural Products Chemistry*, 63, 211-242.

[7.13] Bernaerts, T.M., Gheysen, L., Kyomugasho, C., Kermani, Z.J., Vandionant, S., Foubert, I., Hendrickx, M.E., Van Loey, A.M. 2018. Comparison of microalgal biomasses as functional food ingredients: Focus on the composition of cell wall related polysaccharides. *Algal Research*, 32, 150-161.

[7.14] Bharti, R.K., Srivastava, S., Thakur, I. 2021. Sequestration of carbon dioxide by microorganism and production of value added product. in: *Environmental Microbiology and Biotechnology*, Springer, pp. 235-249.

[7.15] Blommaert, L., Vancaester, E., Huysman, M.J., Osuna-Cruz, C.M., D'hondt, S., Lavaud, J., Lepetit, B., Winge, P., Bones, A.M., Vandepoele, K. and Vyverman, W., 2020. Light regulation of LHCX genes in the benthic diatom Seminavis robusta. *Frontiers in Marine Science*, 7, p.192, https://doi.org/10.3389/fmars.2020.00192

[7.16] Bowler, C., Allen, A.E., Badger, J.H., Grimwood, J., Jabbari, K., Kuo, A., Maheswari, U., Martens, C., Maumus, F., Otillar, R.P. 2008. The

Phaeodactylum genome reveals the evolutionary history of diatom genomes. *Nature*, 456(7219), 239-244.

[7.17] Chen, J., Yang, J., Du, H., Aslam, M., Wang, W., Chen, W., Li, T., Liu, Z., Liu, X. 2021. Laminarin, a major polysaccharide in stramenopiles. *Marine Drugs*, 19(10), 576.

[7.18] Couto, C., Hernández, C.P., Alves Sobrinho, R.C.M., Mendes, C.R.B., Roselet, F., Abreu, P.C.J.J.o.A.P. 2021. Optimization of a low-cost fertilizer-based medium for large-scale cultivation of the coastal diatom Conticribra weissflogii using response surface methodology and its effects on biomass composition. 33(5), 2767-2781.

[7.19] Cui, Y., Thomas-Hall, S.R., Schenk, P.M. 2019. Phaeodactylum tricornutum microalgae as a rich source of omega-3 oil: Progress in lipid induction techniques towards industry adoption. *Food Chemistry*, 297, 124937.

[7.20] Desbois, A.P., Lebl, T., Yan, L., Smith, V.J. 2008. Isolation and structural characterisation of two antibacterial free fatty acids from the marine diatom, *Phaeodactylum tricornutum*. *Applied Microbiology and Biotechnology*, 81(4), 755-764.

[7.21] Desbois, A.P., Mearns-Spragg, A., Smith, V.J. 2009. A fatty acid from the diatom *Phaeodactylum tricornutum* is antibacterial against diverse bacteria including multi-resistant *Staphylococcus aureus* (MRSA). *Marine Biotechnology*, 11(1), 45-52.

[7.22] Eilers, U., Bikoulis, A., Breitenbach, J., Büchel, C., Sandmann, G. 2016a. Limitations in the biosynthesis of fucoxanthin as targets for genetic engineering in *Phaeodactylum tricornutum*. *Journal of Applied Phycology*, 28(1), 123-129.

[7.23] Eilers, U., Bikoulis, A., Breitenbach, J., Büchel, C., Sandmann, G.J.J.o.a.p. 2016b. Limitations in the biosynthesis of fucoxanthin as targets for genetic engineering in *Phaeodactylum tricornutum*. 28(1), 123-129.

[7.24] Erdoğan, A., Karataş, A.B., Demirel, Z., Dalay, M.C.J.J.o.A.P. 2022. Purification of fucoxanthin from the diatom *Amphora capitellata* by preparative chromatography after its enhanced productivity via oxidative stress. 34(1), 301-309.

[7.25] Yang, R. and Wei, D., 2020. Improving fucoxanthin production in mixotrophic culture of marine diatom *Phaeodactylum tricornutum* by LED light shift and nitrogen supplementation. *Frontiers in Bioengineering and Biotechnology*, 8, p.820.

[7.26] Faraloni, C. and Torzillo, G., 2017. Synthesis of antioxidant carotenoids in microalgae in response to physiological stress (pp. 143-157). United Kingdom: IntechOpen, DOI: 10.5772/67843

[7.27] Frank, R.A., Kavanagh, R., Burnison, B.K., Arsenault, G., Headley, J.V., Peru, K.M., Van Der Kraak, G., Solomon, K.R. 2008. Toxicity assessment of collected fractions from an extracted naphthenic acid mixture. *Chemosphere*, 72(9), 1309-1314.

[7.28] Gao, F., Teles, I., Wijffels, R.H., Barbosa, M.J.J.B.t. 2020. Process optimization of fucoxanthin production with *Tisochrysis lutea*. 315, 123894.
[7.29] Gao, F., Woolschot, S., Cabanelas, I.T.D., Wijffels, R.H., Barbosa, M.J.J.B.T. 2021. Light spectra as triggers for sorting improved strains of *Tisochrysis lutea*. *Bioresource Technology*, 321, 124434.
[7.30] Gérin, S., Delhez, T., Corato, A., Remacle, C., Franck, F.J.J.o.A.P. 2020. A novel culture medium for freshwater diatoms promotes efficient photoautotrophic batch production of biomass, fucoxanthin, and eicosapentaenoic acid. *Journal of Applied Phycology*, 32(3), 1581-1596.
[7.31] Gilbert-López, B., Barranco, A., Herrero, M., Cifuentes, A., Ibáñez, E.J.F.R.I. 2017. Development of new green processes for the recovery of bioactives from *Phaeodactylum tricornutum*. *Food Research International*, 99, 1056-1065.
[7.32] Gloser, V., Zwieniecki, M.A., Orians, C.M., Holbrook, N.M. 2007. Dynamic changes in root hydraulic properties in response to nitrate availability. *Journal of Experimental Botany*, 58(10), 2409-2415.
[7.33] Gruber, A., Kroth, P.G. 2017. Intracellular metabolic pathway distribution in diatoms and tools for genome-enabled experimental diatom research. *Philosophical Transactions of the Royal Society B: Biological Sciences*, 372(1728), 20160402.
[7.34] Gruber, A., Weber, T., Bártulos, C.R., Vugrinec, S., Kroth, P.G. 2009. Intracellular distribution of the reductive and oxidative pentose phosphate pathways in two diatoms. *Journal of Basic Microbiology*, 49(1), 58-72.
[7.35] Gu, W., Kavanagh, J.M., McClure, D.D. 2022. Towards a sustainable supply of omega-3 fatty acids: Screening microalgae for scalable production of eicosapentaenoic acid (EPA). *Algal Research*, 61, 102564.
[7.36] Guérin, S., Raguénès, L., Croteau, D., Babin, M., Lavaud, J.J.M.d. 2022. Potential for the Production of Carotenoids of Interest in the Polar Diatom *Fragilariopsis cylindrus*. 20(8), 491.
[7.37] Guo, B., Liu, B., Yang, B., Sun, P., Lu, X., Liu, J., Chen, F.J.M.D. 2016. Screening of diatom strains and characterization of *Cyclotella cryptica* as a potential fucoxanthin producer. *Mar. Drugs*, 14(7), 125.
[7.38] Guo, R., Lu, D., Liu, C., Hu, J., Wang, P., Dai, X. 2022. Toxic effect of nickel on microalgae *Phaeodactylum tricornutum* (Bacillariophyceae). *Ecotoxicology*, 1-15.
[7.39] Gupta, J., Gupta, R. 2020. Nutraceutical status and scientific strategies for enhancing production of omega-3 fatty acids from microalgae and their role in healthcare. *Current Pharmaceutical Biotechnology*, 21(15), 1616-1631.
[7.40] Gupta, A.K., Seth, K., Maheshwari, K., Baroliya, P.K., Meena, M., Kumar, A. and Vinayak, V., 2021. Biosynthesis and extraction of high-value carotenoid from algae. *Frontiers in Bioscience-Landmark*, 26(6), pp.171-190.
[7.41] Hedfi, A., Mahmoudi, E., Boufahja, F., Beyrem, H., Aïssa, P. 2007. Effects of increasing levels of nickel contamination on structure of

offshore nematode communities in experimental microcosms. *Bulletin of Environmental Contamination and Toxicology*, 79(3), 345-349.

[7.42] Hempel, F., Bozarth, A.S., Lindenkamp, N., Klingl, A., Zauner, S., Linne, U., Steinbüchel, A., Maier, U.G. 2011. Microalgae as bioreactors for bioplastic production. *Microbial Cell Factories*, 10(1), 1-6.

[7.43] Huang, J.J., Lin, S., Xu, W., Cheung, P.C.K. 2017. Occurrence and biosynthesis of carotenoids in phytoplankton. *Biotechnology Advances*, 35(5), 597-618.

[7.44] Joseph, M.M., Renjith, K., John, G., Nair, S.M., Chandramohanakumar, N. 2017. Biodiesel prospective of five diatom strains using growth parameters and fatty acid profiles. *Biofuels*, 8(1), 81-8.

[7.45] Kadono, T., Kira, N., Suzuki, K., Iwata, O., Ohama, T., Okada, S., Nishimura, T., Akakabe, M., Tsuda, M. and Adachi, M., 2015. Effect of an introduced phytoene synthase gene expression on carotenoid biosynthesis in the marine diatom *Phaeodactylum tricornutum*. *Marine Drugs*, 13(8), pp.5334-5357, https://doi.org/10.3390/md13085334

[7.46] Kandasamy, S., Zhang, B., He, Z., Bhuvanendran, N., EL-Seesy, A.I., Wang, Q., Narayanan, M., Thangavel, P., Dar, M.A. 2022. Microalgae as a multipotential role in commercial applications: Current scenario and future perspectives. *Fuel*, 308, 122053.

[7.47] Kashyap, M., Chakraborty, S., Kumari, A., Rai, A., Varjani, S. and Vinayak, V., 2023. Strategies and challenges to enhance commercial viability of algal biorefineries for biofuel production. *Bioresource Technology*, p.129551, https://doi.org/10.1016/j.biortech.2023.129551

[7.48] Khan, M.J., Ahirwar, A., Schoefs, B., Pugazhendhi, A., Varjani, S., Rajendran, K., Bhatia, S.K., Saratale, G.D., Saratale, R.G. and Vinayak, V., 2021. Insights into diatom microalgal farming for treatment of wastewater and pretreatment of algal cells by ultrasonication for value creation. *Environmental Research*, 201, p.111550, https://doi.org/10.1016/j.envres.2021.111550

[7.49] Kim, S.-K., Pangestuti, R. 2011. Biological activities and potential health benefits of fucoxanthin derived from marine brown algae. *Advances in Food and Nutrition Research*, 64, 111-128.

[7.50] Kim, S.M., Jung, Y.-J., Kwon, O.-N., Cha, K.H., Um, B.-H., Chung, D., Pan, C.-H.J.A.b., biotechnology. 2012a. A potential commercial source of fucoxanthin extracted from the microalga *Phaeodactylum tricornutum*. 166(7), 1843-1855.

[7.51] Kim, S.M., Kang, S.-W., Kwon, O.-N., Chung, D., Pan, C.-H.J.J.o.t.K.S.f.A.B.C. 2012c. Fucoxanthin as a major carotenoid in *Isochrysis aff. galbana*: Characterization of extraction for commercial application. 55(4), 477-483.

[7.52] Konur, O. 2021. A Review of the Research: Biodiesel and Petrodiesel Fuels. *Biodiesel Fuels*, 63-87.

[7.53] Kroth, P., Chiovitti, A., Gruber, A., Martin-Jezequel, V., Mock, T. 2008a. A Model for Carbohydrate Metabolism in the Diatom *Phaeodactylum*.

[7.54] Kroth, P.G., Chiovitti, A., Gruber, A., Martin-Jezequel, V., Mock, T., Parker, M.S., Stanley, M.S., Kaplan, A., Caron, L., Weber, T. 2008b. A model for carbohydrate metabolism in the diatom *Phaeodactylum tricornutum* deduced from comparative whole genome analysis. *PloS One*, 3(1), e1426.

[7.55] Kroth, P.G., Matsuda, Y. 2022. Carbohydrate Metabolism. in: *The Molecular Life of Diatoms*, Springer, pp. 465-492.

[7.56] Kukurba, K.R., Montgomery, S.B. 2015. RNA sequencing and analysis. *Cold Spring Harbor Protocols*, 2015(11), pdb. top084970.

[7.57] Lepetit, B., Sturm, S., Rogato, A., Gruber, A., Sachse, M., Falciatore, A., Kroth, P.G., Lavaud, J. 2013. High light acclimation in the secondary plastids containing diatom Phaeodactylum tricornutum is triggered by the redox state of the plastoquinone pool. *Plant physiology*, 161(2), 853-865.

[7.58] Li, N., Gao, X., Zheng, L., Huang, Q., Zeng, F., Chen, H., Farag, M.A., Zhao, C. 2022a. Advances in fucoxanthin chemistry and management of neurodegenerative diseases. *Phytomedicine*, 154352.

[7.59] Li, Y., Sun, H., Wang, Y., Yang, S., Wang, J., Wu, T., Lu, X., Chu, Y., Chen, F.J.B.T. 2022b. Integrated metabolic tools reveal carbon alternative in *Isochrysis zhangjiangensis* for fucoxanthin improvement. *Bioresour. Technol.*, 347, 126401.

[7.60] Liang, F., Lindberg, P., Lindblad, P. 2018. Engineering photoautotrophic carbon fixation for enhanced growth and productivity. *Sustainable Energy & Fuels*, 2(12), 2583-2600.

[7.61] Liaqat, F., Khazi, M.I., Bahadar, A., He, L., Aslam, A., Liaquat, R., Agathos, S.N., Li, J. Mixotrophic cultivation of microalgae for carotenoid production. *Reviews in Aquaculture.*

[7.62] Lu, X., Liu, B., He, Y., Guo, B., Sun, H., Chen, F.J.B.t. 2019. Novel insights into mixotrophic cultivation of *Nitzschia laevis* for co-production of fucoxanthin and eicosapentaenoic acid. *Bioresour. Technol.*, 294, 122145.

[7.63] Manfellotto, F., Stella, G.R., Falciatore, A., Brunet, C., Ferrante, M.I. 2020. Engineering the unicellular alga *Phaeodactylum tricornutum* for enhancing carotenoid production. *Antioxidants*, 9(8), 757.

[7.64] Marella, T.K., Tiwari, A.J.B.t. 2020. Marine diatom *Thalassiosira weissflogii* based biorefinery for co-production of eicosapentaenoic acid and fucoxanthin. *Bioresour. Technol.*, 307, 123245.

[7.65] Martino, A.D., Meichenin, A., Shi, J., Pan, K., Bowler, C. 2007. Genetic and phenotypic characterization of *Phaeodactylum tricornutum* (Bacillariophyceae) accessions 1. *Journal of Phycology*, 43(5), 992-1009.

[7.66] Matsuda, Y., Hopkinson, B.M., Nakajima, K., Dupont, C.L., Tsuji, Y. 2017. Mechanisms of carbon dioxide acquisition and CO_2 sensing in marine diatoms: a gateway to carbon metabolism. *Philosophical Transactions of the Royal Society B: Biological Sciences*, 372(1728), 20160403.

[7.67] Matthijs, M., Fabris, M., Obata, T., Foubert, I., Franco-Zorrilla, J.M., Solano, R., Fernie, A.R., Vyverman, W. and Goossens, A., 2017. The transcription factor bZIP14 regulates the TCA cycle in the diatom *Phaeodactylum tricornutum. The EMBO Journal*, 36(11), pp.1559-1576,https://doi.org/10.15252/embj.201696392

[7.68] Mohamadnia, S., Tavakoli, O., Faramarzi, M.A., Shamsollahi, Z.J.A. 2020. Production of fucoxanthin by the microalga *Tisochrysis lutea*: A review of recent developments. 516, 734637.

[7.69] Nymark, M., Valle, K.C., Brembu, T., Hancke, K., Winge, P., Andresen, K., Johnsen, G., Bones, A.M. 2009. An integrated analysis of molecular acclimation to high light in the marine diatom *Phaeodactylum tricornutum. PloS One*, 4(11), e7743.

[7.70] Oudot-Le Secq, M.-P., Green, B.R. 2011. Complex repeat structures and novel features in the mitochondrial genomes of the diatoms *Phaeodactylum tricornutum* and *Thalassiosira pseudonana. Gene*, 476 (1-2), 20-26.

[7.71] Pajot, A., Hao Huynh, G., Picot, L., Marchal, L., Nicolau, E. 2022. Fucoxanthin from algae to human, an extraordinary bioresource: Insights and advances in up and downstream processes. *Marine Drugs*, 20(4), 222.

[7.72] Pan, Y., Hu, F., Yu, C., Li, C., Huang, T., Hu, H. 2020. Amino acid catabolism during nitrogen limitation in *Phaeodactylum tricornutum. Frontiers in Plant Science*, 11, 589026.

[7.73] Parkes, R., Archer, L., Gee, D.M., Smyth, T.J., Gillespie, E., Touzet, N.J.B.P. 2021. Differential responses in EPA and fucoxanthin production by the marine diatom *Stauroneis* sp. under varying cultivation conditions. *Biotechnol. Prog.*, 37(6), e3197.

[7.74] Patel, A.K., Choi, Y.Y., Sim, S.J. 2020. Emerging prospects of mixotrophic microalgae: Way forward to sustainable bioprocess for environmental remediation and cost-effective biofuels. *Bioresource Technology*, 300, 122741.

[7.75] Peng, J., Yuan, J.-P., Wu, C.-F., Wang, J.-H. 2011. Fucoxanthin, a marine carotenoid present in brown seaweeds and diatoms: metabolism and bioactivities relevant to human health. *Marine Drugs*, 9(10), 1806-1828.

[7.76] Peraman, M., Nachimuthu, S. 2019. Identification and quantification of fucoxanthin in selected carotenoid-producing marine microalgae and evaluation for their chemotherapeutic potential. *Pharmacognosy Magazine*, 15(64), 243.

[7.77] Pocha, C.K.R., Chia, W.Y., Chew, K.W., Munawaroh, H.S.H., Show, P.L. 2022. Current advances in recovery and biorefinery of fucoxanthin from *Phaeodactylum tricornutum*. *Algal Research*, 65, 102735.

[7.78] Popovich, C.A., Faraoni, M.B., Sequeira, A., Daglio, Y., Martín, L.A., Martínez, A.M., Damiani, M.C., Matulewicz, M.C., Leonardi, P.I. 2020. Potential of the marine diatom *Halamphora coffeaeformis* to simultaneously produce omega-3 fatty acids, chrysolaminarin and fucoxanthin in a raceway pond. *Algal Research*, 51, 102030.

[7.79] Pudney, A., Gandini, C., Economou, C.K., Smith, R., Goddard, P., Napier, J.A., Spicer, A., Sayanova, O. 2019. Multifunctionalizing the marine diatom *Phaeodactylum tricornutum* for sustainable co-production of omega-3 long chain polyunsaturated fatty acids and recombinant phytase. *Scientific Reports*, 9(1), 1-10.

[7.80] Rai, A., Sirotiya, V., Mourya, M., Khan, M.J., Ahirwar, A., Sharma, A.K., Kawatra, R., Marchand, J., Schoefs, B., Varjani, S. and Vinayak, V., 2022. Sustainable treatment of dye wastewater by recycling microalgal and diatom biogenic materials: Biorefinery perspectives. *Chemosphere*, 305, p.135371.

[7.81] Rai, A.K., Gurung, S.A. 2022. Microalgal Promise to the Next Generation: A Dual Potential Perspective as Cosmeceuticals and Biofuels. in: *Microalgae: Next-generation Feedstock for Biorefineries*, Springer, pp. 55-82.

[7.82] Ramaraj, S., Hemaiswarya, S., Raja, R., Ganesan, V., Anbazhagan, C., Carvalho, I.S. and Juntawong, N., 2015. Microalgae as an attractive source for biofuel production. Environmental Sustainability: *Role of Green Technologies*, pp.129-157, DOI 10.1007/978-81-322-2056-5_8

[7.83] Rebolloso-Fuentes, M., Navarro-Pérez, A., Ramos-Miras, J., Guil-Guerrero, J. 2001. Biomass nutrient profiles of the microalga *Phaeodactylum tricornutum*. *Journal of Food Biochemistry*, 25(1), 57-76.

[7.84] Rehmanji, M., Nesamma, A.A., Khan, N.J., Fatma, T., Jutur, P.P. 2022a. Media engineering in marine diatom *Phaeodactylum tricornutum* employing cost-effective substrates for sustainable production of high-value renewables. *Biotechnology Journal*, 2100684.

[7.85] Rehmanji, M., Nesamma, A.A., Khan, N.J., Fatma, T., Jutur, P.P.J.B.J. 2022b. Media engineering in marine diatom *Phaeodactylum tricornutum* employing cost-effective substrates for sustainable production of high-value renewables. *Biotechnol. J.*, 2100684.

[7.86] Rudenko, N.N., Borisova-Mubarakshina, M.M., Ignatova, L.K., Fedorchuk, T.P., Nadeeva-Zhurikova, E.M., Ivanov, B.N. 2021. Role of plant carbonic anhydrases under stress conditions. *Plant Stress Physiol.*, 4(1), 7-38.

[7.87] Schulze, P.S., Hulatt, C.J., Morales-Sánchez, D., Wijffels, R.H., Kiron, V. 2019. Fatty acids and proteins from marine cold adapted microalgae for biotechnology. *Algal Research*, 42, 101604.

[7.88] Secq, O.-L., Grimwood, J., Shapiro, H., Armbrust, E., Bowler, C., Green, B.R. 2007. Chloroplast genomes of the diatoms *Phaeodactylum tricornutum* and *Thalassiosira pseudonana*: comparison with other plastid genomes of the red lineage. *Molecular Genetics and Genomics*, 277(4), 427-439.

[7.89] Seth, K., Kumar, A., Rastogi, R.P., Meena, M., Vinayak, V. 2021. Bioprospecting of fucoxanthin from diatoms—Challenges and perspectives. *Algal Research*, 60, 102475.

[7.90] Sharma, G., Martin, S.S. and Blumenthal, R.S., 2020. Effects of omega-3 fatty acids on major adverse cardiovascular events: what matters most: the drug, the dose, or the placebo?. *JAMA*, 324(22), pp.2262-2264, doi:10.1001/jama.2020.22387

[7.91] Suganya, T., Varman, M., Masjuki, H., Renganathan, S. 2016. Macroalgae and microalgae as a potential source for commercial applications along with biofuels production: a biorefinery approach. *Renewable and Sustainable Energy Reviews*, 55, 909-941.

[7.92] Sun, Z., Wang, X., Liu, J.J.A.R. 2019. Screening of Isochrysis strains for simultaneous production of docosahexaenoic acid and fucoxanthin. 41, 101545.

[7.93] Szabó, Z., Marosvölgyi, T., Szabó, É., Bai, P., Figler, M., Verzár, Z. 2020. The potential beneficial effect of EPA and DHA supplementation managing cytokine storm in coronavirus disease. *Frontiers in Physiology*, 11, 752.

[7.94] Tachihana, S., Nagao, N., Katayama, T., Hirahara, M., Yusoff, F.M., Banerjee, S., Shariff, M., Kurosawa, N., Toda, T., Furuya, K.J.F.i.b., biotechnology. 2020. High productivity of eicosapentaenoic acid and fucoxanthin by a marine diatom *Chaetoceros gracilis* in a semi-continuous culture. 8, 602721.

[7.95] Taddei, L., Chukhutsina, V.U., Lepetit, B., Stella, G.R., Bassi, R., van Amerongen, H., Bouly, J.-P., Jaubert, M., Finazzi, G., Falciatore, A. 2018. Dynamic changes between two LHCX-related energy quenching sites control diatom photoacclimation. *Plant Physiology*, 177(3), 953-965.

[7.96] Telussa, I., Nurachman, Z.J.A.R. 2019. Dynamics of β-carotene and fucoxanthin of tropical marine *Navicula* sp. as a response to light stress conditions. *Algal Research*, 41, 101530.

[7.97] Tong, S., Xu, D., Wang, Y., Zhang, X., Li, Y., Wu, H., Ye, N. 2021. Influence of ocean acidification on thermal reaction norms of carbon metabolism in the marine diatom *Phaeodactylum tricornutum*. *Marine Environmental Research*, 164, 105233.

[7.98] Vaezi, R., Napier, J.A., Sayanova, O. 2013. Identification and functional characterization of genes encoding omega-3 polyunsaturated fatty acid biosynthetic activities from unicellular microalgae. *Marine Drugs*, 11(12), 5116-5129.

[7.99] Valenzuela, R.W., Sanhueza, J., Valenzuela, A. 2012. Docosahexaenoic acid (DHA), an important fatty acid in aging and the protection of neurodegenerative diseases. *Journal of Nutritional Therapeutics*, 1(1), 63-72.

[7.100] Veaudor, T., Blanc-Garin, V., Chenebault, C., Diaz-Santos, E., Sassi, J.-F., Cassier-Chauvat, C., Chauvat, F. 2020. Recent advances in the photoautotrophic metabolism of Cyanobacteria: biotechnological implications. *Life*, 10(5), 71.

[7.101] Villanova, V., Singh, D., Pagliardini, J., Fell, D., Le Monnier, A., Finazzi, G., Poolman, M. 2021. Boosting biomass quantity and quality by improved mixotrophic culture of the diatom Phaeodactylum tricornutum. *Frontiers in Plant Science*, 12, 642199.

[7.102] Vinayak, V. 2022. Algae as sustainable food in space missions. in: *Biomass, Biofuels, Biochemicals*, Elsevier, pp. 517-540.

[7.103] Vinayak V, Gordon R, Joshi K, Schoefs B. Diafuel. 2018. Trademark application no 3778882. *Trade Marks Journal*. 1846.

[7.104] Vinayak, V., Manoylov, K.M., Gateau, H., Blanckaert, V., Hérault, J., Pencréac'h, G., Marchand, J., Gordon, R., Schoefs, B.J.M.d. 2015. Diatom milking: a review and new approaches. *Mar. Drugs*, 13(5), 2629-2665, https://doi.org/10.3390/md13052629

[7.105] Wang, J.K. and Seibert, M., 2017. Prospects for commercial production of diatoms. *Biotechnology for Biofuels*, 10, pp.1-13, DOI 10.1186/s13068-017-0699-y

[7.106] Wang, S., Verma, S.K., Hakeem Said, I., Thomsen, L., Ullrich, M.S., Kuhnert, N.J.M.c.f. 2018b. Changes in the fucoxanthin production and protein profiles in *Cylindrotheca closterium* in response to blue light-emitting diode light. *Microb Cell Fact.*, 17(1), 1-13.

7.107] Weill, P., Plissonneau, C., Legrand, P., Rioux, V., Thibault, R. 2020. May omega-3 fatty acid dietary supplementation help reduce severe complications in Covid-19 patients? *Biochimie*, 179, 275-280.

[7.108] Wollmann, F., Dietze, S., Ackermann, J.U., Bley, T., Walther, T., Steingroewer, J., Krujatz, F. 2019. Microalgae wastewater treatment: Biological and technological approaches. *Engineering in Life Sciences*, 19(12), 860-871.

[7.109] Wu, C., De Visscher, A., Gates, I.D. 2019. On naphthenic acids removal from crude oil and oil sands process-affected water. *Fuel*, 253, 1229-1246.

[7.110] Wu, S., Gu, W., Jia, S., Wang, L., Wang, L., Liu, X., Zhou, L., Huang, A., Wang, G. 2021a. Proteomic and biochemical responses to different concentrations of CO.

[7.111] Wu, S., Gu, W., Jia, S., Wang, L., Wang, L., Liu, X., Zhou, L., Huang, A., Wang, G. 2021b. Proteomic and biochemical responses to different concentrations of CO2 suggest the existence of multiple carbon metabolism strategies in *Phaeodactylum tricornutum. Biotechnology for Biofuels*, 14(1), 1-17.

[7.112] Wu, S., Gu, W., Jia, S., Wang, L., Wang, L., Liu, X., Zhou, L., Huang, A., Wang, G. 2021c. Proteomic and biochemical responses to different concentrations of CO2 suggest the existence of multiple carbon metabolism strategies in *Phaeodactylum tricornutum. Biotechnology for Biofuels*, 14(1), 235.

[7.113] Xia, S., Wang, K., Wan, L., Li, A., Hu, Q., Zhang, C.J.M.d. 2013b. Production, characterization, and antioxidant activity of fucoxanthin from the marine diatom *Odontella aurita*. 11(7), 2667-2681.

[7.114] Yang, R., Wei, D.J.F.i.B., Biotechnology. 2020. Improving fucoxanthin production in mixotrophic culture of marine diatom *Phaeodactylum tricornutum* by LED light shift and nitrogen supplementation. *Front. Bioeng. Biotechnol.*, 8, 820.

[7.115] Yi, Z., Xu, M., Magnusdottir, M., Zhang, Y., Brynjolfsson, S., Fu, W. 2015. Photo-oxidative stress-driven mutagenesis and adaptive evolution on the marine diatom *Phaeodactylum tricornutum* for enhanced carotenoid accumulation. *Marine Drugs*, 13(10), 6138-6151.

[7.116] Yu, G., Nakajima, K., Gruber, A., Bartulos, C.R., Schober, A.F., Lepetit, B., Yohannes, E., Matsuda, Y., Kroth, P.G. 2022. Mitochondrial phosphoenolpyruvate carboxylase contributes to carbon fixation in the diatom *Phaeodactylum tricornutum* at low inorganic carbon concentrations.

[7.117] Yunus, K., Zuraidah, M., John, A. 2020. A review on the accumulation of heavy metals in coastal sediment of Peninsular Malaysia. *Ecofeminism and Climate Change*, 1(1), 21-35.

[7.118] Zhang, H., Hu, Y., Yang, L., Lin, Z., Zhao, X., Chen, J., Tang, X., Wang, Y. 2021. Transcriptome aberration in marine microalgae *Phaeodactylum tricornutum* induced by commercial naphthenic acids. *Environmental Pollution*, 268, 115735.

[7.119] Zhang, K., Zhou, Z., Li, J., Wang, J., Yu, L. and Lin, S., 2021. SPX-related genes regulate phosphorus homeostasis in the marine phytoplankton, *Phaeodactylum tricornutum. Communications Biology*, 4(1), p.797, https://doi.org/10.1038/s42003-021-02284-x

[7.120] Zhang, W., Wang, F., Gao, B., Huang, L., Zhang, C. 2018. An integrated biorefinery process: Stepwise extraction of fucoxanthin, eicosapentaenoic acid and chrysolaminarin from the same *Phaeodactylum tricornutum* biomass. *Algal research*, 32, 193-200.

[7.121] Zheng, Y., Quinn, A.H., Sriram, G. 2013. Experimental evidence and isotopomer analysis of mixotrophic glucose metabolism in the marine diatom *Phaeodactylum tricornutum. Microbial cell factories*, 12(1), 1-17.

[7.122] Zhu, B.H., Shi, H.P., Yang, G.P., Lv, N.N., Yang, M. and Pan, K.H., 2016. Silencing UDP-glucose pyrophosphorylase gene in *Phaeodactylum tricornutum* affects carbon allocation. *New Biotechnology*, 33(1), pp.237-244, https://doi.org/10.1016/j.nbt.2015.06.003

8

Terraforming Mars with Microalgae, Especially Diatoms

Ira Rai[1‡], Jackson Achankunju[2‡], Richard Gordon[3†] and Vandana Vinayak[4*]

[1]Department of Aerospace Engineering, Indian Institute of Technology, Madras, India
[2]College of Applied and Health Sciences, A'Sharqiyah University, Ibra, Oman
[3]University of Manitoba, Department of Radiology, Alonsa, Manitoba, Canada
[4]Diatom Nanoengineering and Metabolism Laboratory (DNM), School of Applied Sciences, Dr. Harisingh Gour Central University, Sagar, Madhya Pradesh, India

Abstract

During the last century, the space age was initiated, fueled by numerous efforts to extend our knowledge beyond the solar system and our existence beyond Earth. Since the seventies, many telescopes covering a wide range of electromagnetic radiation, space and planet probes, satellites, and crewed orbital missions have been launched, while humans seek to reach and colonize a planet like Mars. Therefore, a crew's survival on a spaceship during long-term missions has become a critical issue requiring extensive research and innovations. Diatoms, the most abundant silica-walled microalgae, are considered the most successful primary producers on Earth, with an extraordinary diversity responsible for 25% of net primary production fixing 10^{13} kg carbon every year. Further, the psychrophilic nature of diatoms makes their adaptability to survive in an ice cold environment well supportive for their survival and a possible life in planet like Mars. Since Mars has a thin atmospheric layer consisting of a pressure of 6–7 mbar, 95%CO_2, 2.8%N_2, 2.1%Ar with only 0.13% O_2 compared to 21% O_2, 78%N_2, 0.04% CO_2 and ~1000 mBar of atmospheric pressure on Earth it has bright possibility of terraforming Mars into planet suitable for life. Therefore, in this chapter we discuss the prospects of simulating life on Mars, which has a bright microalgae of the greatest potential, i.e diatoms.

**Corresponding author*: kapilvinayak@gmail.com
†Corresponding author: DickGordonCan@protonmail.com
‡These authors contributed equally.

Vandana Vinayak and Richard Gordon (eds.) Diatom Cultivation for Biofuel, Food and High-Value Products, (203–230) © 2025 Scrivener Publishing LLC

Keywords: Diatoms, life on Mars, terraformation

8.1 Introduction

Since the beginning of time, when *Homo sapiens* started gazing at the sky, observing the Moon, twinkling stars, and other celestial objects, many questions have arisen which remain unanswered. During humankind's childhood, no one imagined how big the universe was, and most thought that our planet was the core of everything. This idea probably remained until the early sixteenth century, when Nicolaus Copernicus posthumously published his work, "*De revolutionibus orbium coelestium*" (*On the Revolutions of the Heavenly Spheres*) [8.69], which introduced the suggestion that the Earth is not the solar system core, but it is the Sun which centered the entire universe. This initiated the Copernican revolution that inspired Kepler, Galileo, and Newton to set off the basics of modern Astronomy [8.11] [8.20], giving humanity the desire to delve into space, starting with the Moon [8.33].

In the 20th century, humankind's knowledge reached a remarkable turning point, allowing Yuri Gagarin (the first human astronaut) to step up into space for almost two hours, declaring the beginning of the Space Age [8.9]. Later, many space missions were launched, including Sputnik 1 and Apollo, seeking, on the one hand, glory, and on the other hand, answering our ancestors' questions [8.29]. Throughout this era, our understanding of the universe has been simultaneously changing as well as our imagination for living outside Earth. Therefore, it is rational to expand our existence in the solar system and find new opportunities and alternatives. As an example, in the worst-case scenario, we leave Earth temporarily to save our species during catastrophic events on Earth [8.26]. This logic has eventually led to the idea of colonizing space, starting from the potential planets and moons in our solar system [8.46]. Such ideas have attained increasing momentum once we knew the size of the observable universe (estimated diameter of 93 billion light-years) [8.41] and since we discovered more exoplanets in the habitable zone [8.26].

The vastness of our universe also opens the door for further questions on the uniqueness of life on Earth. This has resulted in the emergence of astrobiology as a field of science trying to address the open questions regarding the existence of alien extra-terrestrial life in the expanding universe [8.32]. The probability of finding living organisms somewhere is increasing with the discovery of additional exoplanets in the habitable zone [8.26], various potential biosignatures, and organic molecules in interstellar space [8.15] through space telescopes such as Hubble and James Webb [8.24]. Recently, probable signs of past life on Mars have been reported and are under investigation [8.54]. Such arguments about the presence of alien life are essential, especially when considering colonizing Mars or other planets in the future. On the one hand, this would give us faith in the idea of colonizing other planets (as life might work outside Earth), and on the other hand, gives us concerns about the security of new extra-terrestrial colonies.

To colonize other planets, such as Mars, by transporting humanity there, it is essential to first survive the lengthy space travel so as to get there and start life on Mars. This includes challenges such as microgravity, lack of supplies, and the adaptation to life support systems. Suppose we can surmount the current limitations pertaining to spacecraft stability, the space environment (high cosmic radiation), life support systems, the time required to communicate with Earth, energy sources, and speed restrictions still remain a real challenge. In this circumstance, humans can likely exceed the current realistic limits.

To surmount some of the current limitations confronting crewed long-term missions, developing new concepts and materials for essential crew life applications is necessary. Bio-regenerative life support systems (BLSS) were an extensively proposed concept, where autotrophic organisms (such as plants or algae) have to be integrated for the production of oxygen and food for the artificial ecosystem [8.21]. Here, we focus on diatoms, an important class of microalgae flourishing in many aquatic habitats on Earth. Since diatoms are capable of surviving in harsh environments they are thermophiles, acidophiles, halophiles, shock-proof diatoms, xerophiles, obscurophiles, and cryophiles [8.70]. Researchers have speculated that the infrared spectrum of the Trapezium nebula and galactic center, GC-IRS 7, poses striking similarity with the infrared spectrum of diatoms [8.71]. This is further supported by the fact that diatoms are the most abundant organisms on Antarctic and Arctic ice. They thus may have potential for terraforming planets like Mars. Vinayak V. [3] discussed diatoms and other algae as a potential food and oxygen source for astronauts in space missions due to fact that they alone are responsible for fixing about 25% CO_2 and are thus considered as primary producers on Earth [8.38, 8.72]. Later, Gordon *et al.* [8.14] and Rai *et al.*, [8.38] have suggested diatoms as a primary producer for the BLSS for consuming CO_2, producing O_2, biomass, and valuable metabolites using biowaste generated in space missions, coupling space toilets with algae assisted photosynthetic microbial fuel cells. To turn this into a reality, it is necessary to resolve the survival of diatoms in space and how to incorporate and cultivate them into the BLSS. In this chapter, we intend to initiate discussions on the viability of diatom cultivation in space to encourage this endeavor. Further, the chapter narrates the challenges for cultivation and a list of possible species that could survive and be productive in that Martian environment. The evidences of presence of water in Mars 3 billion years ago during the Noachian and Hesperian eras, has been confirmed. Later Mars lost its ability to host water as it became lost to space [8.73].

8.2 Instrumentation to Artificially Simulate Life on Mars

8.2.1 SpaceQ

The past few years have seen an increase in developing interest to not only mark missions to Mars but also to design instruments and ways to launch rovers to explore the planet via remote sensing [8.74, 8.75]. This is essential since Mars undoubtedly has an extreme environment and temperature, Kurt J. Lesker Company designed a stainless steel cubic-shaped chamber known as SpaceQ (Figure 8.1) for simulating life on Mars and space environments [8.47]. SpaceQ can handle pressures up to 10^{-5} mbar and temperatures between 163 K and 423 K by chilling the plate with liquid nitrogen (LN_2) or heating the walls (using an external heating jacket), respectively. This chamber has an interior volume of 27,000 cm^3 and is a cubical facility composed of stainless steel with an aluminum door (27 L). The working table plate has the following dimensions: 20 cm x 20 cm x 1.8 cm. The chamber includes a number of optical and electrical feedthroughs, including two quartz window viewports, two thermocouple feedthroughs,

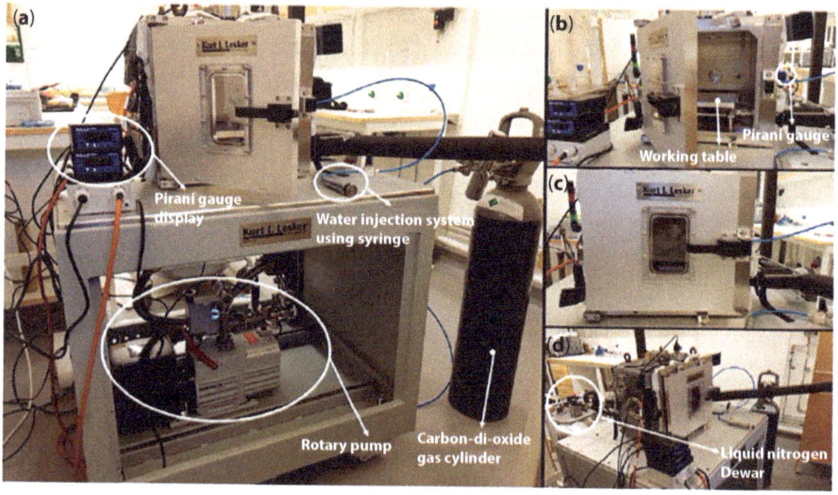

Figure 8.1 Overview of the SpaceQ chamber. (a) This figure shows the entire setup of the chamber, including the external Pirani gauge display, the rotary pump, the CO_2 gas cylinder (to insert a Mars-like atmosphere) and the syringe for water injection. (b) Detail of the experimental table where samples are cooled down to the desired temperature and the KJLC® Cold Cathode/Pirani combination gauge on the right side. (c) Viewport at the front of the chamber, used to take pictures during the experiment. (d) LN_2 Dewar used to cool down the working table. (Copyright [8.47]).

a vacuum-compatible UV lamp, a cold cathode/Pirani combination gauge to measure pressure, two gas inlets, ports for a vacuum pump, connections for a vacuum pump, USB and DB25 ports to read data in real-time, and connectors for the optical cable to spectrometer. Additionally, there is a connector that may be utilized to add more sensors if necessary. A Vaisala HMT334 sensor, which can be operated in vacuum and under Martian conditions, is installed at 10.2 cm above the plate and 5.2 cm from the side-walls to measure the atmospheric temperature and relative humidity inside the chamber, using the legacy from the REMS RH sensor [8.45]. This sensor can measure temperatures in the range of 203.15 K to 453.15 K, with an accuracy of ±2 K, and the relative humidity in the range of 0–100% RH, with an accuracy of ±1% RH. The sensor probe is exposed to the interior environment and is mounted in the chamber using an M22 x 1.5 thread. Vacuum-tight installations have been tested for the thread. The compartment is filled with water using a stainless-steel syringe. It is attached to a Swagelok ¼" connector on the right side of the chamber, which is linked by a tube to a manually driven Swagelok ¼" fitting ball valve, which has a 20 mL capacity. Since the water is connected to the chamber by a valve, it can be injected numerous times during the experiment. When the valve is opened, the water is injected into the chamber's depressurized atmosphere. A cold cathode Pirani gauge, which is intended for the vacuum measurement of gases in the range of $1x10^{9}$ mbar to 1000 mbar, regulates the pressure inside the chamber. In order to lower the pressures to 10^{5} mbar, two phases are needed. First, we use the Pfeiffer 1-phase DUO 10 M rotary pump to create a vacuum level of 10^{3} mbar, and then we couple it with a turbomolecular pump to obtain higher orders of vacuum. In order to manage the flow of the LN_2 supply, a solenoid valve is installed on the working table's liquid nitrogen (LN_2) feedthrough, which is linked to a 50 L Dewar. The thermal control unit receives data from the temperature sensor installed on the working table and uses that information as feedback to cool the table down to the desired temperature. The top and left quartz silica viewports allow us to watch the experiment or start an operation utilizing a certain wavelength on the material. With a viewing surface of 65 mm, the window is constructed of Kodial glass [8.48].

The studies on SpaceQ involved simulating environmental conditions of Mars transitioning from night to day, as seen in Figure 8.2. The temperature on ground (Tg) and air (Ta) were directly measured along with pressure (P) and relative humidity of ice (RHg^{i}) and liquid (RHg^{l}) from Equations 8.1–8.4 [8.31] described below and were derived by using the Formulae 8.1–8.4 [8.47]:

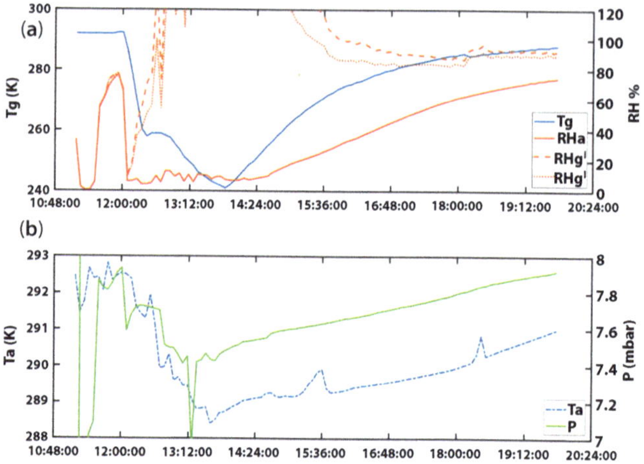

Figure 8.2 Time evolution of the simulated environmental parameters inside the SpaceQ chamber during the transition from a Martian night to a Martian day. Note: In this configuration, the relative humidity on the surface is saturated to allow for frost formation simulating the ground temperature under 260 K. (Copyright [8.47]).

$$RH(T) = \frac{P}{ew_{(T)}} \text{ x } \frac{vmr}{1+vmr} \text{ x100} \quad (8.1)$$

$$ew = 6.112 \text{ x } e\left(17.62 \text{ x } \frac{T-273.14159}{243.12+(T-273.14159)}\right) \quad (8.2)$$

$$ew_{ice(T)} = 6.112 \text{ x } e\left(22.5 \text{x} 1 - \frac{273.14159}{T}\right) \quad (8.3)$$

$$vmr = \frac{W}{1000} \text{ x } \frac{M_d}{M_w} \quad (8.4)$$

where RH(Ta) refers to relative humidity of air, to retrieve volume mixing ratio (vmr) to calculate RHg (Tg). Here, RHi represents the relative humidity with respect to ice, and RHl represents the relative humidity of liquid.

RH: relative humidity in %
P: Pressure in mbar
vmr: volume mixing ratio in parts per million
Tg: table temperature in K
ewliq(Tg): saturation partial pressure over liquid water at a given temperature
ewice(Tg): saturation partial pressure over ice at a given temperature
Mw = 18.0160 (molecular weight of water)
Md = 43.3400 (molecular weight of dry air on Mars)
W = water mass mixing ratio.

When RH_{gi} is above the saturation, atmosphere condenses, when Rh_a is zero the water in Martian atmosphere condenses, while as the surface temperature increases slowly water is lost to the atmosphere of the chamber which increases the chamber pressure P, and raises the RH_a. In such a case, pressure in the simulating chamber SpaceQ is maintained within 8 mbar with the help of a valve. Such simulations thus help in *in-situ* resource utilization (ISRU), which would be required for testing the corrosiveness of materials to be used for Martian architecture as well as to initiate life on Mars by growing our own food. Growing our own food in the form of plants has been researched to be best with the tiniest plants, i.e., microalgae which not only are responsible for consuming and fixing maximum CO_2 but also giving O_2 in return and acting as primary producers [8.76]. Among microalgae, as previously discussed, diatoms stand out best among existing microalgae as they produce 20-30% of the oxygen we breathe.

8.2.2 GraviSat Platform

The microgravity in space may cause varied effects on life in space. It was observed in various space experiments for sustainability of growing microorganisms. It has been observed that *Escherichia coli*, *Bacillus subtilis* and *Salmonella typhimurium* showed varied cell densities in space flights which was higher than the control; however *E. coli* in GeneStat (nanosatellite) and yeast *(Saccharomyces cerevisiae)* in PharmaSat card (nanosatellite) grew slowly in space [8.40]. The nanosatellite with microalgae cells was linked to a low overall rate of atmospheric CO_2 input to the liquid medium, which led to carbon restriction. To make CO_2 available to the algal cells, it must first diffuse through the 50-mm-thick polystyrene barrier and then travel a variety of lengths inside the liquid media. Microalgae strains were cultivated in medium supplemented with sodium bicarbonate as a source of inorganic carbon to postpone the onset of carbon limitation. Some cyanobacteria were inhibited by bicarbonate at the highest tested concentrations (1.0 mM) [8.77]. In contrast to the non-bicarbonate supplemented controls, all cyanobacteria in medium added with 0.1 mM sodium bicarbonate showed faster growth rates, greater photosynthetic efficiencies, and longer growth periods. Higher gas generation was made possible by adding more bicarbonate. The absence of comparable gas generation in control wells with media containing only 0.1 mM sodium bicarbonate suggests that the gas production was caused by photosynthesis rather than dissolved gas nucleating out of the medium. Although gas generation increased when bicarbonate was added to medium, bicarbonate concentrations had little to no impact on the growth or photosynthetic efficiency of green algal strains (*C. vulgaris* and *D. salina*). In terms of the effects of bicarbonate supplementation on growth (F7,112 = 59; p 0.001) and photosynthetic efficiency (F7,112 = 4.1; p 0.001), cyanobacteria (*Synechocystis*) and algae (*C. vulgaris*) differ from one another [8.7].

In yet another experiment, an 85 mL algal culture (the basic media) and two snails were placed in a 120 mL culture container to create the two-element closed aquatic environment (CAES). The guiding concepts for developing this straightforward CAES were that an algal culture (producer) supplies food and oxygen for the snails (consumer), and the algae culture may utilize the waste products and CO_2 created by the snails for photosynthesis and development. In addition, a control chamber was placed in an onboard centrifuge. The culture containers were put into the spacecraft with the CAES chambers enclosed and installed. The culture pots gave the algae light. The transmitted light was proportional to the OD 665 nm of the algae, and it was measured by a light sensor on the other side of the tank. This made it possible to calculate the evolution of the algal population over time. All information was recorded before being downlinked. In January 2001, the Chinese spacecraft Shenzhou-II carried out the space experiment's culture chambers (microgravity group and space-exposed 1g centrifuge group). After entering orbit, the chambers were operated remotely, and the spacecraft used a 12-hour cycle to supply illumination [8.52].

The NASA Johnson Space Center developed the improved suspension culture technique known as the rotating cell culture system (RCCS). The RCCS produced a continuous low-shear by maintaining the cells in a gently fluid orbit [8.52] [8.55].

8.3 Diatoms for Long-Term Space Missions

Diatoms are one of the most diverse and successful classes of unicellular microalgae with nanoporous architecture made up of silica abundantly found on Earth [8.17] [8.42] [8.57]. One of the most outstanding aspects of diatoms is their frustule, a hard shell made of amorphous hydrated silica that encloses the living cell [8.58]. The frustule is made up of two halves that fit together like a petri dish, with unique species-specific geometry and pore patterns [8.42] that give it exceptional mechanical strength [8.19], hydrokinetic characteristics [8.12], and assumed photobiological roles [8.10]. Despite their small size, diatoms have a significant impact on the planet [8.2]. Their vital roles in various aquatic ecosystems were unveiled and became evident through their significant primary production towards organic carbon and oxygen [8.23].

Given the outstanding adaptions of some diatom species to extreme environments and their widespread diversity within a relatively short evolutionary time, diatoms have been suggested as a primary producer for the BLSS in the long-term space missions for multipurpose utilization, as extensively explained in [8.13], [8.38], and [8.14]. Although many projects have suggested or tested green and cyanobacteria microalgal taxa for the BLSS, none of them have, to the best of our knowledge, ever considered diatoms [8.5] [8.38]. Therefore, testing diatom cultivation under space conditions should be considered in future work. Before testing diatoms for the BLSS, several key parameters need to be considered first to successfully grow them or any other microalgae in space, including:

i. **Light:** Like plants, algae need light for photosynthesis. Adequate light sources, such as LEDs or other artificial lighting, are required since Mars is farther from the sun, and therefore receives comparatively less sunlight than Earth, [8.44] as seen in Figure 8.2.
ii. **Nutrients:** Algae need essential macronutrients such as nitrogen, phosphorus, potassium, and trace elements for growth and production of value added products like pigments, lipids, etc. [8.35] [8.59]. The growth medium should be supplemented with these nutrients and controlled to prevent nutrient loss due to stress or microgravity.
iii. **Temperature:** Algae thrive within specific temperature ranges. Maintaining the appropriate temperature is crucial to ensure optimal growth and prevent stress on the algae.

iv. **Carbon Dioxide (CO_2):** Algae, like plants, require CO_2 for photosynthesis. The CO_2 levels should be monitored and maintained at suitable concentrations to support growth.
v. **pH Levels:** Algae tolerate specific pH ranges for optimal growth. The pH levels in the growth medium need to be carefully managed.
vi. **Stirring/Mixing:** Algae require proper agitation or mixing of the growth medium to ensure even distribution of nutrients and prevent settling of algae cells in the cultivation vessel.
vii. **Harvesting Methods:** Developing efficient and nondisruptive methods for harvesting algae biomass is essential for continuous growth and sustainable production.
viii. **Microgravity Adaptation:** Diatoms behave differently in microgravity than on Earth. Researchers should study their behavior and adapt growth systems to accommodate their unique responses to weightlessness.

8.4 Potential Diatoms for the BLSS: Taxa Tolerant to Extreme Conditions

In order to select the potential diatom species (out of the estimated 10^5 species) [8.16] for cultivation within the BLSS, specific criteria should be primarily considered. These criteria include their ability to (i) thrive with minimum cultivation requirements to minimize energy and materials required for the cultivation and maintenance of their cultures and, hence, the payload costs (i.e., they can grow efficiently under low light conditions, low temperature, and low nutrient concentrations) and to (ii) withstand extreme space conditions. Although the BLSS is expected to be well-maintained and protected in terms of temperature, pressure, cosmic radiation, etc., the ability of selected diatom species to overcome such extreme conditions is highly recommended for any accidental malfunctions in the BLSS systems. Thus, the cultivated species should be able to survive under environmental stress conditions of temperature and light. Not all species of diatoms are expected to survive in these extreme conditions. Therefore, identifying the potential species among the available taxa is crucial to prepare a short list for future work.

Among diatoms, certain species have developed exceptional adaptations that enable them to endure the challenging freeze-thawing cycles experienced in these frigid climates, ranging from -80°C to +12°C, for example, *Phaeodactylum tricornutum* [8.3]. Polar and subarctic ecosystems present a unique set of challenges for life to flourish in extreme temperatures, limited light availability during the winter months, and an ever-changing environment characterized by ice formation and melting which are some of the adversities that organisms must confront. Therefore, diatoms inhabiting such habitats have evolved remarkable strategies to survive and thrive in these harsh conditions. One of the key features that make diatoms so well-suited to polar and subarctic ecosystems is their silica-based cell walls, known as frustules. These frustules provide structural integrity to the diatoms besides protecting them against freezing temperatures. When the water temperature drops below freezing, ice crystals may form in the surrounding water. The silica frustules of diatoms act as a natural antifreeze, preventing the cell damage that would be caused by ice crystal formation and expansion [8.78]. This unique property allows them to endure the extreme cold and survive in a suspended state until conditions become favorable again. Furthermore, diatoms are known for their ability to adapt to fluctuations in light availability, which is critical in polar and subarctic regions where periods of complete darkness occur during the winter months. They possess light-harvesting pigments that can efficiently capture and utilize available light energy, enabling them to continue photosynthesizing even under dim light conditions. As the polar day lengthens during the summer, diatoms take full advantage of the increased light availability, rapidly growing and reproducing. In addition to their resilience in freezing temperatures and low light conditions, diatoms have developed robust life strategies to navigate the dynamic freeze-thawing cycles characteristic of polar and subarctic environments. During the warmer months, when temperatures rise above freezing, diatoms flourish and reproduce, building up their populations. When the colder months return and the temperatures drop, some diatoms form specialized spores or resting stages known as auxospores, which protect their genetic material and allow them to withstand adverse conditions. These auxospores remain dormant in the ice or sediment until favorable conditions return, at which point they germinate and restart the reproductive cycle. The ability of diatoms to endure extreme conditions and adapt to the harsh polar and subarctic environments makes them essential players in the ecological dynamics of these regions [8.60]. Their capacity to fix carbon through photosynthesis contributes significantly to primary production, supporting entire food webs. Moreover, as diatoms are sensitive indicators of environmental changes, studying their populations and distributions can provide valuable insights into the impacts of climate change on these delicate ecosystems. By persevering through the challenging environmental conditions, these diatoms contribute to the delicate balance and resilience of polar and subarctic ecosystems.

Further, diatoms adapt to their environment by changing their biovolume, silicification, pore morphology or arrangements. Change in silicification is often observed as a response to external stimuli, such as temperature, This may involve various morphological and physiological changes in their cell.

i. **Biovolume:** Diatoms can adjust their cell size or overall biovolume in response to changes in environmental conditions. For example, when resources are abundant, diatoms may increase their biovolume to take advantage of the available nutrients for growth and reproduction. Conversely, under limited resource conditions, they may decrease their biovolume to conserve energy and nutrients [8.50].
ii. **Silicification:** This refers to the process of depositing silica in their cell walls, forming intricate and sturdy structures called frustules. This silica deposition is essential for the diatom's survival and protection [8.1]. Diatoms can regulate the thickness and composition of their frustules, enabling them to withstand different environmental stresses. Changes in silicification may occur as a response to factors like temperature, nutrient availability, or pH levels [8.25].
iii. **Pore Morphology:** The frustules of diatoms have pores that allow for nutrient uptake and waste excretion. The morphology and arrangement of these pores can be altered in response to environmental cues. By modifying their pore structures, diatoms can regulate the exchange of substances with their surroundings, thus influencing their physiological processes [8.61].
iv. **Cell Arrangements and Assemblages:** Diatoms can change the pattern of their cell arrangements, forming colonies' filaments in various shapes. This adaptation allows them to optimize light exposure, nutrient availability, and protection from predators [8.62]. These changes in silicification could be observed as their response to external stimuli such as temperature. This means that diatom frustules can alter the composition and structure of their silica cell walls depending on changes in temperature.

v. **Temperature:** Temperature fluctuations can directly influence diatom metabolism and growth rates, and as a result, they may modify their frustules to better cope with the changing conditions.

Overall, diatoms exhibit a high level of adaptability through these various mechanisms, allowing them to thrive and persist in diverse aquatic environments. Their ability to adjust their biovolume, silicification, pore morphology, and arrangements is essential for their survival and success in fluctuating environmental conditions.

The diatom *Phaeodactylum tricornutum* Bohlin, a widely used model species cultivated in turbidostat cultures, has shown an increase in the photosynthetic rate per cell. A turbidostat is a type of continuous culture system used in microbiology, where the culture density is maintained at a constant level by regulating the inflow and outflow of culture medium. This characteristic makes it an interesting subject for studying the factors affecting photosynthesis in diatoms. Thus, *P. tricornatum* is widely used as a model organism for studying diatoms due to its well-understood biology, ease of cultivation, and its ability to represent the broader characteristics of diatoms. Unlike some other diatoms that may have a more consistent or uniform cell shape, *P. tricornutum* is a pleimorphic diatom which exhibits a greater range of cell morphologies [8.6] [8.63]. In *P. tricornutum*, the lower concentration of silica in its cell wall contributes to the pleiomorphism, allowing for more flexibility in cell shape and morphology [8.8]. Overall, the unique characteristics of *P. tricornutum*, such as being a widely used model species and its ability to show increased photosynthetic rates per cell in certain culture conditions, make it an important organism for scientific research in the study of diatoms and their ecological significance.

Growth rate and the generated biochemical and molecular data would allow *P. tricornutum* to be considered as a model diatom species for diatom cultivation in space. It is the most promising taxa of diatom which is exploited for producing many biochemicals such as vitamins, nutraceuticals, amino acids, proteins, etc. A total of 12233 coding genes have been deciphered from the genome of this diatom. Analysis of metagenomic data reveals the existence of a temperature correlated transporter gene J43171, which regulates the thermal adaptation and climate resilience of *P. tricornutum* [8.30].

On the offset, psychrophilic diatoms, like those in polar areas viz; Antarctica and Artic regions show a higher content of rubisco as compared

Table 8.1 Different diatoms that thrive under different extreme temperature conditions.

S. no.	Diatom sp.	Optimal temp. range (°C)	Extreme cold tolerance (°C)	Extreme heat tolerance (°C)	References
1	*Thalassiosira antarctica*	0–5	-2	20–25	[8.79]
2	*Pseudo-nitzschia multiseries*	10–15	5	30–35	[8.80]
3	*Chaetoceros muelleri*	15–20	10	35–40	[8.81]
4	*Fragilariopsis cylindrus*	-1–4	-5	18–22	[8.82]
5	*Nitzschia frigida*	0–6	-2	25–30	[8.83]
6	*Thalassiosira weissflogii*	20–25	15	35–40	[8.84]

to temperate diatoms as an adaptation for photosynthesis at low temperature. Thus, the psychrophilic species compensates for the reduced enzyme activity at low temperature by increasing the concentration of rubisco in them [8.56]. Among polar diatoms, *Fragilariopsis cylindrus* is well studied for its adaptation to extremely low temperatures and is known to produce an ice-binding protein (IBP) and extracellular polymeric substance, capable of altering the structure of ice [8.34]. On the other hand, Antarctic species *Thalassiosira gravida* has been reported to contain UV light-absorbing compounds, such as mycosporine-like amino acids (MAAS), which act as photoprotectants in their frustules. Table 8.1 shows different diatoms that survive and thrive under different extreme climate and temperature conditions.

Diatoms are thus incredibly diverse and adaptable organisms, so their temperature preferences can vary based on their geographical location and the conditions they experience in their natural habitats and thus can be a potential organism responsible for terraforming planet Mars.

8.5 Testing Diatom Growth Under Microgravity Conditions

8.5.1 Microgravity and Living Organisms

On Earth, different life forms have evolved and developed under specific gravitational conditions ($g \approx 9.8$ m/s^2), where gravity plays crucial roles in the living organisms. Therefore, the change in the gravitational conditions is expected to alter different biological activities. The satellites or International Space Station (ISS) which travels around the Earth's orbit, experiences freefall conditions, which is conventionally called microgravity. Away from the Earth, e.g., in interplanetary space, the spaceship travels in between celestial objects and does not experience a specific gravity from them if it does not enter their gravitational field. The experience on cultivation of plants and microalgae in space conditions showed microalgae cultivation has been briefly carried out. No doubt, microalgae face problems in culturing under microgravity.

Microgravity is the condition experienced by objects in space, where the force of gravity is significantly reduced, often resulting in a near-weightless environment. This condition is commonly experienced aboard the ISS or during spaceflight missions. Testing microalgae in a microgravity environment can be of scientific interest for several reasons:

i. **Space Exploration:** Understanding how microorganisms like microalgae respond to microgravity is essential for long-duration space missions, such as interplanetary travel or establishing colonies on other celestial bodies. Microalgae could potentially serve as a source of food, oxygen, and life support systems in these environments (see Figure 8.3).
ii. **Bioregenerative Life Support Systems:** Microalgae can be used in BLSS [8.66], where they can help regenerate essential resources like oxygen and remove carbon dioxide from the air. Evaluating their behavior in microgravity can help optimize such systems for space habitats.
iii. **Biofuel Production:** Microalgae have the potential to be a renewable source of biofuels, and studying their growth and lipid production in microgravity may offer insights into their potential as a sustainable energy source for space missions [8.13] [8.64].
iv. **Environmental Applications:** Microalgae play a vital role in carbon capture, absorbing pollutant gases, heavy metals and ions [8.65] and nutrient cycling in Earth's ecosystems.

Understanding how they behave in microgravity can provide insights into how they might contribute to environmental remediation in space or future space habitats.

Overall, studying microalgae in microgravity can offer valuable data and insights that have implications for both space exploration and various other practical applications. These experiments help scientists better understand the behavior of microorganisms under extreme conditions, paving the way for advancements in space research and various terrestrial fields.

Employing the EXPOSE-E launched on 22 October 2008, the European Space Agency allowed long exposures to space conditions and solar UV radiation on ISS. In this facility outside of the ISS, various rock samples with microbial populations on them were kept for 18 months outside the European Columbus laboratory module, one of the nine payloads of the European Technology Exposure Facility (EuTEF). While being subjected to space vacuum and severe temperatures, samples were maintained in darkness. A cyanobacterium (*Gloeocapsa* sp.) and two green algae (*Chlorella* and *Rosenvingiella radicans*) from the microbial community

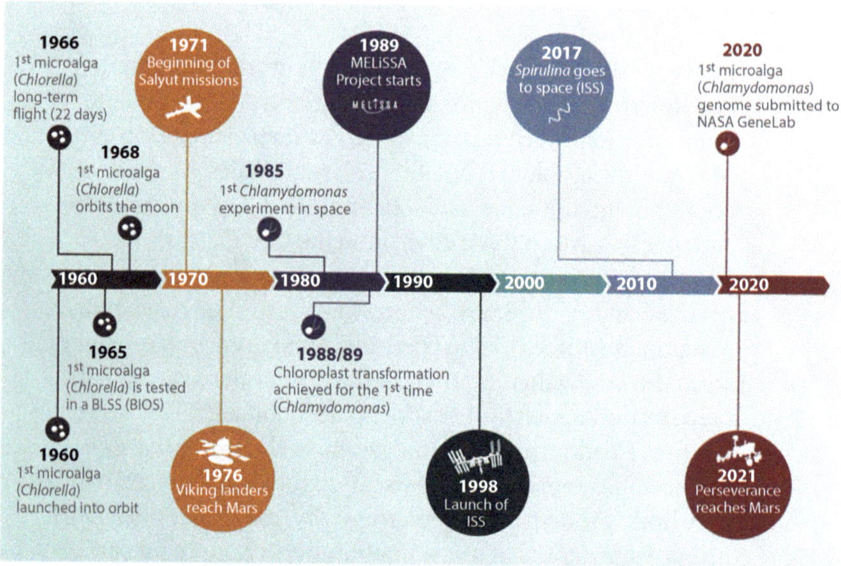

Figure 8.3 List of some microalgae experimented on space missions. Copyright (Mapstone, L.J., Leite, M.N., Purton, S., Crawford, I.A. and Dartnell, L., 2022. Cyanobacteria and microalgae in supporting human habitation on Mars. Biotechnology Advances, 59, p.107946) [8.88].

were reported to have endured the space conditions [8.4]. In yet another study, *Chlamydomonas reinhardtii* cells were carried on the Space Shuttle Endeavour's last trip (STS-134) in 2011 [8.36], both inside the cabin and in a hermetically sealed container in the cargo compartment. Through a transparent cover, the latter exposed a variety of viable algal mutants that were enclosed in a pressurized chamber to space radiation. When compared to the original strain, the mutants showed stronger photosynthetic activity and a faster rate of regrowth, indicating a higher capacity for stress recovery, according to post-flight investigation utilizing PCR, oxygen evolution measurements, and fluorescence measurements [8.51].

The *Chlamydomonas reinhardtii* strain was most notable for being transported on board the Russian Foton-M2 and Foton-M3 missions in 2005 and 2007, the wild type together with four different mutants were tested [8.37]. The D1 protein, crucial for photosynthesis [8.37], has different amino acid changes in the mutations and plays a crucial role in stabilizing and enhancing photosystem II under extreme conditions [8.37]. Photosynthetic performance was measured in flight using fluorescence measurements on cultures that were grown in artificial light. Two mutants showed greater photosynthetic efficiency in space and were able to regrow after returning to Earth, but the wild type and two mutants showed lower photosynthetic activity [8.37]. The researcher then conducted an experiment on the Chinese retrieved satellite on the spacecraft Shenzhou-II in an effort to improve the control system, cultural circumstances, and data gathering system's functionality [8.53]. This spacecraft had freshwater snail (*Bulinus asutralianus*) and *Chlorella pyrenoidosa* on board. Real-time data on the closed aquatic ecosystem (CAES) performance in microgravity was gathered using a variety of sensors to monitor the CAES. To investigate gravity-related issues, an on-board 1 xg centrifuge was included. It was discovered that the main element impacting the CAES performance in space is microgravity. Each day in microgravity, the main producer's biomass changed more than that of the control groups. The mean biomass concentration per day in the microgravity group declined, whereas it grew for a few days before leveling out in the control groups. Microgravity effects that increase the consumer's metabolic rate while decreasing photosynthesis may cause space impacts on a primary producer's biomass [8.52].

In an *in-situ* resource utilization (ISRU) and regeneratives life support system experiment for long-duration human space voyage, the majority of microalgal strains were biocompatible with nanosatellite materials; however, due to CO_2 deprivation, microalgal development was severely restricted in sealed microwells (~1 week) without dissolved bicarbonate. Further microalgae also produced oxygen, which led to the creation of bubbles inside the wells, which tampered with the sensor readings. By reducing light intensities (2–10 mol photons m^{-2} s^{-1}) and temperatures (4–12 °C), researchers were able to extend the development periods of two microalgal strains, *Chlorella vulgaris* and *Dunaliella bardawil*, without producing excessive amounts of oxygen. Though the studies detailed above were carried out to create the GraviSat platform, the outcomes of these studies could be helpful for the introduction of microalgae into other satellite payloads using low-volume microfluidic systems [8.7].

The accepted effects of microgravity on diatom growth are in fact the need of the hour due to the many solid reasons discussed above. Gravity can be important for diatoms, which due to their heavy silica wall, generally sink to the bottom. Microgravity may allow diatoms to float on the surface. Since, diatoms contribute to 25% global O_2 production and CO_2 fixation therefore it is known as chief producer of an ocean. Testing diatom growth during a space mission is limited by the availability and the costs incurred. Alternatively, starting with microgravity simulator experiments on Earth may be the first promising step [8.22].

Since the introduction of the classical clinostat in 1879 by Julius von Sachs, a number of GBFs have been designed to simulate the condition of "weightlessness" or "free fall" in laboratories on Earth [8.28].

8.6 Life Support Systems for Space Missions

One of the most promising life support systems was prepared by NASA, known as the controlled life support system (CELSS) for Mars, which would use controlled environmental agriculture (CEA) to grow plants in a thick-walled pressure hull buried inside the Martian surface. However, simulating the Martian atmosphere on Earth has given different results for different plants; for instance, tomato plant growth was inhibited [8.43]. In another study conducted decades back it was observed that winter rye could germinate at pressure as low as 101 mb but failed to grow at pressure below 31 mb. The advances in bioreactors ever since to simulate Martian atmosphere which is 6 to 7 millibars at its surface has been an important field of study for various scientists and researchers.

The NASA single loop cell culture reactor [8.67] as seen in Figure 8.4a was used to grow three diatoms whose genomes have been sequenced: *Phaeodactylum tricornutum*, *Fragilariopsis cylindrus*, and *Thalassiosira pseudonana*. Among these, *F. cylindrus* is abundant in the Artic and Antarctic, hence a suitable diatom suggested to be tested for Mars atmosphere. In 2007,

Gordon *et al.* [8.87] suggested the fabrication of a single loop cell culture reactor in which the type of diatom cells at a particular stage and live cell types were selected via fluorescence-activated cell sorting (FACS) or cell cytometry. In this study, the first Compustat was examined all diatoms in the chamber but destroyed cells which were farthest away by UV or laser light.

The Cell Specimen Environment Assembly, which holds up to eighteen Cell Specimen Chambers (CSCs, each with 3 to 30 ml capacity), permits up to 60 samples to be drawn (or injected) and stored under preprogrammed ground control, and provides growth medium perfusion, additive delivery, and heat and gas exchange for each chamber individually which has following compartments;

1) Electronics Assembly, which contains all computer and signal conditioning.
2) Video Microscopy Subsystem, which provides 40x and 200x optical magnification views of the specimen cultures to the crew or video downlink to scientists on the ground.
3) Structural Containment Assembly, which supports all the other assemblies and provides interface mounting with the shuttle and ISS host systems.

The temperature was maintained at 4 °C for the psychrophilic diatoms to be cultivated, however before the project could be completed, the cell culture unit (CCU) was modified [8.68] as seen in Figure 8.4b. The SLCC, which was delivered to NASA in 2006, but never flew. The free-swimming *Euglena* was cultured in a closed SLCC having a porous membrane. The light source was 3500–4000 lux which was economical as the system just needed an initial launch cost except filter tapes, nutrient salts from Provasoli-Guillard National Center for Culture of Marine Phytoplankton, and diatom inoculum in fresh chemostat chambers.

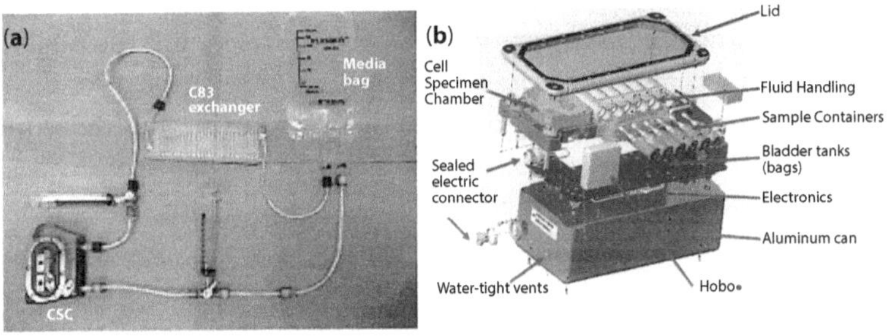

Figure 8.4 (a) Single loop for cell culture and (b) Open-source cell culture chamber. a [8.67], b [8.68].

The payload characteristics are: size: 3.4" x 5" x 10" (2.8 liters); power: each SLCC unit requires 2 Watt (165 mA at 12 VDC) steady-state power, with up to 3.3 Watts of peak power when additional pumps and valves are operating; weight: 5.5 lbs (2.5 kg); it is automated, except at the end of the test, a crew member is needed to help for last sample drying, which takes less than half an hour. An exploded view of the SLCC is also shown.

Researchers have been constantly working on designing ways to produce oxygen on the Red Planet. A recent study showed that production of oxygen on Mars can be done *in situ* by an advanced bioreactor system [8.18]. This system has potential to support life on Mars by converting high-grade iron ores by some biological mechanism and the most advanced seems to be that of a photobioreactor. Recently scientists have shown that iron conductive nanorods as a part of martian dust are of size ca. 9μm and have ability to warm Mars $5x10^3$ times the other gases. These nanorods forward scatter sunlight which reaches earth and blocks the infrared radiation [8.89].

8.7 Management of the Culture Vessel and Elements

Managing the culture vessel and elements is an important aspect of diatom cultivation. It is also important to maintain proper temperature and pH levels within the culture vessel, as diatoms have specific requirements for these parameters. The culture medium, which is the liquid in which the diatoms are grown, must also be regularly monitored and replenished as needed. Nutrients, such as nitrate, phosphate, and silicate, must also be added to the culture medium in appropriate concentrations to support diatom growth. Additionally, the light level and intensity must be controlled to maintain the optimal light intensity for the diatoms to survive and grow. By properly managing the culture vessel and elements, a healthy and thriving diatom culture can be maintained.

Managing biowaste in a space vessel is crucial to ensure the health and safety of astronauts and the proper functioning of the spacecraft. Biowaste refers to biological waste, such as human waste (urine and feces), food scraps, and other organic materials generated during space missions. Efficient biowaste management is essential for minimizing contamination, controlling odors, and recycling valuable resources. Continuous monitoring of waste management systems is essential to ensure their efficiency and safety. Research on advanced biowaste management technologies is ongoing to improve waste handling and resource recycling in space. Effective biowaste management is critical not only for the health and comfort of astronauts but also for the sustainability and success of long-duration space missions. As space exploration advances, innovative and sustainable biowaste management solutions will be essential to support future crewed missions to distant destinations like Mars or beyond [8.74].

8.8 Conclusions

Diatoms, which are thought to be 25% fixators of CO_2 worldwide, have the potential to terraform planets like Mars if diatoms are optimized to grow in microgravity and in space shuttles so as to establish there cultivation for food and sustainability. This is because diatoms may thrive in the CO_2-rich environment of Mars. Concurrently, the oxygen that is released during the process of fixing CO_2 and generating value added products will be the first step towards establishing life for humans and animals on Mars, transforming it into a habitable planet similar to Earth.

Acknowledgments

IR thanks the Indian Institute of Technology Madras, India for support. VV thanks CEFIPRA Indo French Project PPMB 7133/2020 for financial support. The authors thank Mr. Mohamed Ghobara (Department of Physics, Freie Universität Berlin) for discussion and suggestions.

References

[8.1] Ahirwar, A., Meignen, G., Khan, M.J., Khan, N., Rai, A., Schoefs, B., Marchand, J., Varjani, S. and Vinayak, V. (2021) Nanotechnological approaches to disrupt the rigid cell walled microalgae grown in wastewater for value-added biocompounds: commercial applications, challenges, and breakthrough. *Biomass Conversion and Biorefinery*, 1-26.

[8.2] Armbrust, E.V. (2009) The life of diatoms in the world's oceans. *Nature* 459(7244), 185-192.

[8.3] Bernal-Bayard, P., Puerto-Galán, L., Yruela, I., García-Rubio, I., Castell, C., Ortega, J.M., Alonso, P.J., Roncel, M., Martínez, J.I. and Hervás, M. (2017) The photosynthetic cytochrome c 550 from the diatom Phaeodactylum tricornutum. *Photosynthesis Research* 133, 273-287.

[8.4] Cockell, C.S., Rettberg, P., Rabbow, E. and Olsson-Francis, K. (2011) Exposure of phototrophs to 548 days in low Earth orbit: microbial selection pressures in outer space and on early earth. *The ISME Journal* 5(10), 1671-1682.

[8.5] Cycil, L.M., Hausrath, E.M., Ming, D.W., Adcock, C.T., Raymond, J., Remias, D. and Ruemmele, W.P. (2021) Investigating the growth of algae under low atmospheric pressures for potential food and oxygen production on Mars. *Frontiers in Microbiology*, 3171.

[8.6] De Martino, A., Bartual, A., Willis, A., Meichenin, A., Villazán, B., Maheswari, U. and Bowler, C. (2011) Physiological and molecular evidence that environmental changes elicit morphological interconversion in the model diatom Phaeodactylum tricornutum. *Protist* 162(3), 462-481.

[8.7] Fleming, E.D., Bebout, B.M., Tan, M.X., Selch, F. and Ricco, A.J. (2014) Biological system development for GraviSat: A new platform for studying photosynthesis and microalgae in space. *Life Sciences in Space Research* 3, 63-75.

[8.8] Fu, W., Shu, Y., Yi, Z., Su, Y., Pan, Y., Zhang, F. and Brynjolfsson, S. (2022) Diatom morphology and adaptation: Current progress and potentials for sustainable development. *Sustainable Horizons* 2, 100015.

[8.9] Gagarin, Y. (2001) *Soviet Man in Space*. The Minerva Group, Inc.

[8.10] Ghobara, M.M., Ghobara, M.M., Mazumder, N., Vinayak, V., Reissig, L., Gebeshuber, I.C., Tiffany, M.A., Gordon, R. and Gordon, R. (2019) On light and diatoms: A photonics and photobiology review. *Diatoms: Fundamentals and Applications*, 129-189.

[8.11] Gingerich, O. (1993) *The Eye of Heaven: Ptolemy, Copernicus, Kepler*. Springer.

[8.12] Goessling, J.W., González, A.A.S., Raj, V.S.P., Ashworth, M.P., Manning, S.R. and Lopez-Garcia, M. (2020) Biosilica slab photonic crystals as an alternative to cleanroom nanofabrication? *Faraday Discussions* 223, 261-277.

[8.13] Vinayak, V., 2022. Algae as sustainable food in space missions. In *Biomass, Biofuels, Biochemicals*, pp. 517-540, Elsevier, https://doi.org/10.1016/B978-0-323-89855-3.00018-2.

[8.14] Gordon, R., *et al.*, Bubble farming: scalable microcosms for diatom biofuel and the next green revolution. *Diatoms Fundam. Appl.*, 2019: p. 583-654.

[8.15] Guélin, M. and Cernicharo, J. (2022) Organic molecules in interstellar space: Latest advances. *Frontiers in Astronomy and Space Sciences* 9, 787567.

[8.16] Guiry, M.D. (2012) How many species of algae are there? *Journal of phycology* 48(5), 1057-1063.

[8.17] Guiry, M.D., Guiry, G.M., Morrison, L., Rindi, F., Miranda, S.V., Mathieson, A.C., Parker, B.C., Langangen, A., John, D.M. and Bárbara, I. (2014) AlgaeBase: an on-line resource for algae. *Cryptogamie, Algologie* 35(2), 105-115.

[8.18] Gupta, E., Kanu, N.J., Agrawal, M.S., Kamble, A.A., Shaikh, A.N., Vates, U.K., Singh, G.K. and Chavan, S.S. (2021) An insight into numerical investigation of bioreactor for possible oxygen emission on Mars. *Materials Today: Proceedings* 47, 4149-4154.

[8.19] Hamm, C.E. (2005) The evolution of advanced mechanical defenses and potential technological applications of diatom shells. *Journal of Nanoscience and Nanotechnology* 5(1), 108-119.

[8.20] Hanslmeier, A. and Hanslmeier, A. (2014) Zustandsgrößen der Sterne. *Einführung in Astronomie und Astrophysik*, 279-307.

[8.21] Hendrickx, L., De Wever, H., Hermans, V., Mastroleo, F., Morin, N., Wilmotte, A., Janssen, P. and Mergeay, M. (2006) Microbial ecology of the closed artificial ecosystem MELiSSA (Micro-Ecological Life Support System Alternative): reinventing and compartmentalizing the Earth's food and oxygen regeneration system for long-haul space exploration missions. *Research in microbiology* 157(1), 77-86.

[8.22] Herranz, R., Anken, R., Boonstra, J., Braun, M., Christianen, P.C., de Geest, M., Hauslage, J., Hilbig, R., Hill, R.J. and Lebert, M. (2013) Ground-based facilities for simulation of microgravity: organism-specific recommendations for their use, and recommended terminology. *Astrobiology* 13(1), 1-17.

[8.23] Ishak, S., Cheah, W., Waiho, K., Salleh, S., Fazhan, H., Manan, H., Kasan, N. and Lam, S. Aquatic Role of Diatoms: From Primary Producers and Aquafeeds. In: *Diatoms*. CRC Press: 87-104.

[8.24] Kalirai, J. (2018) Scientific discovery with the James Webb space telescope. *Contemporary Physics* 59(3), 251-290.

[8.25] Khan, M.J., Singh, R., Shewani, K., Shukla, P., Bhaskar, P., Joshi, K.B. and Vinayak, V. (2020) Exopolysaccharides directed embellishment of diatoms triggered on plastics and other marine litter. *Scientific reports* 10(1), 18448.

[8.26] Kopparapu, R.K. (2018) The habitable zone: the climatic limits of habitability. *Handbook of Exoplanets*, 58.

[8.27] Kuefner, W., Ossyssek, S., Geist, J. and Raeder, U. (2020) The silicification value: a novel diatom-based indicator to assess climate change in freshwater habitats. *Diatom research* 35(1), 1-16.

[8.28] Kutschera, U. and Niklas, K.J. (2018) Julius Sachs (1868): The father of plant physiology. *American Journal of Botany* 105(4), 656-666.

[8.29] Launius, R.D. (2000) The historical dimension of space exploration: reflections and possibilities. *Space Policy* 16(1), 23-38.

[8.30] Liu, S., Storti, M., Finazzi, G., Bowler, C. and Dorrell, R.G. (2022) A metabolic, phylogenomic and environmental atlas of diatom plastid transporters from the model species Phaeodactylum. *Frontiers in Plant Science* 13, 950467.

[8.31] Martín-Torres, F.J., Zorzano, M.-P., Valentín-Serrano, P., Harri, A.-M., Genzer, M., Kemppinen, O., Rivera-Valentin, E.G., Jun, I., Wray, J. and Bo Madsen, M. (2015) Transient liquid water and water activity at Gale crater on Mars. *Nature Geoscience* 8(5), 357-361.

[8.32] McMahon, S. (2021) Astrobiology (overview). *Oxford Research Encyclopedia of Planetary Science.*

[8.33] Mendell, W.W. (1985) *Lunar bases and space activities of the 21st century.* Lunar and Planetary Institute.

[8.34] Mock, T., Otillar, R.P., Strauss, J., McMullan, M., Paajanen, P., Schmutz, J., Salamov, A., Sanges, R., Toseland, A. and Ward, B.J. (2017) Evolutionary genomics of the cold-adapted diatom *Fragilariopsis cylindrus. Nature* 541(7638), 536-540.

[8.35] Mourya, M., Khan, M.J., Ahirwar, A., Schoefs, B., Marchand, J., Rai, A., Varjani, S., Rajendran, K., Banu, J.R. and Vinayak, V. (2022) Latest trends and developments in microalgae as potential source for biofuels: The case of diatoms. *Fuel* 314, 122738.

[8.36] Niederwieser, T., 2018. Analysis of Factors Affecting the Implementation of an Algal Photobioreactor into a Spacecraft Life Support System (Doctoral dissertation, University of Colorado at Boulder) Corpus ID: 134010772.

[8.37] Niederwieser, T., Kociolek, P. and Klaus, D. (2018) A review of algal research in space. *Acta Astronautica* 146, 359-367.

[8.38] Rai, I., *et al.*, Biowaste recycling strategies for regenerative life support system: An overview. *Sustainable Energy Technologies and Assessments*, 2022. 53: p. 102525.

[8.39] Remmers, I.M., D'Adamo, S., Martens, D.E., de Vos, R.C., Mumm, R., America, A.H., Cordewener, J.H., Bakker, L.V., Peters, S.A. and Wijffels, R.H. (2018) Orchestration of transcriptome, proteome and metabolome in the diatom Phaeodactylum tricornutum during nitrogen limitation. *Algal research* 35, 33-49.

[8.40] Nicholson WL, Ricco AJ. Nanosatellites for Biology in Space: In Situ Measurement of Bacillus subtilis Spore Germination and Growth after 6 Months in Low Earth Orbit on the O/OREOS Mission. Life (Basel). 2019 Dec 29;10(1):1. doi: 10.3390/life10010001. PMID: 31905771; PMCID: PMC7175319

[8.41] Roukema, B. (1996) On determining the topology of the observable Universe via three-dimensional quasar positions. *Monthly Notices of the Royal Astronomical Society* 283(4), 1147-1152.

[8.42] Round, F.E., Crawford, R.M. and Mann, D.G. (1990) *Diatoms: Biology and Morphology of the Genera*. Cambridge University Press.

[8.43] Rule, D. and Staby, G. (1981) Growth of Tomato Seedlings at Subatmospheric Pressures1. *HortScience* 16(3), 331-332.

[8.44] Sagan, C. and Pollack, J.B. (1974) Differential transmission of sunlight on Mars: biological implications. *Icarus* 21(4), 490-495.

[8.45] Shirke, Y.M., Abou-Elanwar, A.M., Choi, W.-K., Lee, H., Hong, S.U., Lee, H.K. and Jeon, J.-D. (2019) Influence of nitrogen/phosphorus-doped carbon dots on polyamide thin film membranes for water vapor/N 2 mixture gas separation. *RSC advances* 9(55), 32121-32129.

[8.46] Smith, C.M. and Davies, E.T.S. (2012) *Emigrating beyond Earth: Human adaptation and space colonization*. Springer.

[8.47] Vakkada Ramachandran, A., Nazarious, M.I., Mathanlal, T., Zorzano, M.-P. and Martín-Torres, J. (2020) Space environmental chamber for planetary studies. *Sensors* 20(14), 3996.

[8.48] Vakkada Ramachandran, A., Zorzano, M.-P. and Martín-Torres, J. (2021) Experimental investigation of the atmosphere-regolith water cycle on present-day Mars. *Sensors* 21(21), 7421.

[8.49] Vinayak, V. (2022) Algae as sustainable food in space missions. In: *Biomass, Biofuels, Biochemicals*. Elsevier: 517-540.

[8.50] Vinayak, V., Bhaskar, P., Pandey, L.K. and Khan, M.J. (2022) Diatoms: the Living Jewels and their Biodiversity, Phycosphere and Associated Phenotypic Plasticity: A Lesson to Learn from the Current Pandemic of Coronavirus. In: *Biodiversity in India: Status, Issues and Challenges*. Springer: 385-429.

[8.51] Vukich, M., Ganga, P.L., Cavalieri, D., Rizzetto, L., Rivero, D., Pollastri, S., Mugnai, S., Mancuso, S., Pastorelli, S. and Lambreva, M. (2012) BIOKIS: a model payload for multidisciplinary experiments in microgravity. *Microgravity Science and Technology* 24(6), 397-409.

[8.52] Wang, G., Liu, Y., Li, G., Hu, C., Zhang, D. and Li, X. (2008) A simple closed aquatic ecosystem (CAES) for space. *Advances in Space Research* 41(5), 684-690.

[8.53] Wang, J., Shanahan, J., Qi, G. and Wang, R.K. (2022) *The Making of The Wandering Earth: A Film Production Handbook*. Routledge India.

[8.54] Webster, C.R., Mahaffy, P.R., Atreya, S.K., Moores, J.E., Flesch, G.J., Malespin, C., McKay, C.P., Martinez, G., Smith, C.L. and Martin-Torres, J. (2018) Background levels of methane in Mars' atmosphere show strong seasonal variations. *Science* 360(6393), 1093-1096.

[8.55] Xiao, Y., Liu, Y., Wang, G., Hao, Z. and An, Y. (2010) Simulated microgravity alters growth and microcystin production in *Microcystis aeruginosa* (cyanophyta). *Toxicon* 56(1), 1-7.

[8.56] Young, J.N., Goldman, J.A., Kranz, S.A., Tortell, P.D. and Morel, F.M. (2015) Slow carboxylation of R ubisco constrains the rate of carbon fixation during A ntarctic phytoplankton blooms. *New Phytologist* 205(1), 172-181.

[8.57] Vinayak, V., Joshi, K.B., Gordon, R., Schoefs, B. 2017. Nanoengineering of Diatom Surfaces for Emerging Applications. in: *Diatom Nanotechnology: Progress and Emerging Applications*, (Ed.) D. Losic, The Royal Society of Chemistry, pp. 0, https://doi.org/10.1039/9781788010160-00055".

[8.58] Ghobara, M.M., Vinayak, V., Smith, D.R., Schoefs, B., Gebeshuber, I.C., Gordon, R. 2015. Diatom frustules as photo-regulators of diatom photobiology. National symposium on horizons of light in molecules, *Materials and Daily Life*. Gour Central University, Sagar, India

[8.59] Mourya, M., Khan, M.J., Ahirwar, A., Schoefs, B., Marchand, J., Rai, A., Varjani, S., Rajendran, K., Banu, J.R., Vinayak, V.J.F. 2022. Latest trends and developments in microalgae as potential source for biofuels: The case of diatoms. 314, 122738 https://doi.org/10.1016/j.fuel.2021.122738

[8.60] Khan, M.J., Mathys, D. and Vinayak, V., 2022. Micro to Nano Ornateness of Diatoms from Geographically Distant Origins of the Globe. Diatom Microscopy, pp.179-219 https://doi.org/10.1002/9781119711568.ch7

[8.61] Vinayak, V., Khan, M.J. and Joshi, K.B., 2023. Nanoengineering diatoms in microfluidic lab on chip devices. In Applications of Multifunctional Nanomaterials (pp. 95-124). Elsevier https://doi.org/10.1016/B978-0-12-820557-0.00030-8

[8.62] Ghobara, M.M., Mazumder, N., Vinayak, V., Reissig, L., Gebeshuber, I.C., Tiffany, M.A. and Gordon, R. (2019) On light and diatoms: A photonics and photobiology review [Chapter 7]. In: Diatoms: Fundamentals & Applications [DIFA, Volume 1 in the series: Diatoms: Biology & Applications, series editors: Richard Gordon & Joseph Seckbach]. J. Seckbach and R. Gordon, (eds.) Wiley-Scrivener, Beverly, MA, USA: 129-190.

[8.63] Vinayak, V., Bhaskar, P.V., Pandey, L.K. and Khan, M.J., 2022. Diatoms: the Living Jewels and their Biodiversity, Phycosphere and Associated Phenotypic Plasticity: A Lesson to Learn from the Current Pandemic of Coronavirus. In Biodiversity in India: Status, Issues and Challenges (pp. 385-429). Singapore: Springer Nature. Singapore.

[8.64] Khan, M.J., Bawra, N., Verma, A., Kumar, V., Pugazhendhi, A., Joshi, K.B., Vinayak, V.J.B.t. 2021. Cultivation of diatom *Pinnularia saprophila* for lipid production: A comparison of methods for harvesting the lipid from the cells. 319, 124129 https://doi.org/10.1016/j.biortech.2020.124129

[8.65] Khan, M.J., Rai, A., Ahirwar, A., Sirotiya, V., Mourya, M., Mishra, S., Schoefs, B., Marchand, J., Bhatia, S.K., Varjani, S.J.B. 2021b. Diatom microalgae as smart nanocontainers for biosensing wastewater pollutants: recent trends and innovations. 12(2), 9531-9549 https://doi.org/10.1080/21655979.2021.1996748

[8.66] Rai, I., Ahirwar, A., Rai, A., Varjani, S. and Vinayak, V., 2022. Biowaste recycling strategies for regenerative life support system: An overview. Sustainable Energy Technologies and Assessments, 53, p.102525 https://doi.org/10.1016/j.seta.2022.102525

[8.67] Vandendriesche, D., Parrish, J., Kirven-Brooks, M., Fahlen, T., Larenas, P., Havens, C., Nakamura, G., Sun, L., Krebs, C., de Luis, J. and Vunjak-Novakovic, G., 2004. Space station biological research project (SSBRP) cell culture unit (CCU) and incubator for international space station (ISS) cell culture experiments. J Gravit Physiol, 11(1), pp.93-103

[8.68] Marchus, C.R., Knudson, J.A., Morrison, A.E., Strawn, I.K., Hartman, A.J., Shrestha, D., Pancheri, N.M., Glasgow, I. and Schiele, N.R., 2022. Low-cost, open-source cell culture chamber for regulating physiologic oxygen levels. HardwareX, 11, p.e00253https://doi.org/10.1016/j.ohx.2021.e00253

[8.69] Copernicus N, Lerner MP, Segonds AP, Verdet JP, Luna C, Savoie D, Toulmonde M. De revolutionibus orbium coelestium. Nueva York: Johnson Reprint Corporation; 1965.

[8.70] Sterrenburg, F.A., *et al.*, Diatoms: living in a constructal environment, in *Algae and cyanobacteria in extreme environments*. 2007, Springer. p. 141-172.

[8.71] Hoover, R.B., *et al.*, *Diatoms on Earth, comets, Europa and in interstellar space.* Astronomical Origins of Life: Steps Towards Panspermia, 2000: p. 197-224.

[8.72] Werner, D., *The biology of diatoms*. Vol. 13. 1977: Рипол Классик.

[8.73] Wright, V., Morzfeld, M. and Manga, M., 2024. Liquid water in the Martian mid-crust. *Proceedings of the National Academy of Sciences*, 121(35), p.e2409983121.

[8.74] Beech, M., Gordon, R. and Seckbach, J. (eds.) (2021) Terraforming Mars [TMRS, Volume in the series Astrobiology Perspectives on Life of the Universe, Eds. Richard Gordon & Joseph Seckbach]. Wiley-Scrivener, Beverly, Massachusetts, USA.

[8.75] Rai. I. 2023. Space™ Vehicles; apparatus for locomotion by land, air or water Trademark No .6143361, 2136. 10.10.2023

[8.76] Goldman JC. Potential role of large oceanic diatoms in new primary production. Deep Sea Research Part I: Oceanographic Research Papers. 1993 Jan 1;40(1):159-68.

[8.77] Chi, Zhanyou, Farah Elloy, Yuxiao Xie, Yucai Hu, and Shulin Chen. Selection of microalgae and cyanobacteria strains for bicarbonate-based integrated carbon capture and algae production system. *Applied Biochemistry and Biotechnology*, 172 (2014): 447-457.

[8.78] Bozarth, A., Maier, U.G. and Zauner, S., 2009. Diatoms in biotechnology: modern tools and applications. *Applied Microbiology and Biotechnology*, 82, pp.195-201.

[8.79] Aletsee L, Jahnke J. Growth and productivity of the psychrophilic marine diatoms *Thalassiosira antarctica* Comber and *Nitzschia frigida* Grunow in batch cultures at temperatures below the freezing point of sea water. *Polar Biology*. 1992 Mar;11(8):643-7.

[8.80] Cho ES, Kotaki Y, Park JG. The comparison between toxic *Pseudo-nitzschia multiseries*(Hasle) Hasle and non-toxic *P. pungens*(Grunow) Hasle isolated from Jinhae Bay, Korea. *Algae*. 2001;16(3):275-85.

[8.81] McGinnis KM, Dempster TA, Sommerfeld MR. Characterization of the growth and lipid content of the diatom *Chaetoceros muelleri*. *Journal of Applied Phycology*. 1997 Feb;9:19-24.

[8.82] Mock T, Hoch N. Long-term temperature acclimation of photosynthesis in steady-state cultures of the polar diatom *Fragilariopsis cylindrus*. *Photosynthesis research*. 2005 Sep;85:307-17.

[8.83] Suzuki Y, Takahashi M. Growth responses of several diatom species isolated from various environments to temperature. *Journal of Phycology*. 1995 Dec;31(6):880-8.

[8.84] Chen J, Guo K, Thornton DC, Wu Y. Effect of temperature on the release of transparent exopolymer particles (TEP) and aggregation by marine diatoms (*Thalassiosira weissflogii* and *Skeletonema marinoi*). *Journal of Ocean University of China*. 2021 Feb;20:56-66.

[8.85] Niederwieser, T., Kociolek, P. and Klaus, D., 2018. A review of algal research in space. *Acta Astronautica*, 146, pp.359-367.

[8.86] [Settles AM, Fisher KE, Zhang J, Bai F, Beisel NS, Hersh H, Tyre KN, Muller BS. Space Algae: Understanding the Genomic Impacts on Microalgae After Growth in the International Space Station. In Plant & Animal Genome Conference (PAG) 30 2023 Jan 13.]

[8.87] Gordon, R. (1996) Computer controlled evolution of diatoms: Design for a compustat. *Nova Hedwigia* 112(Festschrift for Prof. T.V. Desikachary at 75), 215-219.

[8.88] Mapstone, L.J., Leite, M.N., Purton, S., Crawford, I.A. and Dartnell, L., 2022. Cyanobacteria and microalgae in supporting human habitation on Mars. *Biotechnology Advances*, 59, p.107946.

[8.89] Ansari S, Kite ES, Ramirez R, Steele LJ, Mohseni H. Feasibility of keeping Mars warm with nanoparticles. *Science Advances*. 2024 Aug 7;10(32):eadn4650.

9

Diatom: Source of Biofuel and Active Green Anode Material for Advanced Energy Storage Application

Vivek Dalvi[1,2]*, Sumit Dhali[2] and Anushree Malik[2]

[1]Carbon Fix Solutions Pvt Ltd, Mumbai, Maharashtra, India
[2]Applied Microbiology Laboratory, Center for Rural Development & Technology, Indian Institute of Technology, Delhi, Hauz Khas, New Delhi, India

Abstract

Diatoms are single-celled microalgae with silica shells or frustules that can form colonies. More than 100,000 species of diatoms have been discovered to date, with cell sizes ranging from 2 to 200 μm with geometries including circular, triangular, square, or elliptical. They contain both organic and inorganic components and these fractions are evaluated for energy application. The organic fraction is utilized to make biofuels for less polluting transportation. The diatom extracted bio-oil contains both saturated and unsaturated fatty acids, which holds potential to improve biofuel quality and quantity. One of the outstanding complexities of the diatom is the multilayered frustule structure. The frustule consists of amorphous hydrated mesoporous silica with an organized microstructure that is much beyond the reach of synthetic design and fabrication technologies. Research on functional materials used in sophisticated energy storage devices, such as batteries and supercapacitors, can benefit from diatoms' delicate and repeatable nanoscale siliceous structure. This chapter presents the state-of-the-art on diatoms for their utility as biofuel and advanced energy storage systems.

***Keywords*:** Diatom, biofuel, mesoporous silica, frustule, energy storage devices

**Corresponding author*: dalvi.vivek11@gmail.com

Vandana Vinayak and Richard Gordon (eds.) Diatom Cultivation for Biofuel, Food and High-Value Products, (231–244) © 2025 Scrivener Publishing LLC

9.1 Diatoms – Microalgae with Unique Structure and Properties

Diatoms, an unfathomed group of unicellular eukaryotic phytoplankton with their enormous potential and capability in modern science, make them a multi-purpose find for various applications in wastewater treatment [9.1] [9.2] [9.3], drug delivery [9.4], cosmeceuticals [9.5], biosensors [9.6], energy storage [9.7], and many more. They are widely distributed in the aquatic environment and made up of inorganic porous cell walls surrounding the cell structure [9.8]. Diatoms are sometimes called "living factories" because they use the biogeochemical cycling of silica to make shells that are nanostructured, mesoporous, and have a closely packed layered architecture. A diatom's pure and intact silica frustule has two valves that are linked together by girdle bands. The valve consists of a stack of hexagonal compartments, each of which is divided into smaller sections by a thin silica plate. This silica outer shell called frustules have highly ordered hierarchical architecture with three-dimensional pore patterns [9.8]. The well-organized silica frustule (cell wall) structure with unique pore geometry, make it a good choice for diverse industrial applications such as photonics, energy storage (battery), hydrogen storage, etc. [9.5].

The frustules can be obtained in two ways: from diatomaceous earth (diatomite) and by culturing live diatom cells [9.9]. The diatomite is the fossil remains of dead diatom cells whose organic matter decays due to changing environmental conditions. Despite being low cost and available in bulk, they have certain limitations that make them inappropriate for high-end applications. Firstly, the diatomite substructures are often damaged due to harsh environmental conditions experienced over thousands of years. Secondly, the sediments are a mixture of non-uniform shells resulting from distinct species that need further separation and purification [9.10]. Living diatom cells, on the other hand, produce intact, uniform and high-quality silica nanostructures. However, they require cleaning of cultivated diatoms' biomass that involves removal of organic matter from the cells [9.11]. The most commonly used cleaning methods include high temperature baking (400–800 °C), oxidant cleaning and plasma etching. Hydrogen peroxide, sodium dodecyl sulfate (SDS), ethylene diamine tetra acetic acid (EDTA), sulfuric acid and nitric acid are some of the common oxidants employed to clean the organic matter [9.10] [9.12]. Oxygen plasma etching is a nondestructive tool to gently extract the frustule from the diatom cells [9.13].

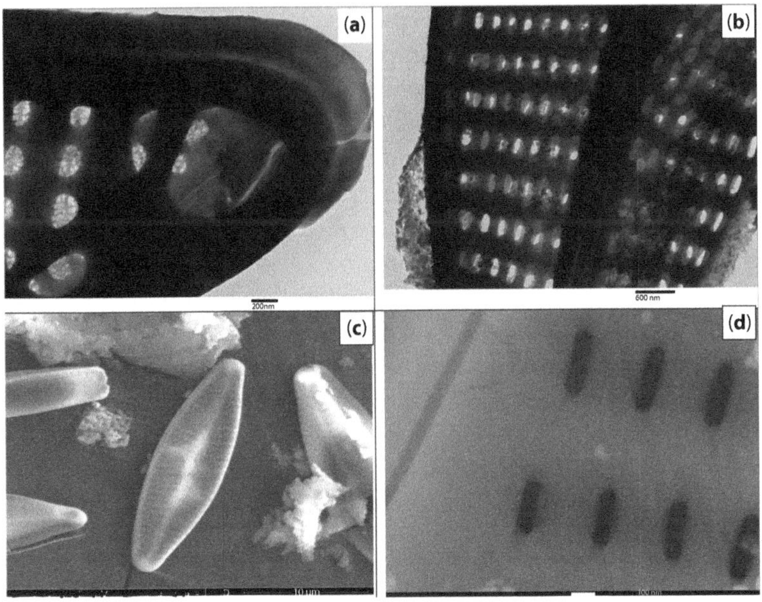

Figure 9.1 Transmission electron microscope (a, b) and Field emission scanning electron microscope (c, d) images of the *Navicula* sp. frustule.

Diatoms have excellent optical scattering capabilities and can be employed as photoelectric devices because their micro- and nanoscale frustule and pore architectures are comparable to the wavelengths of light. In addition, diatoms' photosynthetic receptors are present in chloroplasts close to the silica surface, so the light-channeling and concentrating characteristics of their silica frameworks can aid in the transmission and collection of enhanced light into the photoreceptors. This results in the development of silica frustules that aids in the acceleration of photosynthesis.

On the other hand, the silica shells derived from the diatoms have excellent photonic and mechanical properties that make them an ideal sustainable material for various energy applications like organic solar cells, hydrogen storage, and energy storage. For example, the hydrated, amorphous diatom frustule can be used in lithium-ion batteries. A developed diatom-carbon composite can be used as an anode and enhance its electrochemical performance by high initial coulombic efficiency, brilliant reversible capacity, rate capability, and at the same time mitigate its damage [9.14]. A similar result was reported by another researcher [9.15] in which a carbon black–diatom composite reported a high capacity and durable anode compared to other carbon coatings. Moreover, with its complex 3D structure, frustule can be used to improve light scattering and trapping. For example, a TiO_2 nanocrystal combined with diatom frustule (double layer formation) yields enhanced efficiency (35%) and circuit current density (39%) in quasi-solid-state dye-sensitized solar cells [9.16]. In this review, we will try to compile and represent the latest application of diatoms in advanced energy applications which will help the reader understand its wider applicability in the energy domain.

Diatom silica properties exhibit limitations, such as high resistivity, which are unfavorable for energy conversion and storage and other applications. Numerous studies have been conducted to modify rather than convert silica into other materials, but with preserved diatom configuration with micro- and nanoscale features. Researchers have adopted various metals, semiconductors, and carbon/polymer composites to modify diatom silica.

9.2 Biofuel Application

Diatom consists of an organic and inorganic part. Both of these fractions have individually been studied for energy application. The organic fraction has been used to produce biofuels to shift future transportation industry towards a less polluting system. The oil extracted from diatom generally contains both saturated and unsaturated fatty acids. The saturated fatty acids could contain both mono- and polyunsaturated fatty acids. However, only polyunsaturated fatty acids are present in diatom extracted oil. Diatoms are a good fit in non-arable land for large-scale production, and with their high growth and lipid content, they can take advantage of biodiesel feedstock. For example, biodiesel with least sulfur amount (< 0.0056% w/w) can be produced using *Navicula* sp. along with significantly high heating value of combustion (40.7 MJ/kg) [9.17].

Lipid production in diatoms depends on various factors, such as species, nutrient concentration (nitrate, phosphate, silica), temperature, light (intensity, wavelength) and many more [9.18]. In the case of temperature, lipid production was accelerated with an increase in the surrounding temperature [9.19]. In contrast, a decrease in temperature also reduced the PUFA production in the case of *Cylindrotheca closterium* [9.20]. The characteristics and type of media used for diatom cultivation affect the lipid yield. Also, the species used is important when lipid yield is the prime objective [9.21]. For example, nitrogen-deficient media renders the photosynthetic efficiency of diatom which results in reduced biomass growth and lowered protein content, while carbohydrate and lipid content gets increased [9.21]. Previously, authors have reported *Navicula* sp. can give up to 17% lipid yield in commercial media and wastewater [9.22], which is relatively lower than that grown in agricultural waste rice straw hydrolysate (up to 20% lipid yield) [9.23]. Moreover, silica stress also enhances lipid droplets in diatom cell [9.19]. Lipid production also varies in respect to cultivation mode. For example, mixotrophic cultivation gives an advantage to biomass yield, resulting in increased lipid production by their photoinhibition properties.

On the other hand, the advantage of diatoms for biofuel is that it can be produced from wastewater feedstock. Previously, feedstock like municipal wastewater, industrial wastewater, dairy wastewater, and livestock wastewater, was used to cultivate diatom for wastewater treatment as well as lipid production [9.5] [9.18]. Interestingly, rice straw hydrolysate was also used as a feedstock for diatom cultivation with enhance biomass productivity and lipid production [9.23].

Biodiesel production is also governed by the lipid profile of diatoms. The fatty acid methyl ester (FAME) analysis of lipid can identify the potential component for biodiesel production [9.24]. Although it varies with alteration of environmental parameters, cultivation conditions and stress conditions as well [9.19]. However, their production can vary. Generally, a suitable species can be selected by the "scoring criterion" for mass-scale biodiesel production. This "scoring criterion" was developed based on process score, triacylglyceride (TAG) score and biodiesel score. Suitable species can be selected based on this score for mass-scale biodiesel production. *Halamphora coffeaeformis, Navicula cincta,* and *N. gregaria* showed the best result in this context (overall score 1.65–1.5). Furthermore, lipid productivity can also be enhanced by optimized culture conditions. A moderately alkaline pH (pH 8) and saline water (3% NaCl) with a low concentration of $NaNO_3$ (18.75 mg/L) (source of nitrogen) and NaH_2PO_4 (3.48 mg/L) (source of phosphorus) can enhance growth, biomass, lipid quality and quantity in the case of *Nitzschia* sp. RRSE2. Similarly, biodiesel production can also be accelerated by osmotic shock pretreatment. In the case of *C. muelleri,* a high lipid recovery (72%) and improved methane yield and production (35%) was reported [9.25] [9.26] [9.27] [9.28] [9.29].

Overall, the diatom is shown to be a promising agent for biofuel production. The U.S. Department of Energy (DOE) initiated the Aquatic Species Program (ASP) in 1980–1996 to identify the most suitable microalgae species for use as futuristic biofuel source. The ASP screened more than 3000 strains of microalgae to assess their capability for biomass and lipid production under varying parameters such as media composition, CO_2

injection; and the environmental parameters of wastewaters such as light, pH, and dissolved oxygen (DO). The DOE noted that diatoms can be a promising source of "third generation biofuel" in terms of lipid productivity and ability to survive under challenging environmental conditions,

Table 9.1 Various types of feedstocks and lipid content in diatoms.

Type of feedstock	Algal strain	Biomass yield mg L^{-1}	Lipid content	Reference
30% AQW: ICP-SiO2	*Chaetoceros gracilis*	279.85	16.82[a]	[9.3]
80% AQW: ICP-SiO2	*Thalassiosira weissflogii*	224.56	21.05[a]	[9.3]
MW:SW=2:1	*Phaeodactylum tricornutum*	–	316[b]	[9.14]
MW:SW=1:1	*Phaeodactylum tricornutum*	–	332[b]	[9.14]
Urban domestic wastewater	Diatom consortium	NA	22.2[a]	[9.18]
F/2	*Cylindrotheca closterium*	225	28[a]	[9.20]
F media	*Navicula* sp.	NA	35.25[a]	[9.22]
Rice straw media	*Navicula* sp.	472	18.9[a]	[9.23]
F/2	*Nitzschia* sp.	690	77.5[b]	[9.25]
F/2 (salinity -25PSU)	*Navicula incerta*	123	26.3[a]	[9.26]
F/4 (salinity -25PSU)	*Navicula incerta*	129	24.6[a]	[9.26]
F/8(salinity -25PSU)	*Navicula incerta*	135	20.4[a]	[9.26]

#MW: Municipal wastewater; SW: Sea water; AQW:ICP: Aquaculture wastewater coupled with inductively coupled plasma nanosilica, PSU = Practical salinities unity (g/kg), a: lipid content in percentage of dry weight; b: lipid content in mg L^{-1}.

with 50 out of the tested strains demonstrating exceptional performance [9.30]. In those chosen algal strains, diatoms made up 60% of the population [9.31]. Some examples of diatom feedstocks for lipid production are shown in Table 9.1.

9.3 Diatom Silica: Material for Li-Ion Battery Anode

Currently, lithium-ion batteries (LIBs) are commonly preferred commercial energy storage devices for various day-to-day applications [9.32]. They are effectively employed in electric cars as well as portable electronics like mobile phones, laptops, and cameras. This is due to certain advantages that LIBs possess, such as significantly high energy density, extended cyclability, and low environmental toxicity, compared to conventional lead-acid batteries [9.33]. Additionally, they have made it possible to produce long-range electric vehicles and store energy generated from renewable sources like solar, wind, and hydro power [9.34]. Still, there are a number of obstacles that need to be addressed in the Li-ion battery technology field, such as those pertaining to cost, the availability of resources, and safety concerns regarding the possibility of overheating in the batteries [9.35]. As lithium is being mined and purified, the resultant very high environmental impact and footprint that is left behind by the creation of LIBs is yet another disadvantage of these types of batteries. As a result of the fact that LIBs are dominating the battery market for portable electronics as well as electric vehicle market, it is important to address and solve the issues that have been mentioned above [9.36]. One way to do this is by, for instance, identifying and developing new electrode materials as well as other components that are used in the production of complete batteries [9.37] [9.38].

Designing innovative electrodes is important for developing the next generation of LIBs. Hence, these areas of study are hotspots for enhancing performance, such as enhancing specific energy density, extension of battery life, improving durability, and faster recharging speed, while also reducing capital required. The graphite and hard carbon are commonly used carbonaceous anodes in LIBs. The present carbonaceous anodes, which are based on intercalation processes, are severely constrained because of their low specific capacity (300–400 mA h g^{-1}) and an unavoidable solid electrolyte interface [9.39] [9.40] [9.41]. Theoretically, silicon-based anodes are among those with the most potential, with specific capacity as high as ~4200 mA h g^{-1}, and hence significant research is done using them. However, silicon possesses certain downsides [9.42].

Silicon-based anodes have not been used much commercially because during the lithiation and delithiation processes, the volume violently swells by up to 300–400%, which causes the anode material to break down quickly due to strain and results in poor electrical conductivity [9.43]. Comparatively, the silica anode with significant specific capacity (~1965 mA h g^{-1}) exhibits less volume fluctuation. Even though expansion and contraction cause micron-sized silicon structures to break down, it has been shown that nanostructured Si materials are better able to handle the lithiation and delithiation of Li without breaking down [9.41]. Nanostructured diatom silica in its natural state does not have the physical properties needed for a battery electrode, but it could be redesigned in a way that uses various thermos-catalytic processes to have the battery electrode needed for desired properties [9.42].

Following is a general interpretation of how the process of transformation interactions between silica and Li^+ ions work [9.44]:

$$SiO_2 + 4Li^+ + 4e^- \rightarrow 2Li_2O + Si \tag{9.1}$$

$$2SiO_2 + 4Li^+ + 4e^- \rightarrow 2Li_4SiO_4 + Si \tag{9.2}$$

$$5SiO_2 + 4Li^+ + 4e^- \leftrightarrow 2Li_2Si_2O_5 + Si \tag{9.3}$$

$$Si + xLi^+ + xe^- \rightarrow Li_xSi \tag{9.4}$$

where reaction (9.1) and (9.2) are irreversible reactions. During reaction (9.1) and (9.2) the side reaction of electrolyte degradation and the development of the solid electrolyte interface occur concurrently (SEI). On the other hand, reaction (9.3) and (9.4) are reversible reactions. The formation of the metastable $Li_2Si_2O_5$ structure takes place during reaction (9.3). While during reaction (9.4), the alloying and dealloying of Si and Li+ ions take place.

It is essential to enhance the efficiency of the silica-based anode by adding new nanoscopic morphologies with more surface area and increasing the pore volume, such as nanorods, nanowires, nanosheets, and nanospheres. This will increase the electrical conductivity, charge storage, and ion transport across membrane, as well as stop the particles from breaking up [9.45].

Nowak *et al.* used different structures of silicon to achieve higher discharge capacities, such as three-dimensional silica particles (2500–2600 mA h g^{-1}), silica nanotube arrays (1700–1800 mA h g^{-1}), and silica nanotubes

(900–1000 mA h g^{-1}). Si-based anodes could also be made to work better by coating them with polymers, like silicohydrides. But the conditions for their reactions are hard, take a long time, and require complex steps like elevated heat or pressure [9.46].

In their recent study, Nowak *et al.* showed that diatom silicon layered with carbon and changed into a different form can be used as Li-ion battery electrode material. The diatom silica was reduced by employing a magnesiothermic technique to synthesize diatom silicon and then carbonizing it with polyacrylic acid (PAA). The NanoSi synthesized using diatoms had a remarkable specific surface area of 162.6 cm^2 g^{-1} [9.47]. Using diatom-synthesized porous Si/C composite anode material, Wang *et al.* also demonstrated that high specific charge-discharge capacity is achievable (1700 mA h g^{-1}) at low current density (0.1 A g^{-1}). Additionally, the anode has a substantially higher cyclic stability (520 mA h g^{-1}) than that of pure porous silicon (205 mA h g^{-1}) [9.48]. Findings suggested that limiting particle pulverization was the key to enhancing the functionality of lithium-ion batteries with silica anodes.

9.4 Conclusion

Diatoms, microalgae with unique symmetrical structural architecture with size ranging from 2–200 μm, exhibit the presence of significant organic and inorganic fraction. The oil extracted from the organic fraction possesses huge potential to be applied as a source of future biofuel for internal combustion engines. On the other hand, the inorganic fraction, which exhibits silica multilayered frustule structure, can have material for application in various fields. After employment of thermo-catalytic processes, the synthesized nanosilica material exhibits good charge storage property, thus can be used as green anode material for lithium-based energy storage devices. However, the literature suggests that there is a dire need to take up research activities that address how to scale up production and improve the economics of diatom-based energy generation (biofuel) and storage (battery) applications, which will be fruitful for societal energy needs.

Acknowledgment

The authors gratefully acknowledge this project's sponsor under Biotechnology Ignition Grant (BIG) scheme by the Biotechnology Industry

Research Assistance Council (BIRAC), Government of India, for their financial support (BIRAC/FITT0824/BIG-17/20).

The authors thank the Ministry of Education, Government of India, for awarding the "Prime Minister Research Fellowship" to Mr Sumit Dhali.

References

[9.1] Li, J., Sang, H., Guo, H., Popko, J. T., He, L., White, J. C., Parkash Dhankher, O., Jung, G., & Xing, B. (2017). Antifungal mechanisms of ZnO and Ag nanoparticles to *Sclerotinia homoeocarpa. Nanotechnology*, 28(15), 155101. https://doi.org/10.1088/1361-6528/aa61f3

[9.2] Reid, A., Buchanan, F., Julius, M., & Walsh, P. J. (2021). A review on diatom biosilicification and their adaptive ability to uptake other metals into their frustules for potential application in bone repair. *Journal of Materials Chemistry B*, 9(34), 6728-6737.

[9.3] Saxena, A., Kumar Singh, P., Bhatnagar, A., & Tiwari, A. (2022). Growth of marine diatoms on aquaculture wastewater supplemented with nanosilica. *Bioresource Technology*, 344, 126210. https://doi.org/10.1016/j.biortech.2021.126210

[9.4] Terracciano, M., De Stefano, L., & Rea, I. (2018). Diatoms green nanotechnology for biosilica-based drug delivery systems. *Pharmaceutics*, 10(4), 242.

[9.5] Sharma, N., Simon, D. P., Diaz-Garza, A. M., Fantino, E., Messaabi, A., Meddeb-Mouelhi, F., ... & Desgagné-Penix, I. (2021). Diatoms biotechnology: Various industrial applications for a greener tomorrow. *Frontiers in Marine Science*, 8, 636613.

[9.6] Selvaraj, V., Thomas, N., Anthuvan, A. J., Nagamony, P., & Chinnuswamy, V. (2018). Amine-functionalized diatom frustules: a platform for specific and sensitive detection of nitroaromatic explosive derivative. *Environmental Science and Pollution Research*, 25(21), 20540-20549.

[9.7] Ramachandra, T. V., Mahapatra, D. M., & Gordon, R. (2009). Milking diatoms for sustainable energy: biochemical engineering versus gasoline-secreting diatom solar panels. *Industrial & Engineering Chemistry Research*, 48(19), 8769-8788.

[9.8] Richthammer, P., Börmel, M., Brunner, E., & Van Pée, K. H. (2011). Biomineralization in Diatoms: The Role of Silacidins. *ChemBioChem*, 12(9), 1362–1366. https://doi.org/10.1002/cbic.201000775

[9.9] Sardo, A., Orefice, I., Balzano, S., Barra, L., & Romano, G. (2021). Mini-Review: Potential of Diatom-Derived Silica for Biomedical Applications. *Applied Sciences*, 11(10), 4533. https://doi.org/10.3390/app11104533

[9.10] Van Eynde, E., Lenaerts, B., Tytgat, T., Verbruggen, S. W., Hauchecorne, B., Blust, R., & Lenaerts, S. (2014). Effect of pretreatment and temperature on the properties of *Pinnularia* biosilica frustules. *RSC Advances*, 4(99), 56200–56206. https://doi.org/10.1039/c4ra09305d

[9.11] Jain, R., Nigam, H., Mathur, M., Malik, A., & Arora, U. K. (2021). Towards green thermal power plants with blowdown water reuse and simultaneous biogenic nanostructures recovery from waste. *Resources, Conservation and Recycling*, 168, 105283.

[9.12] Saad, E. M., Pickering, R. A., Shoji, K., Hossain, M. I., Glover, T. G., Krause, J. W., & Tang, Y. (2020). Effect of cleaning methods on the dissolution of diatom frustules. *Marine Chemistry*, 224(June 2019), 103826. https://doi.org/10.1016/j.marchem.2020.103826

[9.13] Jeffryes, C., Campbell, J., Li, H., Jiao, J., & Rorrer, G. (2011). The potential of diatom nanobiotechnology for applications in solar cells, batteries, and electroluminescent devices. *Energy and Environmental Science*, 4(10), 3930–3941. https://doi.org/10.1039/c0ee00306a

[9.14] Wang, X.-W., Huang, L., Ji, P.-Y., Chen, C.-P., Li, X.-S., Gao, Y.-H., & Liang, J.-R. (2019). Using a mixture of wastewater and seawater as the growth medium for wastewater treatment and lipid production by the marine diatom *Phaeodactylum tricornutum*. *Bioresource Technology*, 289, 121681. https://doi.org/10.1016/j.biortech.2019.121681

[9.15] Norberg, A.N., Wagner, N.P., Kaland, H., Vullum-Bruer, F., Svensson, A.M., 2019. Silica from diatom frustules as anode material for Li-ion batteries. *RSC Adv.* 9, 41228–41239. https://doi.org/10.1039/C9RA07271C

[9.16] Bandara, T.M.W.J., Furlani, M., Albinsson, I., Wulff, A., Mellander, B.-E., 2020. Diatom frustules enhancing the efficiency of gel polymer electrolyte-based dye-sensitized solar cells with multilayer photoelectrodes. *Nanoscale Adv.* 2, 199–209. https://doi.org/10.1039/C9NA00679F

[9.17] Marella, T. K., Saxena, A., & Tiwari, A. (2020). Diatom mediated heavy metal remediation: A review. *Bioresource Technology*, 305, 123068.

[9.18] Marella, T. K., Datta, A., Patil, M. D., Dixit, S., & Tiwari, A. (2019). Biodiesel production through algal cultivation in urban wastewater using algal floway. *Bioresource Technology*, 280, 222-228.

[9.19] Tang, C. C., Tian, Y., Liang, H., Zuo, W., Wang, Z. W., Zhang, J., & He, Z. W. (2018). Enhanced nitrogen and phosphorus removal from domestic wastewater via algae-assisted sequencing batch biofilm reactor. *Bioresource Technology*, 250(September 2017), 185–190. https://doi.org/10.1016/j.biortech.2017.11.028

[9.20] Almeyda, M. D., Scodelaro Bilbao, P. G., Popovich, C. A., Constenla, D., & Leonardi, P. I. (2020). Enhancement of polyunsaturated fatty acid production under low-temperature stress in *Cylindrotheca closterium*. *Journal of Applied Phycology*, 32(2), 989-1001.

[9.21] Fu, W., Wichuk, K., & Brynjólfsson, S. (2015). Developing diatoms for value-added products: Challenges and opportunities. *New Biotechnology*, 32(6), 547–551. https://doi.org/ 10.1016/j.nbt.2015.03.016

[9.22] Fimbres-Olivarría, D., López-Elías, J. A., Martínez-Córdova, L. R., Carvajal-Millán, E., Enríquez-Ocaña, F., Valdéz-Holguín, E., & Miranda-Baeza, A. (2015). Growth and biochemical composition of *Navicula* sp. cultivated at two light intensities and three wavelengths. *The Israeli Journal of Aquaculture-Bamidgeh.* DOI: 10.46989/001c.20715

[9.23] Dhali, S., Jain, R., Malik, A., Sharma, S., & Raliya, R. (2022). Cultivation of *Navicula* sp. on rice straw hydrolysate for the production of biogenic silica. *Bioresource Technology*, 360, 127577.

[9.24] Tan, X., Lam, M., Uemura, Y., Lim, J., Wong, C., & Lee, K. (2018). Cultivation of microalgae for biodiesel production: A review on upstream and downstream processing. *Chinese Journal of Chemical Engineering*, 26(1), 17–30. https://doi.org/10.1016/j.cjche.2017.08.010

[9.25] Harini, A.B., Rajkumar, R., Takriff, M.S., 2020. Enhanced Production of Lipid as Biofuel Feedstock from the Marine Diatom *Nitzschia* sp. by Optimizing Cultural Conditions 20. *BioRes.*

[9.26] González-González, L.M., Astals, S., Pratt, S., Jensen, P.D., Schenk, P.M., 2021. Osmotic shock pre-treatment of *Chaetoceros muelleri* wet biomass enhanced solvent-free lipid extraction and biogas production. *Algal Res.* 54, 102177. https://doi.org/10.1016/j.algal.2020.102177

[9.27] Kalwani, M., Devi, A., Patil, K., Kumari, A., Dalvi, V., Malik, A., ... & Pabbi, S. (2022). Microalgae-mediated wastewater treatment and enrichment of wastewater-cultivated biomass for biofuel production. In *Expanding Horizon of Cyanobacterial Biology* (pp. 259-281). Academic Press.

[9.28] Dalvi, V., Patil, K., Nigam, H., Jain, R., Pabbi, S., Malik, A. (2021). Environmental Resilience and Circular Agronomy Using Cyanobacteria Grown in Wastewater and Supplemented with Industrial Flue Gas Mitigation. In: Rastogi, R.P. (eds) *Ecophysiology and Biochemistry of Cyanobacteria*. Springer, Singapore. https://doi.org/10.1007/978-981-16-4873-1_14

[9.29] Dalvi, V., Chawla, P., & Malik, A. (2021). Year-long performance assessment of an on-site pilot scale (100 L) photobioreactor on nutrient recovery and pathogen removal from urban wastewater using native microalgal consortium. *Algal Research*, 55, 102228. https://doi.org/10.1016/j.algal.2021.102228

[9.30] Dhanker, R., Kumar, R., Tiwari, A., & Kumar, V. (2022). Diatoms as a biotechnological resource for the sustainable biofuel production: a state-of-the-art review. *Biotechnology and Genetic Engineering Reviews*, 38(1), 111-131.

[9.31] Guiry, M. (2012). How many species of algae are there? *Journal of Phycology*, 48(5), 1057–1063. https://doi.org/10.1111/j.1529-8817.2012.01222.x

[9.32] Armand, M., & Tarascon, J. M. (2008). Building better batteries. *Nature*, 451(7179), 652-657.

[9.33] Choi, J. W., & Aurbach, D. (2016). Promise and reality of post-lithium-ion batteries with high energy densities. *Nature Reviews Materials*, 1(4), 1-16.

[9.34] Tarascon, J. M., & Armand, M. (2001). Issues and challenges facing rechargeable lithium batteries. *Nature*, 414(6861), 359-367.

[9.35] Whittingham, M. S. (2004). Lithium batteries and cathode materials. *Chemical Reviews*, 104(10), 4271-4302.

[9.36] Lee, K. T., Jung, Y. S., & Oh, S. M. (2003). Synthesis of tin-encapsulated spherical hollow carbon for anode material in lithium secondary batteries. *Journal of the American Chemical Society*, 125(19), 5652-5653.

[9.37] Park, C. M., Choi, W., Hwa, Y., Kim, J. H., Jeong, G., & Sohn, H. J. (2010). Characterizations and electrochemical behaviors of disproportionated SiO and its composite for rechargeable Li-ion batteries. *Journal of Materials Chemistry*, 20(23), 4854-4860.

[9.38] Si, Q., Hanai, K., Ichikawa, T., Phillipps, M. B., Hirano, A., Imanishi, N., ... & Takeda, Y. (2011). Improvement of cyclic behavior of a ball-milled SiO and carbon nanofiber composite anode for lithium-ion batteries. *Journal of Power Sources*, 196(22), 9774-9779.

[9.39] Lee, J. I., & Park, S. (2013). High-performance porous silicon monoxide anodes synthesized via metal-assisted chemical etching. *Nano Energy*, 2(1), 146-152.

[9.40] Lee, J. I., Lee, K. T., Cho, J., Kim, J., Choi, N. S., & Park, S. (2012). Chemical-assisted thermal disproportionation of porous silicon monoxide into silicon-based multicomponent systems. *Angewandte Chemie*, 124(11), 2821-2825.

[9.41] Vu, A., Qian, Y., & Stein, A. (2012). Porous electrode materials for lithium-ion batteries–how to prepare them and what makes them special. *Advanced Energy Materials*, 2(9), 1056-1085.

[9.42] Nowak, A. P., Sprynskyy, M., Wojtczak, I., Trzciński, K., Wysocka, J., Szkoda, M., Buszewski, B., & Lisowska-Oleksiak, A. (2020). Diatoms Biomass as a Joint Source of Biosilica and Carbon for Lithium-Ion Battery Anodes. *Materials* (Basel, Switzerland), 13(7), 1673. https://doi.org/10.3390/ma13071673

[9.43] Wang, Z., Zhao, J., Liu, S., Cui, F., Luo, J., Wang, Y., ... & Yang, X. (2021). Cultured diatoms suitable for the advanced anode of lithium ion batteries. *ACS Sustainable Chemistry & Engineering*, 9(2), 844-852.

[9.44] Norberg, A. N., Wagner, N. P., Kaland, H., Vullum-Bruer, F., & Svensson, A. M. (2019). Silica from diatom frustules as anode material for Li-ion batteries. *RSC advances*, 9(70), 41228–41239. https://doi.org/10.1039/c9ra07271c

[9.45] He, Q., Xu, C., Luo, J., Wu, W., & Shi, J. (2014). A novel mesoporous carbon@ silicon–silica nanostructure for high-performance Li-ion battery anodes. *Chemical Communications*, 50(90), 13944-13947.

[9.46] Nowak, A. P., Lisowska-Oleksiak, A., Wicikowska, B., & Gazda, M. (2017). Biosilica from sea water diatoms algae—electrochemical impedance spectroscopy study. *Journal of Solid State Electrochemistry*, 21(8), 2251-2258.

[9.47] Nowak, A. P., Sprynskyy, M., Brzozowska, W., & Lisowska-Oleksiak, A. (2019). Electrochemical behavior of a composite material containing 3D-structured diatom biosilica. *Algal Research*, 41, 101538.

[9.48] Wang, W., Gutu, T., Gale, D. K., Jiao, J., Rorrer, G. L., & Chang, C. H. (2009). Self-assembly of nanostructured diatom microshells into patterned arrays assisted by polyelectrolyte multilayer deposition and inkjet printing. *Journal of the American Chemical Society*, 131(12), 4178-4179.

Part III

LOW-VALUE PRODUCTS

10

Milking of Diatoms: A Realistic Approach to Serve the Biorefinery Concept

Mrinal Kashyap[1,2]

[1]*Department of Biosciences and Biomedical Engineering, Indian Institute of Technology, Indore, Madhya Pradesh, India*
[2]*Porter School of Environment and Earth Sciences, Tel Aviv University, Tel Aviv, Israel*

Abstract
The algae-based biodiesel industry is moving forward with time, but the production cost is still a limiting step in the process. Diatoms are unicellular eukaryotic, photosynthetic microorganisms having a siliceous cell wall. They are known as the source of diafuel (fuel from diatoms), and their application in the sustainable energy field is the current hot topic. High-value compounds in these unicellular cells make them crucial for biorefinery. The robustness of the diatoms cell wall makes extracting the valuable compounds unprofitable. High-end instruments like ultrasonication, high-pressure pretreatment and chemical treatments are needed to extract beneficial compounds from the cells. Due to the production cost, economic pressure is a significant barrier to diafuel industrialization. Different approaches to milk diatoms for diafuel and other intracellular compounds are discussed in the current chapter. The new emerging tool for successful milking is nanotechnology. The interdisciplinary interaction of diatoms and nanoparticles can develop future tools for nano-biorefinery. The present chapter sheds light on the need for process development for milking diatoms and the potential of available processes in this direction.

***Keywords*:** Milking, nano-biorefinery, diatoms, nanotechnology, diafuel

Email: mkashyap610@gmail.com

Vandana Vinayak and Richard Gordon (eds.) Diatom Cultivation for Biofuel, Food and High-Value Products, (247–266) © 2025 Scrivener Publishing LLC

10.1 Introduction

Microalgae are better at utilizing sunlight and nutrients than vascular plants. Therefore, they are a reliable source for the sustainable production of value-added compounds [10.1]. The productivity of these compounds is considered one of the significant parameters for estimating the economics of algal bio-refinery [10.1]. The unpropitious advancement of global warming and climate change has led the way to making microalgae an alluring candidate for biofuels [10.2]. The use of algae for biodiesel production has several advantages. They have great photosynthetic activity (12.6%), good metal handling capacity, recovery of nutrients from wastewater, and carbon sequestration capability [10.2] [10.3]. Although several species can produce various value-added compounds, certain microalgae are better than others [10.1]. Diatoms can fix up to 20% of the carbon dioxide in the surroundings and have better productivity than many other microalgae [10.4]. Along with that, diatoms are flexible, which allows them to survive harsh environmental conditions. Thus, they can be a better candidate for the algal bio-refinery approach [10.5] [10.1] [10.4]. Diatoms are unicellular photosynthetic microorganisms in the aquatic environment [10.6]. They have various value-added compounds and contribute 20–25% of net primary production [10.7] [10.8] [10.9]. The diatoms' outer cell wall contains silica, and the two valves are joined together by girdle bands. This structure looks like a Petri dish [10.10]. This silica wall is rough and rugged, limiting intracellular compounds' extraction from the diatoms [10.11]. Therefore, even though the diatoms are a potential candidate for value-added compounds, their commercialization is still limited. The advent of algal nanotechnology has helped to grow the field of applied diatomology by providing insights into new applications like dye-sensitized solar cells (DSSCs) [10.12], nanoparticle biosynthesis [10.13], and milking of cells [10.14].

The use of nanotechnology with diatoms has opened new dimensions of research. Nanotechnologists have studied the nanopores on the diatom cell's surface for numerous applications. However, producing such nanostructures with clarity and precision is a complicated task for nanotechnologists [10.15]. The intricately arranged cell wall of diatoms consists of nano-silica, and these structures are species-specific, further demonstrating the genetic basis of their formation. The fascinating morphogenesis in constructing these nano-silica structures is of great interest to nanotechnologists [10.16]. The frustules of diatoms having nano-silica can have different sizes, shapes, and functionalities depending upon the species [10.17] [10.18]. The bio-silica from diatoms has shown great possibilities in the pharma and medicine industry, which needs more focus [10.18]. Nanoparticle (NP) biosynthesis is another essential dimension of diatom nanotechnology, where diatoms are explored as a potential candidate for nanoparticle manufacturing. The nano-silica material is hydrophobic, which promotes the use of diatom frustules in biomedicine. The biosynthesis of nanoparticles through extracts prepared from diatoms can further aid in the investigation of new tools for medicinal use as well as its use in biofuel industry.

Nowadays, nanotechnology has become one of the crucial units of applied sciences, and diatom nanotechnology is an emerging field in this sector. Issues with chemical and physical methods of nanoparticle biosynthesis led researchers to look for an environment-friendly process. Therefore, nanoparticle biosynthesis came into existence. The literature shows the use of diatoms in synthesizing gold and silver nanoparticles for antibacterial and many other applications [10.19] [10.20] [10.21]. Diatoms are also considered an excellent renewable energy source and can be vital for algal biorefinery [10.22]. Diatoms can produce 13–21 carbon fatty acid methyl esters (FAME). These FAME biomolecules are primarily saturated and monounsaturated fatty acids (MUFAs). The MUFAs provide more energy than polyunsaturated fatty acids (PUFAs) upon oxidation.

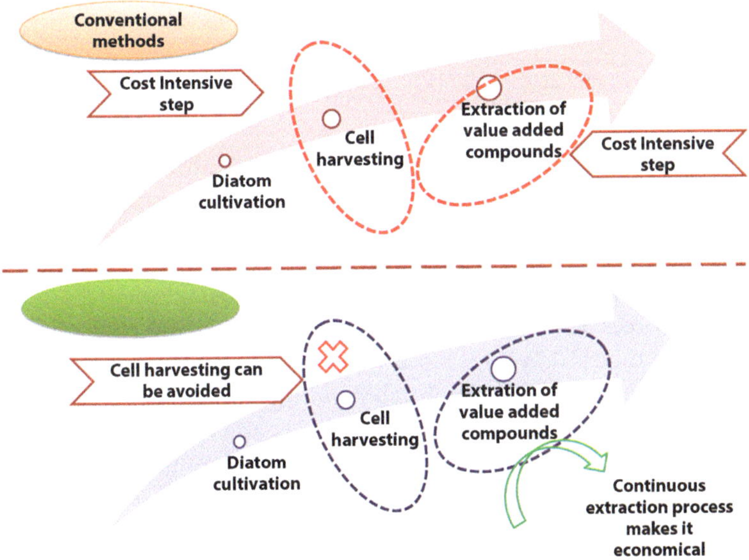

Figure 10.1 Schematic representation of conventional and milking approach in diatom biomass processing.

Therefore, diatoms can be an essential source of biofuels and play an important role in algal biorefinery establishments [10.23]. However, the major challenge is the cell wall disruption of the diatoms and the energy involved in the process [10.11]. Thus, this issue of the high energy investment process in cell disruption can be overcome by using innovative concepts like cell milking [10.24] [10.25]. The current chapter discusses different cell disruption methodologies available at present. The concept of milking has been discussed in detail and Figure 10.1 illustrates the use of the milking approach to address the cost-intensive steps in the cell harvesting and extraction of biomolecules. It sheds light on the economic perspective of the algal biorefinery approach and cell disruption processes. Lastly, it discusses the milking approach's challenges and future perspectives.

10.2 Cell Disruption Methods

The lipid productivity can be enhanced using an appropriate cell disruption method in conventional microalgae species cultivated for biofuel production. In diatoms, the nano-silica cell frustule has 750 μN mechanical strength, which limits the process of cell wall disruption [10.26] [10.27]. Cell wall disruption is necessary to extract valuable compounds from microalgae. However, the cost involved in the cell disruption processes overpowers the economics of extracted biomolecules. The following cell disruption methods have been used in the last two decades by various global researchers to address cell wall disruption problems:

10.2.1 Ultrasonication of Cells to Extract Value-Added Compounds

Acoustic cavitation can be produced by interacting ultrasound waves and bubbles in the aquatic environment [10.28]. A frequency of 20 Hz can generate shear forces with acoustic cavitation. The cavitation generated can cause cell rupture [10.29]. In a study, the extent of cell rupture by ultrasonication was measured through cell counting and optical measurements at 200 nm. It was shown that the extracellular polysaccharides (EPS) that limit the lipid extraction in *Navicula* sp. can be overcome by removing EPS using ultrasonication [10.30]. In another study performed using diatoms, 40% of lipid per dry weight was extracted using ultrasonication. The ultrasonication was performed for 20 minutes in an ultrasonic bath with 20 mL of n-hexane as an extraction solvent [10.31]. In a reported study, the ultrasonication process was compared to other methods of microalgal cell disruption. A Bandelin SONOPULS HD 2070 ultrasonic homogenizer was used in the study, and it was found that the ultrasonic treatment resulted in 9 mg FAME/L [10.32]. The extraction efficiency of ultrasonication was found to be better than the microwave process but lower than the thermal and hydrothermal processes of lipid extraction [10.32]. The ultrasonication processes can be used for biofuel industrialization; however, the economics of it is still challenging [10.33] [10.34]. Therefore, it is necessary to keep looking for more sustainable and economically suitable processes for Diafuel and biofuel commercialization. Although the energy consumption could be higher using ultrasonication, ultrasonication at an industrial scale is widely accepted in various industries. Therefore, parameter optimization to lower the energy consumption for cell rupture could be viable in algal biomass processing.

10.2.2 Microwave-Assisted Cell Wall Disruption Method

The cell disruption by microwave involves using non-ionizing electromagnetic waves with a frequency range of 300 MHz to 300 GHz [10.35]. The calorific energy is generated from electromagnetic waves and used for cell disruption [10.36] [10.37]. However, more technical knowledge is required in the field to understand the mechanism involved. The cell disruption also depends on the type of cell, its structure, and its size [10.37] [10.38]. One of the studies compared it to the ultrasonication pretreatment process for lipid extraction from microalgae. The results showed that the microwave-assisted extraction resulted in 2 mg FAME/L, and ultrasonicated cells released 3 mg FAME/L [10.32]. In another study, no significant difference ($p>0.05$) was observed between ultrasonicated cells ($24.6 \pm 1.3\%$) and microwave-assisted extraction ($24.2 \pm 0.9\%$) [10.35]. The use of microwave for cell disruption is limited to lab-scale production and is barely feasible for scale-up processes.

10.2.3 High-Pressure Homogenization

Solvent is needed to extract intracellular compounds from the cells by using their polarity and miscible properties. However, for the solvent to gain access to the cells, it must counter the cell wall. Therefore, cell disruption is essential in biomass processing to extract intracellular compounds. High-pressure homogenization (HPH) can increase the free fatty acids, but it could also have a negative impact on the oil quality [10.39]. The HPH treatment causes an increase in lipolysis; microalgal cell integrity is vital for studying the lipid stability during wet storage of the biomass after harvesting it [10.40]. A study used HPH to treat *Navicula* sp. cells after treatment with acids (HCL, pH = 1.5) [10.41]. A benchtop HPH was used at 20 °C temperature, and a pressure of 390 bar was used to treat cells; 25% of *Navicula* sp. cells were reported to be ruptured after thermal pretreatment along with HPH at 123 °C [10.41]. This study further suggests that even HPH requires pretreatments for better yields. The extraction efficiency can be increased using HPH by disrupting cell walls and making cells susceptible to solvents [10.39].

10.2.4 Chemical Methods

Traditionally, many toxic chemicals are used as solvents to extract value-added products from microalgae [10.42]. Similarly, some proposed algal biorefinery approaches used volatile organic compounds (VOC) [10.43]. The conventionally used VOCs can be replaced by ionic liquid solvents (ILs), considered a new class of green solvents. The ILs consists of inorganic anions and organic cations; they are a cheap source of non-flammable and highly stable solvent for extracting value-added compounds [10.43]. A research group uses ethyl-3-methylimidazolium dibutyl phosphate (EMIM DBP) to extract astaxanthin from *Haematococcus pluvialis* [10.44]. The extraction efficiency of ≥70% was obtained for astaxanthin extraction in the same study [10.44]. In another study, a water-miscible ionic liquid solvent was used to extract lipids from *Thalassiosira lundiana* CSIR-CSMCRI 001. The ILs used in the study were 1-ethyl-3-methylimidazoliumacetate ([EMIM][Ac]), and a total lipid yield of 91.1 mg/g DW of lipid was obtained [10.43]. Thus, as advancement and more research are taking place in this direction, the cells can be milked for value-added compounds with intact cells. However, the current method needs more research input to validate the process.

10.2.5 Pulsed Electric Field

Exposure of cells to pulsed electric fields (PEF) leads to the introduction of an additional transmembrane voltage (TMV) [10.45]. This additional potential can help in cell rupture, and the expression for this TMV can be obtained for a spherical cell using the Schwan equation in a spherical coordinate system [10.45]. In a study, electroporation was used to extract proteins from the diatom *Phaeodactylum tricornutum*, and a 20 kV/cm electric field was applied to treat the cells. In this study, the PEF showed the least effect on the extraction of biomolecules compared to ultrasonication and high voltage electric discharges (HVED) [10.46]. In another study, the PEF was demonstrated to be helpful in carbohydrate extraction in brown macroalga *Alaria esculenta*. The electric field strength of the pulses was 7.5 kV/cm with 1.2 Hz frequency and 720 pulses. The treatment using PEF resulted in the extraction of 21.4 ± 1.8 mg GE g^{-1} DW of carbohydrate [10.47]. Extracting biomolecules using PEF also implies stress on the cells during the initial exposure period. The exposure of cells to PEF can increase the production of desired biomolecules and their extraction [10.12]. In a recent study, PEF is also proposed as an alternative to conventional methods to break microalgae cell walls for milking pigments [10.12]. The parameters used in PEF for cell wall disruptions make it a reversible or irreversible cell disruption process [10.45]. The milking of biomolecules is only possible if the process is reversible. The cells can heal themselves after the extraction of desired biomolecules and can be used for another extraction cycle using a reversible PEF process.

10.2.6 Milking of Diatoms

Ramachandra and his group, along with Richard Gordon, proposed the milking of diatoms. The article they published together mentioned the role of diatoms in the biofuel industry. The same report also discusses the use of milking and its possibilities. Using genomically modified diatoms that can secrete high amounts of lipids on their own could be one of the possibilities [10.24]. They proposed a new solar bar approach to milk diatoms for lipids [10.24]. The advent of the milking method is a new dimension of research in the field of algal biorefinery; therefore, optimizing the processes crucial to milking should be validated. The concept of milking the cells revolves around the techniques which can lead to the extraction of value-added compounds (VAC) from the cells without cell sacrifice [10.11] [10.48]. In a recent article, the nanotechnology application is proposed to be used for milking diafuels from diatoms [10.11].

Similarly, the role of PEF has also been discussed in the milking concept recently [10.14]. The milking idea is an effective process of cell pore creation using various methods so that the cells are leaky and the intracellular compounds can leak out. The leaky cells can then be rehealed once the appropriate nutrition conditions are provided to the cells [10.11] [10.48] [10.25]. It is a relatively new technique, and process optimization requires more research input. However, it is one of the most sustainable processes proposed in the last two decades for the advent of the algal biorefinery approach.

10.3 Concept of Milking Cells for Value-Added Compounds

In 2004, Hejazi and Wijffels published an article where they reported the use of the milking process of microalgae [10.49]. This study proposed continuous production of value-added compounds from microalgae cultures. The study demonstrates the harvesting of β-carotene from *Dunaliella salina* using the milking approach [10.49]. In the milking method developed to extract β-carotene from *Dunaliella salina,* the cells were stressed to overexpress β-carotene. Once the cells reach this stage, the cells are exposed to a biocompatible organic solvent to extract β-carotene. Since the cells continue to grow and produce the β-carotene, a continuous extraction cycle can be maintained [10.50] [10.51]. The capacity of cells to perform a constant endo- and exocytosis mechanism can also be vital to the milking process [10.49]. The milking process is directly or indirectly related to the production pathway. The relationship between the production of biomolecule production and milking is crucial to understand the optimization of the milking process [10.51]. The biosynthesis of a biomolecule can be enhanced by a stress factor, thus modulating its production pathway. Therefore, it is necessary to understand the production pathway to implement the milking process.

Nanotechnology has been recently proposed as a medium of milking diatoms for a feasible biorefinery approach [10.11]. The use of a pulsed electric field has also been proposed in a similar manner for the milking of cells [10.14]. The significant difference between the two processes is the former works on the photothermal effect and the latter on the electric field strength. The goal is to disrupt the cell wall so that the cells are not sacrificed (Figure 10.2). It means that cellular leakage should be controlled, leading to the extraction of desired biomolecules. In recent reviews, the milking approach using nanoparticles was proposed for microalgae [10.11] [10.48]. The article suggests using wastewater as the growth media for microalgae, and then the cells grown were treated with biosynthesized metallic nanoparticles [10.11] [10.48].

10.3.1 Advancements in the Milking Approach

The silica cell wall of diatoms makes it challenging to break, and thus the cost involved in the cell disruption processes remains a limitation for diafuel production [10.52]. *Diadesmis confervaceae* is one of the diatom species existing in nature that oozes out oil naturally [10.53]. Identifying such strains from

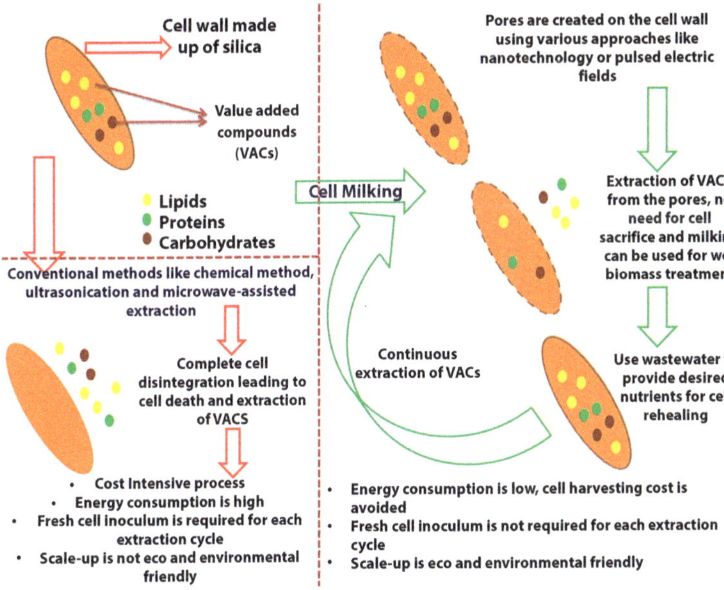

Figure 10.2 Schematic representation of the milking concept.

nature can aid in developing a milking approach for algal biorefinery [10.54]. Suppose the diatoms can be milked for the lipids or any other desired products without cell sacrifice. In that case, it has two significant advantages: 1) the cost of harvesting and drying the cells can be avoided, and 2) the continuous extraction of desired biomolecules can be achieved. In the past, lots of work proposing the use of technologies like PEF and nanotechnology has been published on the milking of algae [10.11] [10.14] [10.48]. In one of the reported studies in which *Nitzschia palea* was used as the dye-sensitized solar cell (DSSC) component, oozing of lipids in the supernatant was reported [10.12]. In a related work by Ramachandra *et al.*, the milking of diatoms coupled with solar panels using genomically modified diatoms was hypothesized [10.24]. OriginOil, a biofuel-based company based in Los Angeles, USA, patented technology for microalgae cell electrical modulation. The electrical modulation of cells proposed resulted in the enhancement of cell permeability. Thus, this technology can produce oil without cell sacrifice [10.55]. However, little work has been done in this field, especially in diatoms. Therefore, exploring the mechanism and hypothesis proposed in the previously reported results is necessary.

Both nanotechnology and PEF approaches can be implemented simultaneously for better yields. However, the PEF and nanotechnology approach has not been proposed or used anywhere. Furthermore, the PEF is a pretreatment method for extracting proteins and other value-added compounds [10.56]. Therefore, it is possible to use both processes in configuration, making the feasibility of the nano-biorefinery approach a reality. However, in this case, the nanotechnology approach of treating cells with nanoparticles will be used as a pretreatment step, and then the cells are subjected to PEF. This is because nanoparticle exposure can also induce stress in the cells. The stress-induction can further increase the synthesis of the value-added compounds, and the chances of them oozing out.

Further exposure to PEF under optimized conditions can cause cell disruption, and the value-added compounds can be extracted. Once the cells are extracted for one cycle, the cells can be given appropriate nutrients for their healing and growth, and a continuous extraction cycle can be obtained (Figure 10.3). The conjugation of PEF and nanoparticles is schematically shown in Figure 10.3. Since PEF is one of the green and sustainable methods to extract biomolecules from algae without using any chemicals, the combination of PEF with biosynthesized nanoparticles can be more beneficial.

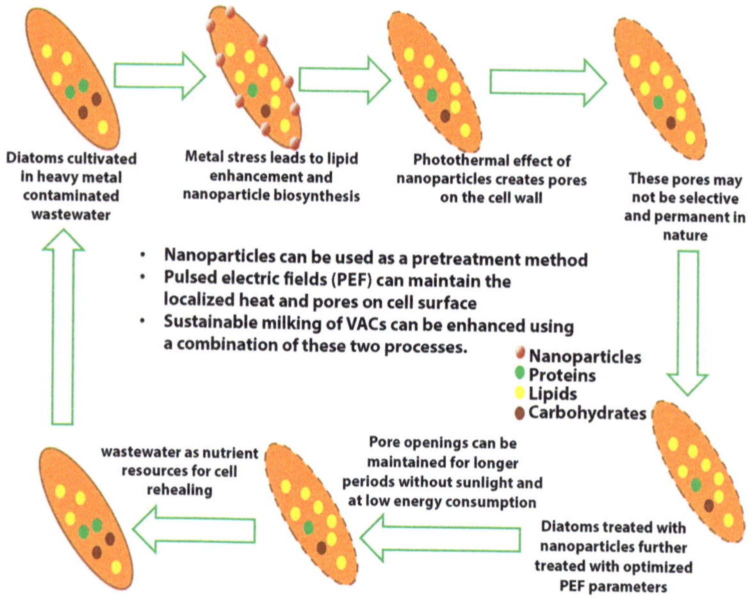

Figure 10.3 Schematic illustration of proposed cycle for sustainable extraction of VACs using milking approach channelized by nanoparticles and pulsed electric fields.

10.4 Economic Perspectives of Biofuels and Cell Disruption

The carbon-neutral prospect of biofuels has put them at the center of renewable and sustainable energy sources [10.57]. The usage of biofuels releases carbon fixed by the plant and microbes, making the net carbon release zero [10.57]. This carbon neutrality is a significant aspect of biofuel refinery, but it has been limited by the cost involved in biofuel production. Many companies (Shell, BP, and Chevron) have globally invested in microalgae to make it feasible and readily available [10.58]. Harvesting is the most cost-intensive step in biofuel production [10.59]. Once the cells have accumulated the lipids to be extracted, the cells should be separated from the aqueous bulk [10.60]. Since the size of the unicellular microalgae is small, their terminal velocity is relatively low. Therefore, mechanical harvesting is a limitation for small-sized algae [10.61]. Other processes like thermal drying and centrifugation are also energy inefficient and require lots of energy input [10.62]. Once the cells are harvested, the cell wall must be disrupted to extract lipids. Different methods were used in the past for cell wall disruption, as discussed in Section 10.2. The milking approach has been demonstrated in a few reported studies to be successful [10.50] [10.51]. The milking process saves the energy involved in harvesting and drying. Harvesting and drying are significant steps in algal biomass processing to extract biofuels. However, they are also the most cost-incurring steps in the algal biorefinery approach. This indicates that if we can avoid the actions of harvesting and drying and use wet biomass to extract value-added compounds, it would be more feasible from an economic point of view. The milking approach is based on the cell permeability being increased to the extent that the desired biomolecules can be oozed out. The oozing of lipids or other biomolecules does not require cells to be separated from the aqueous environment. However, the system should be able to put the cells in contact with a green solvent compatible with the extraction of lipids. The recirculation of the solvent is maintained to extract desired biomolecules [10.51].

The other method is to genetically modify a species to ooze out lipids on their own continuously [10.24]. However, conflicts between native and foreign species could be a challenge. Thus, the safest way would be to integrate wastewater treatment or metal recovery methods with the milking approach [10.11]. Therefore, more emphasis should be given to multiple product synthesis in a biorefinery approach. Developing simultaneous products from the same species would help keep the process's economics in check [10.63]. Strategies like harvesting; drying and cell disruption must be taken care of to develop an eco-friendly biorefinery system. The milking approach concept helps in avoiding the steps of harvesting and drying and the cell disruption is done to create pores in the algal cell wall. Thus, the energy requirements in the process are minimized, leading to the development of the biorefinery approach. Table 10.1 discusses various methods of cell disruption used and their energy consumption/conditions and the species used. However, it was clear from the literature there are not many reports on diatoms using techniques like PEF and ultrasonication, unlike most of the studies that use solvent extraction methods.

Table 10.1 Cell disruption methods with lipid yields of diatoms.

Species name	Technique	Energy consumed/ conditions	Lipid yield	Reference
Chaetoceros calcitrans	• Ultrasonication • Microwave assisted extraction	18.7 GJ Kg^{-1} 13.0 GJ Kg^{-1}	24.6 ± 1.3% 24.2 ± 0.9%	[10.35]
Pinnularia saprophilla	• PEF • Centrifugation	Electric field strength= 110kV/27mA Centrifugation RCF= 13000–15000 g	2.86% 3.39–3.49%	[10.64]
Phaeodactylum tricornutum	Microwave assisted extraction	2.45 GHz, 90 °C for 5 minutes	32% (w/w) glycerides	[10.65]
Phaeodactylum tricornutum	Subcritical n-butane	Extraction at 40 °C with 15 bar pressure	Omega 3 poly unsaturated fatty acids (PUFA)= 313 mg/g Omega 6 (PUFA)= 267 mg/g	[10.66]
Nitzchia palea	Resonating microfluidic chamber	Treatment time= 40 minutes with 220 KHz frequency, mask design 6 with 100 μm width micropillars	Lipids in cell supernatant= 4.35% Lipids in cell pellet= 13.14%	[10.67]
Diatoms/green microalgae	Nanoparticle assisted (Hypothesized)	To be optimized according to species cell wall structure	Not reported	[10.11]

10.5 Prospects and Challenges of the Milking Process

The milking process needs optimization in protocol design and the pre-treatment conditions. The working process reported till now usually uses an organic phase. However, the chemicals used as an organic phase could be a green solvent, but again it will raise the question of cost-effectiveness during process development. Therefore, avoiding using costly and environmentally unsustainable chemicals or using the solvent recovery approach is necessary. The other reported studies which reported the hypothesized process of milking need validation. The methodology discussed in earlier sections seems feasible and accessible, but various factors govern the process of using nanotechnology or PEF for milking diatoms. Diatoms exposed to biosynthesized nanoparticles in the presence of sunlight could experience the local heat on the surface. The regional heat experienced by the diatoms is due to the photothermal property of nanoparticles. The extent of the heat dissipation will decide if the pores are developed for the extraction of lipids or not.

Similarly, the milking process will be governed by parameters like electric field strength, the number of pulses, frequency, etc., used during the electroporation of cells through PEF. Therefore, it is crucial to estimate the dissipated heat in the nanotechnology approach and optimize the concentration of nanoparticles in use. Furthermore, in the PEF approach, validating the parameters for maximum lipid extraction from the cells is necessary. Thus, the hyphenated milking approach proposed in the current chapter should validate critical parameters in both PEF and nano-algal biorefinery approaches. The parameters mentioned above and their validations make it challenging to develop the milking process using PEF and nanotechnology. The studies done on nanoparticles as "milking molecules" are none to a few in the literature. Therefore, it is necessary to have more research inputs in this direction. More emphasis should be on combining the PEF and nanotechnology approaches to develop a sustainable milking process. The economics of the process is also a critical factor while designing the process protocol. However, integrated strategies like using wastewater/seawater to cultivate diatoms and utilizing the other value-added compounds recovered can solve this issue. Heavy metal bioremediation can also be incorporated into the process, as shown in Figure 10.2. Methodologies like simultaneous production of nanoparticles and lipids are an example of integrated approaches. More work is needed in the direction of multiple product biosynthesis to make the milking process more sustainable and cost-effective.

10.6 Conclusions

Milking of diatoms is possible using a combination of PEF and nanoparticles. The sustainability of the process depends on process optimization and multiple product biosynthesis. Integrating various techniques, like utilizing wastewater treatment and nanoparticle biosynthesis with bioremediation, is possible during the milking process. The proposed hypothesis can reduce the overall production cost of biofuels from diatoms. Therefore, it can be concluded from the discussion in the current chapter that the milking of lipids is feasible, and it needs validation by more research input in this direction. The cell disruption methods add to the cost of biofuel production, and milking can help reduce this cost. Furthermore, in order for the milking process to be used as a process more research input is required.

References

[10.1] Hildebrand, M., Davis, A. K., Smith, S. R., Traller, J. C., & Abbriano, R. (2012). The place of diatoms in the biofuels industry. *Biofuels*, 3(2), 221–240. https://doi.org/10.4155/BFS.11.157

[10.2] Saranya, G., & Ramachandra, T. V. (2020). Novel biocatalyst for optimal biodiesel production from diatoms. *Renewable Energy*, 153, 919–934. https://doi.org/10.1016/j.renene.2020.02.053

[10.3] Yiitolu, M., & Temoçin, Z. (2010). Immobilization of Candida rugosa lipase on glutaraldehyde-activated polyester fiber and its application for hydrolysis of some vegetable oils. *Journal of Molecular Catalysis B: Enzymatic*, 66(1–2), 130–135. https://doi.org/10.1016/j.molcatb.2010.04.007

[10.4] Tréguer, P., Nelson, D. M., Van Bennekom, A. J., Demaster, D. J., Leynaert, A., & Quéguiner, B. (1995). The silica balance in the world ocean: A reestimate. *Science*, 268(5209), 375–379. https://doi.org/10.1126/science.268.5209.375

[10.5] Field, C. B., Behrenfeld, M. J., Randerson, J. T., & Falkowski, P. (1998). Primary production of the biosphere: Integrating terrestrial and oceanic components. *Science*, 281(5374), 237–240. https://doi.org/10.1126/science.281.5374.237

[10.6] Gordon, R., Losic, D., Tiffany, M. A., Nagy, S. S., & Sterrenburg, F. A. S. (2009). The Glass Menagerie: diatoms for novel applications in nanotechnology. *Trends in Biotechnology*, 27(2), 116–127. https://doi.org/10.1016/j.tibtech.2008.11.003

[10.7] Mann, D. G., & Droop, S. J. M. (1996). *Biodiversity, biogeography and conservation of diatoms BT - Biogeography of Freshwater Algae: Proceedings of the Workshop on Biogeography of Freshwater Algae, held during the*

Fifth International Phycological Congress, Qingdao, China, June 1994 (J. Kristiansen (ed.); pp. 19–32). Springer Netherlands. https://doi.org/10.1007/978-94-017-0908-8_2

[10.8] Street-Perrott, F. A., & Barker, P. A. (2008). Biogenic silica: a neglected component of the coupled global continental biogeochemical cycles of carbon and silicon. *Earth Surface Processes and Landforms*, 33(9), 1436–1457. https://doi.org/https://doi.org/10.1002/esp.1712

[10.9] Werner, D. (1977). *Introduction with a note on taxonomy. The Biology of Diatoms, Hrsg.: D. Werner*. Blackwell Scientific Publications, Oxford.

[10.10] Round, F. E., Crawford, R. M., & Mann, D. G. (1990). *Diatoms: biology and morphology of the genera*. Cambridge university press.

[10.11] Ahirwar, A., Meignen, G., Khan, M. J., Khan, N., Rai, A., Schoefs, B., Marchand, J., Varjani, S., & Vinayak, V. (2021). Nanotechnological approaches to disrupt the rigid cell walled microalgae grown in wastewater for value-added biocompounds: commercial applications, challenges, and breakthrough. *Biomass Conversion and Biorefinery*, 0123456789. https://doi.org/10.1007/s13399-021-01965-1

[10.12] Gautam, S., Kashyap, M., Gupta, S., Kumar, V., Schoefs, B., Gordon, R., Jeffryes, C., Joshi, K. B., & Vinayak, V. (2016). Metabolic engineering of TiO2 nanoparticles in *Nitzschia palea* to form diatom nanotubes: An ingredient for solar cells to produce electricity and biofuel. *RSC Advances*, 6(99), 97276–97284. https://doi.org/10.1039/c6ra18487a

[10.13] Gupta, S., Kashyap, M., Kumar, V., Jain, P., Vinayak, V., & Joshi, K. B. (2018). Peptide mediated facile fabrication of silver nanoparticles over living diatom surface and its application. *Journal of Molecular Liquids*, 249, 600–608. https://doi.org/10.1016/j.molliq.2017.11.086

[10.14] Ahirwar, A., Khan, M. J., Sirotiya, V., Mourya, M., Rai, A., Schoefs, B., Marchand, J., Varjani, S., & Vinayak, V. (2022). Pulsed Electric Field–Assisted Cell Permeabilization of Microalgae (*Haematococcus pluvialis*) for Milking of Value-Added Compounds. *Bioenergy Research*, 0123456789. https://doi.org/10.1007/s12155-022-10414-4

[10.15] Gordon, R., Sterrenburg, F. A. S., & Sandhage, K. H. (2005). A special issue on diatom nanotechnology. *Journal of Nanoscience and Nanotechnology*, 5(1), 1–4. https://doi.org/10.1166/jnn.2005.017

[10.16] Kröger, N., & Poulsen, N. (2008). Diatoms - From cell wall biogenesis to nanotechnology. *Annual Review of Genetics*, *42*, 83–107.

[10.17] Bradbury, J. (2004). Nature's nanotechnologists: Unveiling the secrets of diatoms. *PLoS Biology*, 2(10). https://doi.org/10.1371/JOURNAL.PBIO.0020306

[10.18] Rabiee, N., Khatami, M., Jamalipour Soufi, G., Fatahi, Y., Iravani, S., & Varma, R. S. (2021). Diatoms with invaluable applications in nanotechnology, biotechnology, and biomedicine: Recent advances. *ACS Biomaterials Science and Engineering*, 7(7), 3053–3068. https://doi.org/10.1021/acsbiomaterials.1c00475

[10.19] Borase, H. P., Patil, C. D., & Suryawanshi, R. K. (2017). Mechanistic approach for fabrication of gold nanoparticles by *Nitzschia* diatom and their antibacterial activity. *Bioprocess and Biosystems Engineering*, 40(10), 1437–1446. https://doi.org/10.1007/s00449-017-1801-3

[10.20] Feurtet-Mazel, A., Mornet, S., Charron, L., Mesmer-Dudons, N., Maury-Brachet, R., & Baudrimont, M. (2016). Biosynthesis of gold nanoparticles by the living freshwater diatom *Eolimna minima*, a species developed in river biofilms. *Environmental Science and Pollution Research*, 23(5), 4334–4339. https://doi.org/10.1007/s11356-015-4139-x

[10.21] Jena, J., Pradhan, N., Dash, B. P., Panda, P. K., & Mishra, B. K. (2015). Pigment mediated biogenic synthesis of silver nanoparticles using diatom *Amphora* sp. and its antimicrobial activity. *Journal of Saudi Chemical Society*, 19(6), 661–666. https://doi.org/10.1016/j.jscs.2014.06.005

[10.22] Graham, J. M., Graham, L. E., Zulkifly, S. B., Pfleger, B. F., Hoover, S. W., & Yoshitani, J. (2012). Freshwater diatoms as a source of lipids for biofuels. *Journal of Industrial Microbiology and Biotechnology*, 39(3), 419–428. https://doi.org/10.1007/s10295-011-1041-5

[10.23] Levitan, O., Dinamarca, J., Hochman, G., & Falkowski, P. G. (2014). Diatoms: A fossil fuel of the future. *Trends in Biotechnology*, 32(3), 117–124. https://doi.org/10.1016/j.tibtech.2014.01.004

[10.24] Ramachandra, T. V., Mahapatra, D. M., Karthick, B., & Gordon, R. (2009). Milking diatoms for sustainable energy: Biochemical engineering versus gasoline-secreting diatom solar panels. *Industrial and Engineering Chemistry Research*, 48(19), 8769–8788. https://doi.org/10.1021/ie900044j

[10.25] Vinayak, V., Manoylov, K. M., Gateau, H., Blanckaert, V., Hérault, J., Pencréac'H, G., Marchand, J., Gordon, R., & Schoefs, B. (2015). Diatom milking? A review and new approaches. *Marine Drugs*, 13(5), 2629–2665. https://doi.org/10.3390/md13052629

[10.26] Hamm, C. E., Merkel, R., Springer, O., Jurkojc, P., Maier, C., Prechtel, K., & Smetacek, V. (2003). Architecture and material properties of diatom shells provide effective mechanical protection. *Nature*, 421(6925), 841–843. https://doi.org/10.1038/nature01416

[10.27] Khan, M. J., Bawra, N., Verma, A., Kumar, V., Pugazhendhi, A., Joshi, K. B., & Vinayak, V. (2021a). Cultivation of diatom *Pinnularia saprophila* for lipid production: A comparison of methods for harvesting the lipid from the cells. *Bioresource Technology*, 319(August 2020), 124129. https://doi.org/10.1016/j.biortech.2020.124129

[10.28] Li, W., Gamlath, C. J., Pathak, R., Martin, G. J. O., & Ashokkumar, M. (2019). Ultrasound–The Physical and Chemical Effects Integral to Food Processing, Ref. Modul. *Food Sci.*

[10.29] Yatipanthalawa, B. S., Ashokkumar, M., Scales, P. J., & Martin, G. J. O. (2022). Ultrasound-Assisted Extracellular Polymeric Substance Removal from the Diatom *Navicula* sp.: A Route to Functional

Polysaccharides and More Efficient Algal Biorefineries. *ACS Sustainable Chemistry and Engineering*, 10(5), 1795–1804. https://doi.org/10.1021/acssuschemeng.1c06290

[10.30] Yatipanthalawa, B., Li, W., Hill, D. R. A., Trifunovic, Z., Ashokkumar, M., Scales, P. J., & Martin, G. J. O. (2021). Interplay between interfacial behaviour, cell structure and shear enables biphasic lipid extraction from whole diatom cells (*Navicula* sp.). *Journal of Colloid and Interface Science*, 589, 65–76. https://doi.org/10.1016/j.jcis.2020.12.056

[10.31] Neto, A. M. P., Sotana de Souza, R. A., Leon-Nino, A. D., da Costa, J. D. ar. A., Tiburcio, R. S., Nunes, T. A., Sellare de Mello, T. C., Kanemoto, F. T., Saldanha-Corrêa, F. M. P., & Gianesella, S. M. F. (2013). Improvement in microalgae lipid extraction using a sonication-assisted method. *Renewable Energy*, 55, 525–531. https://doi.org/10.1016/j.renene.2013.01.019

[10.32] Passos, F., Carretero, J., & Ferrer, I. (2015). Comparing pretreatment methods for improving microalgae anaerobic digestion: Thermal, hydrothermal, microwave and ultrasound. *Chemical Engineering Journal*, 279, 667–672. https://doi.org/10.1016/j.cej.2015.05.065

[10.33] Halim, R., Danquah, M. K., & Webley, P. A. (2012). Extraction of oil from microalgae for biodiesel production: A review. *Biotechnology Advances*, 30(3), 709–732. https://doi.org/10.1016/J.BIOTECHADV.2012.01.001

[10.34] Pernet, F., & Tremblay, R. (2003). Effect of Ultrasonication and Grinding on the Determination of Lipid Class Content of Microalgae Harvested on Filters. *Lipids*, 38(11), 1191–1195. https://doi.org/10.1007/s11745-003-1178-6

[10.35] Nogueira, D. A., Da Silveira, J. M., Vidal, É. M., Ribeiro, N. T., & Veiga Burkert, C. A. (2018). Cell Disruption of *Chaetoceros calcitrans* by Microwave and Ultrasound in Lipid Extraction. *International Journal of Chemical Engineering*, 2018. https://doi.org/10.1155/2018/9508723

[10.36] Cravotto, G., Boffa, L., Mantegna, S., Perego, P., Avogadro, M., & Cintas, P. (2008). Improved extraction of vegetable oils under high-intensity ultrasound and/or microwaves. *Ultrasonics Sonochemistry*, 15(5), 898–902. https://doi.org/10.1016/J.ULTSONCH.2007.10.009

[10.37] Lee, J., Yoo, C., Jun, S., Ahn, C., & Oh, H. (2010). Bioresource Technology Comparison of several methods for effective lipid extraction from microalgae. *Bioresource Technology*, 101(1), S75–S77. https://doi.org/10.1016/j.biortech.2009.03.058

[10.38] Ma, Y. A., Cheng, Y. M., Huang, J. W., Jen, J. F., Huang, Y. S., & Yu, C. C. (2014). Effects of ultrasonic and microwave pretreatments on lipid extraction of microalgae. *Bioprocess and Biosystems Engineering*, 37(8), 1543–1549. https://doi.org/10.1007/s00449-014-1126-4

[10.39] Balduyck, L., Bruneel, C., Goiris, K., Dejonghe, C., & Foubert, I. (2018). Influence of High Pressure Homogenization on Free Fatty Acid Formation in *Nannochloropsis* sp. *European Journal of Lipid Science and Technology*, 120(4), 1–6. https://doi.org/10.1002/ejlt.201700436

[10.40] Balduyck, L., Stock, T., Bijttebier, S., Bruneel, C., Jacobs, G., Voorspoels, S., Muylaert, K., & Foubert, I. (2017). Integrity of the microalgal cell plays a major role in the lipolytic stability during wet storage. *Algal Research*, 25(August 2016), 516–524. https://doi.org/10.1016/j.algal.2017.06.013

[10.41] Spiden, E. M., Scales, P. J., Yap, B. H. J., Kentish, S. E., Hill, D. R. A., & Martin, G. J. O. (2015). The effects of acidic and thermal pretreatment on the mechanical rupture of two industrially relevant microalgae: *Chlorella* sp. and *Navicula* sp. *Algal Research*, 7, 5–10. https://doi.org/10.1016/j.algal.2014.11.006

[10.42] Miazek, K., Kratky, L., Sulc, R., Jirout, T., Aguedo, M., Richel, A., & Goffin, D. (2017). Effect of organic solvents on microalgae growth, metabolism and industrial bioproduct extraction: A review. *International Journal of Molecular Sciences*, 18(7). https://doi.org/10.3390/ijms18071429

[10.43] Krishna, P. M., Polisetti, V., Damarla, K., Mandal, S. K., & Kumar, A. (2021). Improved biorefinery pathways of marine diatoms using a water miscible ionic liquid and its colloidal solution: efficient lipid extraction and *in situ* synthesis of fluorescent carbon dots for bio-imaging applications. *RSC Advances*, 11(35), 21207–21215. https://doi.org/10.1039/d1ra01425k

[10.44] Desai, R. K., Streefland, M., Wijffels, R. H., & Eppink, M. H. M. (2016). Novel astaxanthin extraction from *Haematococcus pluvialis* using cell permeabilising ionic liquids. *Green Chemistry*, 18(5), 1261–1267. https://doi.org/10.1039/c5gc01301a

[10.45] Golberg, A., Sack, M., Teissie, J., Pataro, G., Pliquett, U., Saulis, G., Stefan, T., Miklavcic, D., Vorobiev, E., & Frey, W. (2016). Energy-efficient biomass processing with pulsed electric fields for bioeconomy and sustainable development. *Biotechnology for Biofuels*, 9(1), 1–22. https://doi.org/10.1186/s13068-016-0508-z

[10.46] Zhang, R., Lebovka, N., Marchal, L., Vorobiev, E., & Grimi, N. (2020). Comparison of aqueous extraction assisted by pulsed electric energy and ultrasonication: Efficiencies for different microalgal species. *Algal Research*, 47(March). https://doi.org/10.1016/j.algal.2020.101857

[10.47] Einarsdóttir, R., Þórarinsdóttir, K. A., Aðalbjörnsson, B. V., Guðmundsson, M., Marteinsdóttir, G., & Kristbergsson, K. (2022). Extraction of bioactive compounds from *Alaria esculenta* with pulsed electric field. *Journal of Applied Phycology*, 34(1), 597–608. https://doi.org/10.1007/s10811-021-02624-8

[10.48] Kashyap, M., & Kiran, B. (2021). Milking microalgae in conjugation with nano-biorefinery approach utilizing wastewater. *Journal of Environmental Management*, 293(February), 112864. https://doi.org/10.1016/j.jenvman.2021.112864

[10.49] Hejazi, M. Amin, & Wijffels, R. H. (2004). Milking of microalgae. *Trends in Biotechnology*, 22(4), 189–194. https://doi.org/10.1016/j.tibtech.2004.02.009

[10.50] Hejazi, M. A., De Lamarliere, C., Rocha, J. M. S., Vermuë, M., Tramper, J., & Wijffels, R. H. (2002). Selective extraction of carotenoids from the microalga *Dunaliella salina* with retention of viability. *Biotechnology and Bioengineering*, 79(1), 29–36. https://doi.org/10.1002/bit.10270

[10.51] Hejazi, M. A., Holwerda, E., & Wijffels, R. H. (2004). Milking Microalga *Dunaliella salina* for β-Carotene Production in Two-Phase Bioreactors. *Biotechnology and Bioengineering*, 85(5), 475–481. https://doi.org/10.1002/bit.10914

[10.52] Vinayak, V., Khan, M. J., Varjani, S., Saratale, G. D., Saratale, R. G., & Bhatia, S. K. (2021). Microbial fuel cells for remediation of environmental pollutants and value addition: Special focus on coupling diatom microbial fuel cells with photocatalytic and photoelectric fuel cells. *Journal of Biotechnology*, 338, 5–19. https://doi.org/10.1016/J.JBIOTEC.2021.07.003

[10.53] Vinayak, V., Gordon, R., Gautam, S., & Rai, A. (2014). Discovery of a Diatom That Oozes Oil. *Advanced Science Letters*, 20(7), 1256–1267. https://doi.org/10.1166/asl.2014.5591

[10.54] Rai, A., Khan, M. J., Ahirwar, A., Deka, R., Singh, N., Schoefs, B., Marchand, J., Varjani, S., & Vinayak, V. (2022). Hydrogen economy and storage by nanoporous microalgae diatom: Special emphasis on designing photobioreactors. *International Journal of Hydrogen Energy*, 47, 100, 42099–42121. https://doi.org/10.1016/j.ijhydene.2022.01.057

[10.55] Malcata, F. X. (2011). Microalgae and biofuels: A promising partnership? *Trends in Biotechnology*, 29(11), 542–549. https://doi.org/10.1016/j.tibtech.2011.05.005

[10.56] Postma, P. R., Pataro, G., Capitoli, M., Barbosa, M. J., Wijffels, R. H., Eppink, M. H. M., Olivieri, G., & Ferrari, G. (2016). Selective extraction of intracellular components from the microalga *Chlorella vulgaris* by combined pulsed electric field-temperature treatment. *Bioresource Technology*, 203, 80–88. https://doi.org/10.1016/j.biortech.2015.12.012

[10.57] Yadugiri, A. V. T. (2009). *Milking diatoms – a new route to sustainable energy Published by : Current Science Association Milking diatoms - a new route to sustainable energy*. 97(6), 748–750.

[10.58] Mascarelli, A. L. (2009). Gold rush for algae. *Nature*, 461(7263), 460–461. https://doi.org/10.1038/461460a

[10.59] Chaudry, S., Bahri, P. A., & Moheimani, N. R. (2016). Selection of an Energetically More Feasible Route for Hydrocarbon Extraction from Microalgae – Milking of *B. braunii* as a Case Study. In *Computer Aided Chemical Engineering* (Vol. 38). Elsevier Masson SAS. https://doi.org/10.1016/B978-0-444-63428-3.50262-9

[10.60] Amaro, H. M., Guedes, A. C., & Malcata, F. X. (2011). Advances and perspectives in using microalgae to produce biodiesel. *Applied Energy*, 88(10), 3402–3410. https://doi.org/10.1016/J.APENERGY.2010.12.014

[10.61] Brennan, L., & Owende, P. (2010). Biofuels from microalgae—A review of technologies for production, processing, and extractions of biofuels and co-products. *Renewable and Sustainable Energy Reviews*, 14(2), 557–577. https://doi.org/10.1016/J.RSER.2009.10.009

[10.62] Danquah, M. K., Ang, L., Uduman, N., Moheimani, N., & Forde, G. M. (2009). Dewatering of microalgal culture for biodiesel production: Exploring polymer flocculation and tangential flow filtration. *Journal of Chemical Technology and Biotechnology*, 84(7), 1078–1083. https://doi.org/10.1002/jctb.2137

[10.63] Kashyap, M., Samadhiya, K., Ghosh, A., Anand, V., Lee, H., Sawamoto, N., Ogura, A., Ohshita, Y., Shirage, P. M., & Bala, K. (2021). Synthesis, characterization and application of intracellular Ag/AgCl nanohybrids biosynthesized in *Scenedesmus* sp. as neutral lipid inducer and antibacterial agent. *Environmental Research*, 201(February), 111499. https://doi.org/10.1016/j.envres.2021.111499

[10.64] Khan, M. J., Bawra, N., Verma, A., Kumar, V., Pugazhendhi, A., Joshi, K. B., & Vinayak, V. (2021b). Cultivation of diatom *Pinnularia saprophila* for lipid production: A comparison of methods for harvesting the lipid from the cells. *Bioresource Technology*, 319(September 2020), 124129. https://doi.org/10.1016/j.biortech.2020.124129

[10.65] Lohman, E. J., Gardner, R. D., Halverson, L., Macur, R. E., Peyton, B. M., & Gerlach, R. (2013). An efficient and scalable extraction and quantification method for algal derived biofuel. *Journal of Microbiological Methods*, 94(3), 235–244. https://doi.org/10.1016/j.mimet.2013.06.007

[10.66] Feller, R., Matos, Â. P., Mazzutti, S., Moecke, E. H. S., Tres, M. V., Derner, R. B., Oliveira, J. V., & Junior, A. F. (2018). Polyunsaturated ω-3 and ω-6 fatty acids, total carotenoids and antioxidant activity of three marine microalgae extracts obtained by supercritical CO2 and subcritical n-butane. *The Journal of Supercritical Fluids*, 133, 437–443. https://doi.org/10.1016/J.SUPFLU.2017.11.015

[10.67] Vinayak, V., Kumar, V., Kashyap, M., Joshi, K. B., Gordon, R., & Schoefs, B. (2017). Fabrication of resonating microfluidic chamber for biofuel production in diatoms (Resonating device for biofuel production). *2016 3rd International Conference on Emerging Electronics, ICEE* 2016, 4–9. https://doi.org/10.1109/ICEmElec.2016.8074628

11

Dissection of Gene Expression Pattern and Metabolic Profile Under Enhanced Oil Production Conditions in Diatoms

Geetanjali Kumawat[1], Pallavi Vyas[1], Sandhya Deora[1], Sneha Sabu[1], Amit Kumar Gupta[1], Mukesh Meena[1], Ashwani Kumar[2], Vandana Vinayak[3] and Harish[1]*

[1]*Department of Botany, Mohanlal Sukhadia University, Udaipur, Rajasthan, India*
[2]*Department of Botany, University of Allahabad (A Central University), Prayagraj, (UP) India*
[3]*Diatom Nanoengineering and Metabolism Laboratory (DNM), School of Applied Sciences, Dr. Harisingh Gour Central University, Sagar, Madhya Pradesh, India*

Abstract

In this era of high energy demand, all we need is a sustainable and renewable biofuel or biodiesel source. As we all know, third-generation biofuel (algae-based) is preferable to the first- (crop-based) and second-generation (Petro-crop based) biofuels. However, if we increase lipid formation in algae by limiting nutrients like nitrogen and phosphorus, it will directly reduce the amount of obtained biomass. To address this issue, metabolomic techniques have been created in which we directly target the biosynthesis pathway or the genes implicated in it. Upregulation and downregulation of these genes, such as *ACCase* (acetyl-CoA carboxylase), *MAT* (malonyl-CoA:ACP-transacylase), *GPAT* (glycerol phosphate acyltransferase), *LPAAT* (lysophosphatidic acid acyltransferase), *DGAT* (diacylglycerol acyltransferase), *TE* (thioesterase), etc., regulate the lipid biosynthetic pathway. Accumulation of lipid, whether in the form of total lipid or TAG, which provides carbon for conversion to a variety of types of fuel, is a key goal for the development of biofuels. The genetic and metabolic engineering approaches for enhanced oil production are highlighted in this chapter.

**Corresponding author*: harish@mlsu.ac.in

Vandana Vinayak and Richard Gordon (eds.) Diatom Cultivation for Biofuel, Food and High-Value Products, (267–322)

***Keywords*:** Diatoms, lipid, ACCase, DGAT, DAG, stress, gene expression, rDNA technology

Abbreviations

AAD	Acyl ACP desaturase
AAT	Acyl ACP thioesterase
ACCase	Acetyl CoA carboxylase
ACP	Acyl carrier protein
ACS	Acetyl-CoA synthase
AFDW	Ash free dry weight
AGP	ADP-glucose pyrophosphorylase
ASP	Aquatic Species Program
BC	Biotin carboxylase
CO_2	Carbon dioxide
CRISPR	Clustered regularly interspaced short palindromic repeats
CT	Carboxyl transferase domain
CV	Chrysolaminarin vacuole
DAG	Diacylglycerol
DCW	Dry cell weight
DECL	Declining phase
DGAT	Diglyceride acyltransferase
DHA	Docosahexaenoic acid
EMP	Embden–Meyerhof–Parnas
EPA	Eicosapentaenoic acid
EXP	Exponential phase
FAME	Fatty acid methyl esters
FAS	Fatty acid synthase
GL	Glycerol lipids
GPAT	Glycerol phosphate acyl transferase
GR	Glutathione reductase
GSH	Glutathione
GST	Glutathione S-transferase
GPX	Glutathione Peroxidase
H_2O_2	Hydrogen peroxide

KOH	Potassium hydroxide
LPA	Lysophosphatidate
LPAAT	Lysophosphatidic acid acyltransferases
MAT	Malonyl CoA ACP transacylase
miRNA	MicroRNA
MUFA	Monounsaturated fatty acid
NaCl	Sodium chloride
NADPH	Nicotinamide adenine dinucleotide phosphate
NGS	Next generation sequencing
PA	Phosphatidate
PAP	Phosphatidic acid phosphatase
POME	Palm oil mill effluent
PPM	Parts per million
SFA	Saturated fatty acid
Si NP	Silicon nanoparticle
SiO_2 NP	Silicon dioxide nanoparticle
siRNA	Small Interfering RNA
STAT	Stationary phase
TAG	Triacylglycerol
TALENs	Transcription activator-like effector nucleases
TE	Thioesterase
TFA	Trifluoroacetic acid
ZFN	Zinc finger nuclease

11.1 Introduction

In order to compete with the limited supply of fossil fuels, we must work to find alternative energy sources. Biofuels (e.g., bioalcohol, biodiesel, biogas, and biohydrogen) are considered a good alternative, sustainable source of energy because they may be produced from biological sources such as plants, bacteria, animals, and trash [11.13] [11.59] [11.77]. As a blend of fatty-acid alkyl esters, biodiesel is a liquid biofuel that can replace existing fossil fuels with sustainable energy. Due to its low toxicity, environmental friendliness (reduced emissions of CO_2, CO, hydrocarbons, and organic particle matter), ability to mix with other fuels, and ease of storage due to its high flash point, biodiesel is recommended [11.4] [11.15] [11.201].

11.1.1 Why Algae Over Other Sources?

Microalgae are a special class of single organisms with the potential to address a range of transportation industry requirements through a variety of options [11.108] [11.120]. From the lipid bodies in their cells, all microalgae have the ability to produce energy-dense bio-oils [11.23] [11.24] [11.108]. Compared to terrestrial sources of biomass used for biofuels, microalgae have a smaller impact on the environment [11.23] [11.24] [11.108].

11.1.2 Basic Cell Structure of Diatoms

There are about 8,000 different species that have been identified as thriving in lakes and oceans all over the world, but according to several researchers [11.10] [11.52] [11.88] [11.137], the number of species that are still alive is thought to be between 20,000 and 200,000 [11.52]. They are one of the most significant primary producers in the oceans and often play a significant role in the global oxygen, carbon, and silica cycles in aquatic environments [11.68] [11.69] [11.79] [11.213].

Diatoms are thought to be one of the primary producers of crude oil [11.9] [11.55] [11.82] [11.229] and first appeared on our planet approximately 150 million years ago [11.117] [11.210] [11.229]. Diatoms are single-celled eukaryotic creatures that are distinguished by their surrounding silica cell wall and yellow-brown coloration [11.18] [11.114]. It is believed that silicon cell walls contribute to the development and production of diatoms [11.114] [11.258]. The silica-based cell wall of diatoms, commonly known as the frustule, is its most noticeable characteristic. Their nanostructured silica cell wall, which is duplicated faithfully over generations by genetically controlled assembly processes, is their most distinguishing cellular feature. A diatom frustule is made up of two valves (a larger epivalve and a smaller hypovalve) that are joined by a girdle made of components that resemble bands. It is believed that a diatom's maximum lipid potential is species-specific and heritable [11.69] [11.79].

Diatoms acquired chloroplasts through secondary endosymbiosis; as a result, new diatom research has linked chloroplasts to that of red algae. Diatoms have chloroplasts that contain chlorophylls a, c1, and c2 [11.114] [11.126]. Diatom plastids have four membranes, two of which are part of the organelle itself and the other two of which are the plastid endoplasmic reticulum. Diatoms are typically split into two subgroups: centric diatoms, which are radially symmetric (Centrales), and pennate diatoms, which are bilaterally symmetric (Pennales) [11.114] [11.214]. It is possible for these photosynthetic autotrophs to coexist with other photosynthetic organisms

[11.114] [11.126]. Unicellular photosynthetic organisms with distinctive silica cell walls (frustules) are named diatoms (Bacillariophyta).

Diatoms are typically between some few and a few hundred microns in size, with the bulk of species falling between 10 and 50 μm (micro meter). Diatoms are part of the Stramenopile or heterokont class of algae, which are very distinct from green algae and vascular plants and have a more complex evolutionary history [11.11] [11.94]. According to calculations of productivity per unit of agricultural carbon [11.94] [11.224], diatoms are also better at fixing carbon than other coexisting types of microalgae. Due to their increased capacity for storing nutrients, cells with larger vacuoles may take longer than cells with smaller ones to reach nutrient limitation in terms of lipid productivity. In terms of storing carbohydrates, diatoms are distinct from other kinds of algae. Diatoms store fixed carbon as chrysolaminarin, a complex carbohydrate that has an 11:1 ratio of β(1-3) and β(1-6) linked glucose units instead of starch [11.19] [11.94]. Because it is soluble, chrysolaminarin is preserved in the chrysolaminarin vacuole (CV). The CV is fairly substantial [11.42] [11.94], and the imaging studies conducted in association with Robyn Roth at Washington University (MO, USA) show that it can make up as much as 50% of the cell volume. Diatoms can be found in a wide range of habitats, such as general habitats like marine, freshwater, and soil, as well as more specialized severe or variable habitats for salinity, temperature (such as ice or hot springs), or pH. Diatoms are well acclimated to turbulence and in contrast to other groups of marine microalgae, are more prolific in high turbulence environments [11.94] [11.141]. Although some studies suggest that diatoms are particularly sensitive to shear during cultivation, there appears to be some apparent discrepancy in the literature regarding this subject [11.33] [11.48] [11.94]. In diatoms, certain haptophytes, brown algae, and other organisms, fucoxanthin is a significant pigment that performs as a light-harvesting pigment by interacting with the fucoxanthin-chlorophyll binding protein [11.106] [11.229]. Diatoms can form complex clusters in the wild known as "spring blooms" [11.109] [11.153] [11.229], which satellites can spot [11.67] [11.229]. Diatoms are eukaryotic, photosynthetic microalgae that are not only present in groups other than the Bacillariophyta family, with which we are primarily concerned [11.187] [11.232].

11.1.3 Why Diatoms?

Diatoms are the main producers in both freshwater and seawater because they can fix up to 20% of the carbon dioxide in the atmosphere [11.110] [11.114]. Carbon dioxide is necessary for the growth and metabolism of diatoms. Although the production of storage compounds also needs carbon. They include significant levels of sterols, lipids, carotenoids, isoprenoids, and other compounds, and some of these substances have been linked to toxins and metabolites [11.114] [11.214]. Diatoms produce lipids to keep them buoyant because lipids are lighter than their other chemical components and allow them to exist suspended in the photic zone [11.114] [11.214]. In diatoms, lipids serve as the principal form of storage [11.1] [11.114].

Diatoms are secondary endosymbionts that belong to the heterokont group, along with other silica-forming algae, including brown algae, bolidophytes, chrysophytes, heliozoans, silicoflagellates, and synurophytes. In comparison to another silica-forming heterokont, nonmineralized bolidophytes, a genus of biflagellated unicells, are closely related to diatoms [11.87] [11.227]. Diatoms have a unique metabolic profile thanks to this special characteristic that sets them apart from other algae [11.12] [11.227].

The most dominating characteristic of diatoms, which is absent in dinoflagellates and coccolithophores, is their huge storage vacuole [11.198] [11.227]. At a pace of two to four divisions per day, diatoms can outcompete other algae of the same size [11.155] [11.227]. They are dominant over other eukaryotic algae when subjected to strong turbulence and mixing circumstances, which makes them an ideal model for mass culture under various mixing conditions [11.227].

Diatoms serve as conduits in the biogeochemical cycles of carbon, silicon and contribute 40% of the ocean's total primary productivity [11.22] [11.139]. Diatoms have been carefully designed and optimized for boosting lipid productivity as well as usage as a suitable biodiesel feedstock because of their high lipid content and generation of a wide variety of TAGs [11.22] [11.72] [11.139]. Growth and culture conditions, nutrient deficiency, light intensity, and temperature fluctuations all have a significant impact on lipid accumulation and partitioning into fatty acids [11.22] [11.41]. This further shifts the carbon flux, which reduces chrysolaminarin, a storage starch of diatoms, and increases production of TAG [11.22] [11.94].

To date, more than 15,000 phytochemicals, such as carotenoids, fatty acids, sterols, antioxidants, etc., have been produced and identified from different classes of microalgae. However, only a few species are currently being investigated and produced industrially [11.194] [11.255]. Surprisingly, diatoms, the primary producers in an aquatic food chain, are the least explored aquatic species that need the utmost attention [11.194] [11.228]. Researchers are focusing on the replacement of dirty sources of energy as a result of rising fuel costs brought on by the depletion of renewable fossil fuel resources and growing atmospheric CO_2 generation

[11.148] [11.150] [11.162] [11.172] [11.235]. Microalgae are regarded as a plentiful source of biofuel among the options for switching to cleaner energy [11.143] [11.148] [11.240].

Due to their distinctive distribution of C14, C16, and C20 fatty acids compared to other green algae and land plants, diatoms are widely regarded as viable feedstocks for the generation of biofuels [11.169] [11.175]. Several international initiatives on the production of biofuel from diatoms have been undertaken, including the biofuel and biochemical potential of the marine diatom *Amphora coffeaeformis* in open raceways [11.174] [11.175], quality evaluation of biodiesel from marine diatom *Navicula cincta* [11.169] [11.175], and exploration of the biodiesel prospects of the diatom *Halamphora coffeaeformis* [11.142] [11.175]. Diatoms have been raised for aquaculture purposes for many years. They are well known for their value-added products, accepted omics, and genetic manipulation techniques, with enormous potential in the nutraceutical and biofuel industries. However, the true potential of these "jewels of the sea" is still untapped, which was the main driving force behind researchers around the world [11.94] [11.227].

Comparatively, *C. gracilis* had a considerably higher lipid content than *T. weissflogii* under adaptive culture techniques. Under various culture conditions, the lipid content (percent DCW) in *Chaetoceros* sp. and *Thalassiosira* sp., both belonging to the same group, was 22–47 and 29–52% respectively [11.22] [11.195]. Additionally, total lipid content reported in several species, which ranges from 20 to 69% DCW, supports the idea that diatoms are the dominant source of high lipid production under various culture conditions. These species include *Amphora exigua, Fistulifera solaris, Navicula lyra, Phaeodactylum tricornutum, Skeletonema costatum,* and *Chaetoceros* [11.194] [11.196].

The successful evolution of diatoms on our planet can be attributed to their silica cell walls [11.178] [11.227]. They are the least studied species currently, although having significant potential in several fields of algae biotechnology. Therefore, an effort has been made in this chapter to investigate the potential of diatoms for algal biotechnology. It is crucial to research how the lipid content and profile of the researched species connect with growth under various environmental circumstances and nutritional loads in order to maximize the lipid yield. Additionally, excessive fertilizer use in agriculture encourages dead zones and unwelcome algae growth. As a result, fertilizers and clean water are not required for the cultivation of microalgae. Diatom cultivation in wastewater serves as a medium for feedstock production in addition to the cleaning of polluted water, but this aspect of using diatoms for phycoremediation of polluted water was not studied until recently. Growing diatoms in wastewater utilizes leftover nutrients and transfer oxygen, thus acting as a photosynthetic catalyst to promote the growth of aerobic bacteria uniformly in the deposited layer of eutrophic water bodies. Wastewater cultivation of diatoms as a growth medium has been much sought-after as a solution to these problems. The development of high-value products and the growth of diatom algae on wastewater reduce not only the need for clean water and fertilizer but also turn "waste to money." Diatoms have a significant role in the enormous biomass production and reduction of greenhouse emissions in the ocean [11.213] [11.227]. They are also fixing 40% of the annual output of oceanic biomass and nearly 20% of the photosynthetic CO_2 that is produced, which is comparable to the photosynthetic activity of all rainforests combined [11.37] [11.177] [11.227].

11.1.4 Percentage Lipid Extraction as Per Dry Cell Weight

The marine diatom *Chaetoceros muelleri,* for instance, can store lipids up to more than 80% (dry mass basis) and has a maximum growth rate of four doublings per day [11.36] [11.144], while the green alga *Botryococcus braunii* can acquire 75% lipid per dry weight [11.36] [11.43]; although maximum lipid levels within species can range from less than 15% to more than 80% for both diatom and green algae [11.36] [11.98].

Assimilated carbon is stored in the cells of many diatomaceous species as triacylglycerol (TAG), which takes about 25–45% of the dry weight of the cells [11.43] [11.52] [11.229] and 40–78% of the weight of the silica shell [11.209] [11.229]. Only a few of the 20,000–2,000,000 species of diatoms that have been researched as potential biofuel producers [11.10] [11.87] [11.137] [11.229] include *Phaeodactylum tricornutum, Thalassiosira pseudonana, Thalassiosira weissflogii, Cyclotella cryptica* [11.52], *Fistulifera solaris* [11.220], *Chaetoceros gracilis, C. muelleri* [11.2] [11.144] [11.229], *Cylindrotheca* and *Nitzschia* [11.43] [11.229].

Thalassiosira sp. had the highest total lipid content as assessed by gravimetric analysis as DCW (dry cell weight) of the collected biomass (52%), followed by *Skeletonema* sp. (44%) and *Chaetoceros* sp. (22% DW of extract). According to Marella and Tiwari, the lipid percent DCW of *T. weissflogii* under both the modified (30.1%) and standard f/2 Si (19.5%) medium was lower than that found in this investigation [11.22] [11.139]. The earlier study on indoor mass cultivation found that *Thalassiosira* sp. had the second-highest total lipid percent as dry weight at 29.66%, followed by *Chaetoceros* sp. at 44.33% [11.194] [11.193]; and 10–20% of the total lipid was found in a study on *Thalassiosira* sp. bulk culture [11.194] [11.237].

Similar to this, the lipid percent DW in *Chaetoceros* sp. and *Skeletonema* sp. was higher than that in *S. costatum* reported by Gao, Renaud and colleagues, 17 to 24.4% and 17% and 13.3%, respectively [11.22] [11.72] [11.179]. According to the fatty acid profile of *S. costatum*, which was recorded under control in a previous study [11.22] [11.72], *Skeletonema* sp. had the highest concentration of neutral lipid specific fatty acids with 64.45% SFA and 17.61% MUFA, followed by *Thalassiosira* sp. with 52.73% SFA and 21.48% MUFA, and *Chaetoceros* sp. with *Thalassiosira* sp. had the highest lipid percentage as dry cell weight (52%) followed by *Skeletonema* sp. (44%) and *Chaetoceros* sp. (22%). They have an inherent capacity to dominate, even under low light and nutrition conditions. This makes them highly advantageous over other classes of microalgae [11.139] [11.194].

11.1.5 Aquatic Species Programme (ASP)

Out of more than 3000 candidates, 50 microalgal strains were chosen by the Aquatic Species Program in 1998 as having the greatest potential for producing biofuels [11.94]. Diatoms made up 60% of the chosen strains, which were chosen in accordance with standards including rapid growth and lipid yields, resistance to adverse environmental conditions, and effectiveness in large-scale cultures. Algae have a number of benefits as a feedstock for biodiesel, including better photosynthetic efficiency (12.6%), effective CO_2 sequestration, the ability to bioremediate contaminated streams and non-arable areas [11.25] [11.191] [11.261], as well as higher oil content and algal biomass production. Triglycerides, a high-quality and abundant raw material, are the form in which algae store fats and lipids. When stressed, most oleaginous microalgae acquire up to 50% more total lipids than their dry cell weight (which is typically 20%) [11.43] [11.191]. The contribution of the Aquatic Species Program (ASP) is of utmost significance when considering diatoms as a source of lipids (oil) for the synthesis of biofuels. With careful separation of approximately 3000 microalgal species, particularly from challenging environmental circumstances like saline and brackish waters, and subsequent screening of algae for species with higher lipid content, it was one of the pioneering efforts in algal biotechnology. The outcomes of this program offered insightful information on the physiology and biochemistry of algae, systems biology, and concepts for mass-scale algae production using open ponds, as well as a timeline of research activities from the 1980s to the 1990s, outdoor testing and system analyses, and cost-benefit analysis. Out of the total number of isolated algae, the close-out report on ASP found that 50 species, 60% of which were diatoms, were better lipid accumulators. These figures best capture the significance of diatoms as a substantial source of lipids for biodiesel generation, among other microalgae [11.96] [11.122] [11.175] [11.176]. Diatoms are the fastest-growing eukaryotic unicellular microalgae in the world, accounting for around one-fifth of all photosynthesis [11.139] [11.194]. Other significant elements crucial to the development and acclimation of diatoms are ideal temperature, pH, light intensity, and adaptable growing techniques [11.194] [11.193]. Identifying the most promising naturally occurring fuel for the synthesis of biodiesel was the focus of the Aquatic Species Program. This program demonstrated that, under specific development conditions, diatoms can synthesize up to 60% of their cellular mass as TAG [11.114] [11.259]. But not all species of microalgae accumulate lipid equally, even within a taxa [11.36] [11.83] [11.98].

The goal of the Aquatic Species Program of the US Department of Energy has been to identify the optimum species for the production of biofuel [11.204] [11.242]. Thousands of species were investigated in the lab and put to the test in the field as part of the project, which cost millions of dollars. It was an initiative that resulted in a great deal of extremely helpful scientific data, and some breeds are currently being researched [11.242]. They have taken over as the dominant primary producers in the ocean and are thought to be responsible for 20% of the world's carbon fixation [11.69] [11.94] [11.230].

11.2 Generalized Pathway for Lipid Biosynthesis in Diatoms

One distinction in the glycolysis pathway is that *T. pseudonana* and *P. tricornutum* do not encode a gene for glucokinase [11.94] [11.118]. More unexpectedly, the lower half of the glycolysis process, from triose phosphate isomerase to pyruvate kinase to the mitochondria, is located there [11.94] [11.118]. According to research, diatoms have the unusual capacity to complete process glyceraldehyde-3-phosphate into pyruvate inside of the mitochondria. It's unknown what this incomplete pathway does in the mitochondrion, however it might be used to direct resources in the direction of the citric acid cycle. A glycolysis process that is localized in the mitochondria could also eliminate the requirement for the malate-aspartate shuttle to import the reducing equivalents required for the production of ATP via oxidative phosphorylation. Saturated FAs are more desirable because they raise the fuel's ignition quality (cetane number), and shorter chain lengths are generally preferred to improve the cold-flow features of biofuels [11.94] [11.116] [11.173].

Lipid production in algae follows a biosynthetic pathway that is almost identical to that of plants (Fig. 11.1). Plastids carry out the first two stages of synthesis, and the smooth endoplasmic reticulum handles the last stage. Acetyl-CoA Carboxylase (ACCase), which is present in plastids and cytoplasm and is in charge of the first step of fatty acid biosynthesis, is essential [11.125] [11.148]. The primary enzyme that catalyzes the production of malonyl-CoA is thought to be ACCase. Malonyl-CoA is a precursor for fatty acid elongation and secondary metabolites in the cytosol and catalyzed *de novo* fatty acid biosynthesis in plastids [11.64] [11.91] [11.148]. Fatty acid synthase (FAS) generates fatty acids as a result of the condensation of malonyl-ACP (acyl carrier protein) and acetyl-ACP. After that, separate acyl-ACP thioesterase (AAT) terminates the extension of fatty acids by hydrolyzing acyl-ACP in free fatty acids (FFAs) [11.148] [11.199]. Acetyl-CoA synthase (ACS) binds to fatty acids and cofactor coenzyme A (CoA)

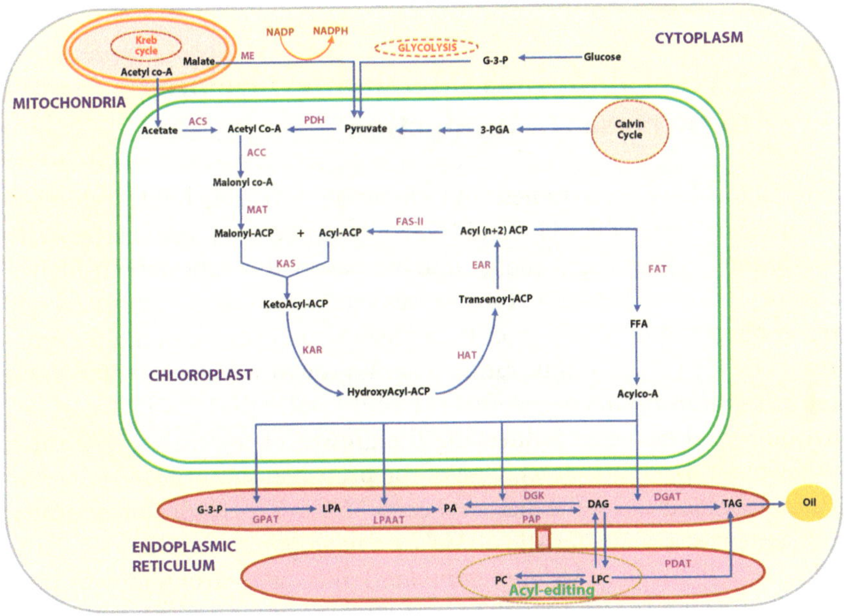

Figure 11.1 Schematic representation of lipid biosynthesis pathway in diatoms: ME- Malic enzyme, ACS- Acetyl-CoA synthase, PDH- Pyruvate dehydrogenase complex, ACC- Acetyl-CoA carboxylase, MAT- Malonyl acyl transferase, KAS- Keto acyl reductase, KAR- Ketoacyl-ACP-reductase, FAS-II- Fatty acid synthase, HAT- 3-hydroxyacyl-ACP reductase, EAR- Enoyl-ACP reductase, FAT- fatty acyl transferase, GPAT- Glycerol-3-phosphate acyltransferase, LPAAT- Lysophosphatidic acid acyltransferase, DGK- Diacylglycerol kinase, PAP- Phosphatidate phosphatase, DGAT- Diacylglycerol acyltransferase, PDAT- Phospholipid:diacylglycerol acyltransferase.

after leaving the plastids to create acyl-CoA [11.138] [11.148]. All cytosolic fatty acids are transferred by acyl-CoA to the smooth endoplasmic reticulum, where they are further lengthened and desaturated [11.112] [11.148] [11.151]. With the aid of the enzyme glycerol-3-phosphate acyltransferase (GPAT), acyl-CoA transfers one fatty acid to the Sn-1 position of glycerol-3-phosphate and creates lysophosphatidate (LPA). Lysophosphatidate acyltransferase (LPAAT) adds another fatty acid to the Sn-2 position in the synthesis of phosphatidate (PA), and phosphatidic acid phosphatase (PAP) adds another fatty acid to the Sn-3 position in the formation of diacylglycerol (DAG) from phosphatidate (PA). Diacylglycerol acyltransferase, which forms TAG from DAG, is the final step (DGAT). Diatoms also possess the Embden-Meyerhof-Parnas (EMP) route, which necessitates carbon transport both within and outside of the plastid membrane [11.75] [11.148] [11.238] and the chloroplast [11.148] [11.211] [11.222]. As a result, the mitochondria also operate a second EMP route of glycolysis, which generates pyruvate [11.118] [11.128] [11.148].

11.3 Stress Conditions (Metabolites) Helping to Increase Oil Production

Many factors have been studied with a focus on increasing lipid production in diatom species (Tables 11.1 and 11.2). Compared to *Chlorella vulgaris, P. tricornutum* converts light energy into biomass with a substantially higher efficiency [11.115] [11.227]. Thus, it was established that diatoms exhibit greater photosynthetic efficiency than green algae, even in low light conditions [11.145] [11.227]. Integrating diatom growth with wastewater is the best method for producing biofuel [11.140] [11.226] [11.227]. One of the environmental elements influencing the growth of microalgae and their lipid content is salinity stress. One of the probable candidates for the raw material for biodiesel is the diatom species *Melosira* sp. High lipid and fatty acid content is a characteristic of *Melosira moniliformis* [11.133] [11.216]. *Aulacoseira granulata* (*Melosira granulata*) also has a strong growth rate, high lipid content, and a quick rate of deposition, all of which improve the harvesting process [11.65] [11.216]. Lipid content can be impacted by environmental factors and microalgae growing media [11.164] [11.216]. It is further noted that these environmental parameters also include salinity level, nitrogen content in the growing medium, temperature, light intensity, lighting time, and nutrition [11.133] [11.188] [11.216].

The *Phaeodactylum tricornutum* study found that in nitrate-starved circumstances, carbon fluxes into all major sinks drastically decline, with

Table 11.1 Impact on lipid level in different diatomic species using various stress factors.

S. no.	Algal species	Stress condition	Impact	References
1.	*Chaetoceros gracilis*	Silicon and NaCl restricted media	Lipid level increase up to 73%	[11.114]
2.	*Fragilaria capucina*	Nitrogen and silicon-starved condition	Lipid % increased up to 32.8	[11.36]
3.	*Phaeodactylum tricornutum*	0.25 m H_2O_2 with nitrogen depletion	Amount of lipids increase up to 92%	[11.30]
4.	*Haslea ostrearia*	Cu and Zn stress	Results in induced lipid bodies	[11.76] [11.113] [11.163]
5.	*Asterionella glacialis*	Hg and Cd stress	Increased in lower fatty acids and sterol levels	[11.73] [11.76]
6.	*Phaeodactylum tricornutum*	Nitrogen and phosphorus starvation	Lipid content increase up to 70%	[11.32] [11.44] [11.234]
7.	*Chlorella* (TISIR 8990)	Nitrogen minimal media with 475.5 mg/l KNO3	Highest lipid content found 31.14%	[11.258]
8.	*P. tricornutum*	Nitrogen-starved f/2 medium 0.5 mM NaNO3	Results in increased lipid content	[11.251] [11.258]
9.	*Chaetoceros gracilis*	1% w/v sea salt in medium	TAG production is increased	[11.229]

(*Continued*)

Table 11.1 Impact on lipid level in different diatomic species using various stress factors. (*Continued*)

S. no.	Algal species	Stress condition	Impact	References
10.	*C. gracilis*	Urea as a nitrogen source	TAG production and cell growth both increased	[11.229]
11.	*Cylindrotheca fusiformis*	Si deficient condition	Lipid accumulation increased up to 57%	[11.94] [11.207]
12.	*Cyclotella cryptica*	Si-deprived condition	Lipid content increased up to 17-20%	[11.94] [11.207]
13.	*Amphiprora* sp.	Silicon dioxide (SiO_2) NP	Greater lipid content (0.42%) than standard medium (0.15%)	[11.108] [11.232]
14.	*Chaetoceros muelleri*	Nitrogen-starved condition	Total lipid content increased up to 7-fold	[11.144] [11.148]
15.	*Thalassiosira pseudonana*	Nitrogen and silica-starved condition	Natural lipid content was increased	[11.72] [11.148]
16.	*Skeletonema marinoi*	Phosphate stress	Increased TAG production	[11.148] [11.260]
17.	*Navicula pelliculosa*	Silicate restriction	Lipid content increased up to 36%	[11.56] [11.148] [11.189]
18.	*Cyclotella cryptica*	Silicon insufficiency	Lipid content increased up to 20–28%	[11.148] [11.185]

(*Continued*)

Table 11.1 Impact on lipid level in different diatomic species using various stress factors. (*Continued*)

S. no.	Algal species	Stress condition	Impact	References
19.	*Diadesmis gallica* (CCALA 765, CCALA 766) *Navicula atomus* (CCALA 383)	Si NP 2mM	Increased lipid accumulation	[11.148]
20.	*Amphiprora*	2% SiNP	Lipid percentage increased up to 52.94	[11.108] [11.148]
21.	*Chaetoceros* sp.	Temperature 25 °C	Lipid content increased from 12–16.8%	[11.148] [11.182]
22.	*Phaeodactylum tricornutum*	Temperature 25–10 °C	EPA and PUFA increased by 120%	[11.119] [11.148]
23.	*Nitzschia closterium*	Light intensity 300 mol photons/m2/s	TAG accumulation increased up to 9.2%	[11.148] [11.154]

proteins experiencing the largest decline and lipids experiencing the smallest. Equilibrium between growth rate and lipid content is the main barrier to industrial biodiesel production at prices competitive with fossil fuels [11.85] [11.112]. It has been investigated how temperature affects the fatty acid composition of *P. tricornutum* and *Chlamydomonas muelleri*. In addition to these fatty acids, *C. muelleri* also possessed the fatty acids 18:1(n9) t and 20:4(n6). The fatty acids 14:0, 16:0, 16:1, 16:2, 16:3, 18:0, 18:1(n9) c, 18:2(n6), and 20:5(n3) were found in *P. tricornutum*. *C. muelleri* was more affected than *P. tricornutum* by a 2-hour temperature increase over the optimal development temperature in terms of fatty acid content and degree of saturation. However, protein assay analysis revealed that *P. tricornutum*, not *C. muelleri*, was under stress, even though both species' lipid levels were impacted [11.85] [11.114]. Several investigations have shown that diatoms are capable of accumulating fat even in nutrient-restricted environments. The ability of *Thalassiosira weissflogii*, *Phaeodactylum tricornutum*, and *C. muelleri* to accelerate the build-up of lipids following nitrogen deprivation has been demonstrated [11.114]. *Chaetoceros gracilis* can store lipids up to 73% when silicon and sodium chloride are restricted [11.114]. According to Bozarth *et al.*, oleaginous microalgae store carbon as lipid [11.26] [11.36], and lipid accumulation typically rises when nutritional conditions are less favorable than ideal, such as when there is insufficient nitrogen (N) [11.36] [11.83] or silica (Si) for diatoms [11.36] [11.38] [11.49] [11.144]. The maximum lipid concentration of *F. capucina* was 32.8%, while reducing N and Si caused a spike in lipid accumulation. The maximum lipid content was found at 10 mol L^{-1} Si and 100 or 10 mol L^{-1} N concentrations at 10 and 20 °C; 10 mol/L Nitrogen and 10 mol/L Silicon prevented lipid build-up at 30 °C. It is noteworthy that, generally speaking, the fat content rose faster as Si was reduced than as Nitrogen was reduced [11.36].

Abiotic stress is known to increase the lipid synthesis of microalgae [11.30] [11.98]. Abiotic stressors frequently have a detrimental impact on microalgal growth rates and reduce biomass productivity [11.30] [11.160] [11.254], but if lipid synthesis increases overall after adjusting for biomass loss, generating stress may be a useful method. This proves that adding H_2O_2 (0.25 mM) to *P. tricornutum* cultures during their early stationary phase causes the amount of neutral lipids in the separated biomass to grow by up to 156%. The amount of neutral lipids increased by 92%, from 0.29 mg/mg to 0.56 mg/mg, when nitrogen depletion and H_2O_2 therapy were combined. The neutral lipid content of *P. tricornutum* was improved by peroxide treatment at levels of 5% and 10% nitrogen supplementation, but not at levels of 25% nitrogen supplementation, according to further tests examining levels of nitrogen-limitation. When treated with H_2O_2, the total amount of neutral lipids harvested was enhanced by up to 50%, from 64 mg/L to 96 mg/L, indicating its potential efficacy as a pre-harvest technique to improve the *P. tricornutum*'s potential as a feedstock for biofuels. Analysis of many stressors shows that hydrogen peroxide therapy enhances neutral lipid synthesis under nitrogen-depleted settings but not under phosphorus-depleted conditions, indicating that the effects of hydrogen peroxide on lipid synthesis are regulated by ambient nitrogen levels [11.30].

Additionally, when stress conditions like low nutrients, high temperatures, low silica, high or low light, etc., are applied to diatoms, the concentration of their oil is said to increase. As a result, lipid metabolism can involve everything from lipid storage to the creation of intracellular plasma membranes [11.76] [11.194]. An examination of the prior literature revealed that just one diatom, *Diadesmis confervacea*, produced the most oil among all diatoms; nonetheless, the apparent oil content was relatively low [11.76] [11.239]. Diatoms naturally ooze oil due to their high lipid content, which eliminates the costly stage in energy production where oil must be extracted from other diatoms. The diatom *Stephanodiscus minutulus* cell size decreased noticeably when nutrients were few, and N-limited cells were shown to be substantially smaller than Si or P-limited cells [11.76] [11.135].

A soft metal with strong peroxidation potential for sulfhydryl-ligand affinities is cadmium (Cd). With the exception of oceanic diatoms, Cd is therefore typically thought to be a hazardous element. Certain diatom species undergo numerous metabolic pathway reprogramming as a result of nitrogen shortage, such as a decrease in primary carbon metabolism and an increase in lipid content [11.76] [11.247]. These findings show that the core carbon metabolism response of other eukaryotic photoautotrophs to nitrogen deprivation differs dramatically from *Thalassiosira pseudonana* nitrogen starvation and is comparable to cyanobacterial reactions. Another proteomic test comprehensively investigated molecular alterations in *Phaeodactylum tricornutum* in response to neutral lipid buildup during nitrogen shortage. The findings demonstrated that nitrogen deprivation led to a genetic environment that was more regulated for nitrogen and fatty acid biosynthetic pathway, followed by less regulated genes in the process of photosynthesis and lipid catabolism, changing the diatom metabolic network from a carbon flux to a lipid accumulation [11.76] [11.96]. According to reports, the concentration of oil rises when diatoms are subjected to stress factors, including high levels of heavy metals (Pb and Se) [11.73] [11.76], lack of nutrients, high temperature, lack of silica, high or low level of light, etc. In comparison to *Nitzschia* sp., *Diadesmis confervacea* showed the highest lipid content (14.6% of fatty acid methyl esters, or FAME), 6% of the total. In terms of nutrient stress, lipid bodies are more prevalent, although it is unknown what would happen under high metal stress. The diatom *Haslea ostrearia* was observed to have larger lipid contents under Cu stress in contrast to the treatments in laboratory studies [11.76] [11.113] [11.163], which reported induction of lipid bodies recognized under Cu and Zn stress in the cultured population of the phytoplankton in the laboratory. Lower fatty acid and sterol levels were observed in Hg and Cd stress in the coastal diatom *Asterionella glacialis* [11.73] [11.76].

One of the candidates in this strategy is algae oil because it has a great potential and superior benefits to other fuels made from energy crops [11.43] [11.44]. Multiple researchers have noted that under various nutritional stresses, such as nitrogen or phosphate deprivation, oleaginous cells, such as algae (*P. tricornutum*), can store up to 70% lipids in their cells. Lack of a nitrogen source impacts not only cell development but also other processes [11.32] [11.44] [11.234]. At 504 hours of cultivation, the maximum lipid concentration and algal lipid productivity were 38.08, 2.91 mg/L and

0.08, 0.01 mg/L hr, respectively. These values were comparable to those achieved without providing $NaNO_3$ (36.23 2.68 mg/L and 0.07 0.00 mg/L hr). Previously, when *Chlorella* sp. TISTR 8990 was grown photoautotrophically in nitrogen-minimal media, compared to the control condition, the nitrogen concentration at 475.5 mg/L KNO_3 produced the largest lipid content (31.14%) [11.258]. When *P. tricornutum* (CCMP2561) was grown in f/2 medium with 0.5 mM $NaNO_3$, the high lipid accumulation in this study exhibited comparable tendencies. As acetate is digested to form an intracellular carbon through the glyoxylate cycle in such a stress situation, storage lipids (TAGs) are observed to accumulate; consequently, the Kennedy route is used to produce TAGs from intermediate metabolites of carbohydrates [11.251] [11.258].

11.3.1 Salt Stress

Different seawater dilutions were used to grow *Chaetoceros gracilis* cells. Typically, 4% (w/v) concentrations of sea salt were employed. Growth rate, cell density during the fourth and tenth days of cultivation, and TAG content of *C. gracilis* were unaffected when sea salt concentrations were decreased to 2 or 3% (w/v). Growth rate was significantly slower (DT, 12.4 hr) when *C. gracilis* was cultured in a medium containing 1% sea salt, but TAG production was noticeably higher. Growth rate was reduced and TAG production decreased to less than half the previous rates when the NaCl concentration was raised from 3 to 4%. On the other hand, as previously stated, as NaCl concentrations were below 3% in *C. gracilis,* the growth properties were improved. Even so, the marine diatom *C. gracilis* is known to be adaptable to a broad range of salinities, flourishing well and depositing significant levels of TAG at nearly all of the salinity conditions tested in this work (0–3% NaCl) [11.229]. Outstanding results have been achieved by optimizing the growth of multiple diatom species (including *P. tricornutum, Navicula saprophila, Nitzschia,* etc.) under various culture conditions (heterotrophy and mixotrophy) [11.194] [11.238]. When both nitrogen and salinity are present, algae maintain osmotic equilibrium with their culture media. However, when all of the nitrogen is used up, the stress from the salinity causes the intracellular components to slow down, which stimulates the synthesis of lipids, in line with the results of the prior study [11.194] [11.205]. Even so, the presence of high salt is unrelated to the process underlying high lipid levels brought on by nitrogen deprivation; rather, high salinity contributed to an improvement in lipid purity [11.6] [11.194]. Studies on *C. vulgaris* have previously shown that the growth declines as salt rises. According to Church and his colleagues, regulating the salinity period leads to an increase in total lipid productivity; however, in the present investigation, no influence on growth was seen despite salinity [11.47].

11.3.2 Urea as a Nitrogen Source

In F/2-enriched seawater, *C. gracilis* developed significantly more quickly with urea than nitrate (DTs, 8.9 and 10.3 hr, respectively), and urea also produced more TAG and had higher cell density. Growth and TAG buildup were modestly inhibited when the medium also contained 70 ppm urea and nitrate. Growth was accelerated but TAG generation was decreased when 70 ppm urea was added to artificial saltwater (Marine Art SF-1) mixed with Daigo IMK medium [11.229].

11.3.3 Nutrient Stress

It was shown that *C. gracilis* also produced large amounts of TAG with no nitrogen or silica deprivation, which is routinely applied to many other microalgae to stimulate lipid formation [11.2] [11.229] [11.256].

Under light irradiation without CO_2 enrichment, the time it took for the cell population to double was about 10 hours. Triacylglycerols (TAGs) and fucoxanthin managed to accumulate in significant amounts under a variety of salinity and nutrient conditions, reaching concentrations of about 200 and 18.5 mg/L, respectively [11.229].

Diatoms naturally bloom in nutrient-rich environments, which results in the maximum growth rate to density possible and is perfect for just a biofuel production system. For diatoms under N limitation, Shifrin and Chisholm published values of 12–45% of dry cell weight (DCW) for total lipids [11.94] [11.207]. An analysis of the literature of 55 micro-algal species from different classes revealed that diatoms as a class performed above average, with total lipid content under nutrient-rich conditions reaching 27.4% (19% better than average) and under N-deficient conditions reaching 36.1% (13% better than average) [11.83] [11.94]. The average lipid content for diatoms restricted for Si in N deficient conditions was 41%, which was 28% higher than all species and 13% higher when compared to other species [11.83] [11.94].

In terms of neutral lipid (TAG), it was observed that *Cylindrotheca fusiformis* contained 57% in Si-deficient cultures and 17–20% in Si-replete cultures for *C. cryptica* in Si-deficient cultures, compared to 32% in Si-replete cultures. A study of the Si-induced lipid accumulation in *C. cryptica* shows that the total lipid based on DCW was 42% [11.94] [11.207], while the total lipid based on AFDW was 54.1% [11.94] [11.186]. Protein content is often lower in N-deprived microalgae; for *P. tricornutum*, N depletion resulted in a decline in RuBisCO abundance, particularly the small subunit [11.74] [11.94]. Acetyl-CoA carboxylase activity, the first committed step in FA biosynthesis, must increase by *de novo* protein synthesis [11.185] [11.259]. Under N limitation, a diatom and *Dunaliella tertiolecta* showed increased general protease activity [11.21] [11.83] [11.94].

11.3.4 Light Stress

Chlorella vulgaris and *P. tricornutum* both had comparable rates of light conversion into biomass under nonfluctuating light settings; however, under varying light conditions, *P. tricornutum* had a rate nearly twice as high as *C. vulgaris* [11.94] [11.241]. During one investigation, TFA increased in *Nitzschia closterium* from 7.6 to 12.6% of dry weight while TAGs grew from 1.7 to 9.2% when light conditions were 30 and 300 mol photons/m^2/s at 17 °C and 1% CO_2 [11.148] [11.154]. Furthermore, it was discovered through research on the diatom *Thalassiosira pseudonana* that the total lipid content of TAG increased from 7 to 10% in log phase to 47% during stationary phase at a light intensity of 100 mol photons/m^2/s [11.29] [11.148].

11.3.5 Nanoparticle Stress

Since silica makes up the cell wall of diatoms, two different types of media were created, with the SiO_2-NP medium receiving the majority of attention. The growth rate and biomass of the microalga were increased by the use of nanosized particles [11.108]. The microalga *Amphiprora* sp. accurately represents the overall lipid content. It demonstrated that the modified medium containing silicon dioxide (SiO_2) nanomaterial had a greater lipid content than the standard F2 medium for *Amphiprora* sp. As the first investigation into employing the nanomaterial as a culture medium for microalgae, this approach is regarded as unique. Additionally, it is well known that the lipid content of dry microalgae is strongly influenced by its biomass [11.108] [11.232]. The microalga *Amphiprora* sp. has had its nutrient optimization done in such a way that it was determined that the medium containing nanomaterials would increase the lipid content of the microalga. The maximum dry cell weights were noted on the 29th day for all the light intensities. The dry cell weight declined even though the light intensity was increased to 30 mol/m2/s, making 24 mol/m2/s the best suited light intensity out of all those measured. This light intensity showed cell dry weight of 0.2567 0.003 g/L. This resulted from the photoinhibition problem the microalgal cells had. Previous research revealed that photoinhibition happened when the cells were exposed to the most excess excitation energy in the visible range. By altering the photosystem II activity center, this could cause damage to the microalgal cells [11.78] [11.108] [11.202]. According to a study, the total lipid yield percent was calculated to be 3.5%. The microalga *Amphiprora* sp. produced some differences in total lipid levels, when it was cultivated under five different light intensities. The maximum total lipid content was 52.94% at a light intensity of 24 mol/m^2/s, and the lowest level was 6 mol/m^2/s. The highest light intensity, 30 mol/m^2/s, exhibited a lower total lipid content than 24 mol/m^2/s1. This only depended on the photoinhibition factor, which constrained the capacity of *Amphiprora* sp. to produce biomass. For instance, the nutritional stress was induced by changing the microalga (diatom) growing medium and adding nanomaterials. Due to the easy availability of phytohormones (auxins and cytokinins) in their metabolic pathways, microalgae are able to thrive in harsh environments. Despite the harsh environment, these hormones can cause a continuing growth influence on the secondary metabolites produced by the microalgae, including lipids, carotenoids, proteins, and carbohydrates [11.108] [11.184]. It was observed that the maximum methyl esters of 1.25% were created when employing 24 mol/m^2/s of light intensity for microalgal cultivation, while the lowest methyl ester value of 1.14% was recorded during 6 mol/m^2/s of light intensity. The maximum light intensity, however, showed methyl ester concentrations below 24 mol/m^2/s [11.108]. It was found that at concentrations of silica nanoparticles (SiNPs) below 2 mM, all three strains of the diatoms *Diadesmis gallica* (CCALA 765 and CCALA 766) and *Navicula atomus* (CCALA 383) showed an increase in lipid accumulation [11.148]. *Amphiprora* sp. cultivated in modified f/2 media (52.94%) responded significantly to the influence of 2% SiNP [11.108] [11.148]. It demonstrates that employing potassium hydroxide (KOH) as a catalyst and microalgal oil from *Amphiprora* sp. produced outstanding results (75.25%). When the catalyst concentration was 2.5%, it was somewhat greater than the methyl ester content (71.51%), and when the concentration was 1.5%, it was slightly lower (62.85%) [11.108]).

11.3.6 Nitrogen Stress

When there is a nitrogen shortage, the nitrogen taken in through catabolic processes during the urea cycle mechanism aids in recycling intracellular nitrogen, which ultimately serves as a precursor for the production of biofuel [11.5] [11.93] [11.148]. After nitrogen was starved from *Chaetoceros muelleri*, the total lipid content increased seven-fold [11.144] [11.148].

Under nitrogen depletion, the lipid content of *Thalassiosira pseudonana* increased from 13 to 20% of its dry weight [11.111] [11.148]. When *T. pseudonana* cells were starved of nitrogen and silicate, according to research by Hildebrand and colleagues [11.94] [11.148], the natural lipid content increased [11.72] [11.148]. In addition to nitrogen depletion, diatoms showed an increase in ammonium transporters [11.99] [11.148] [11.166]. In a recent study on *Chaetoceros muelleri*, the reaction of the cells lipid and cell density machinery was examined at various nitrogen concentrations (0.18 mM, 0.44 mM, 1.76 mM, 3.53 mM) [11.32] [11.148] [11.256]. When the cells were taken on the second day of either of these nitrogen depletions, it was discovered that *C. muelleri* did not exhibit any lipid analysis methods in cell density. Without losing biomass, the cells grown in 0.18 mM nitrogen-depleted media displayed lipids and carbohydrates that were, respectively, 1.3 and 1.6 times greater than those in the control [11.58] [11.148].

11.3.7 Phosphorus Stress

According to dry weight, *Amphora subtropica* has a 19% lipid content; however, when under phosphate stress, that percentage rises to 53% [11.46] [11.148]. It has been discovered that in *Skeletonema marinoi*, phosphate stress causes a greater response in TAG pathway-related genes for lipid synthesis than nitrogen stress does [11.148] [11.260]. Phosphorus restriction aids in raising the total lipid content of the diatoms *Phaeodactylum tricornutum* and *Chaetoceros* sp. from 8 to 19%, particularly total fatty acids up to 46% [11.99] [11.148] [11.182]. Importantly, the lipid buildup rises in phosphate-limited culture media in numerous old cultures of *Pavlova viridis*, *Chaetoceros* sp., and *P. tricornutum* [11.34] [11.147] [11.148]. In a two-way culture, it was also observed that the brackish oleaginous diatom *Navicula phyllepta* MACC8 increased lipid production. During the second stage of the silica stress, this further increased to 28.78% of weight on day 18 [11.148] [11.189]. Additionally, the diatom displayed a notable rise in lipid to 32.13% of weight when the media, which was devoid of silica and phosphate, was applied [11.148].

11.3.8 Silicon Stress

Research from the past and the present demonstrates that the silicate restriction on the diatom *Navicula pelliculosa* raised the lipid content by 36% to reach 34% of dried lipid content [11.56] [11.148] [11.189]. Similar findings have been supported by earlier research, such as that done by Taguchi and colleagues, who found that on silicate fasting, lipid content increased in *Chaetoceros gracilis* (46% lipid by dry weight), *Cyclotella* sp. (61% lipid by dry weight), and *Hantzschia* sp (54% lipid of dry weight) [11.148] [11.218]. After 12 hours of silicon insufficiency, the lipid content of *Cyclotella cryptica* increased from 20 to 28% of its dry weight [11.148] [11.186].

11.3.9 Temperature Stress

In the *Chaetoceros* sp., total lipid as a percentage of dry weight increased from 12 to 16.8% as temperature decreased from 35 to 25 °C [11.148] [11.183]. In a study by Jiang and Gao, EPA and PUFA increased by 120% in the diatom *Phaeodactylum tricornutum* after being exposed to a temperature change from 25 to 10 °C for 12 hours [11.119] [11.148]. When the temperature was changed from 25 °C to 10 °C for 12 hours, the percentages of EPA and PUFA increased from 13% to 24% and 22% to 32%, respectively [11.54] [11.148].

11.3.10 POME-Based Biofuel

Diatoms are a type of microalgae that are sources or prospective suppliers of a variety of products [11.45] [11.107] [11.157] [11.161] [11.242], including edible oils. Palm oil mill effluent, commonly known as POME, is produced during the manufacturing of palm oil and is a nutrient-rich wastewater. The ability of microalgae to absorb nutrients from POME can lessen the amount of pollution it can cause [11.107] [11.129] [11.157] [11.158]. Diverse microalgae, including diatoms like *Chaetoceros affinis* [11.158], *Gyrosigma* sp. [11.81], *Amphora copulata* [11.80], and *Amphora* sp. [11.108], have grown well on POME-based culture media. Diatoms are particularly interesting because they produce valuable lipids [11.52] [11.81] [11.80] [11.161] [11.192] [11.242]. In oil bodies, diatoms commonly assemble neutral lipids such triacylglycerols [11.136] [11.157]. *Thalassiosira pseudonana* also needs selenium for growth in addition to silicon [11.157] [11.170]. Lack of selenium prevents cell division and alters the ultrastructure of the mitochondrial and chloroplast membrane systems [11.60] [11.157]. Under nutritional stressors that inhibit growth, *T. pseudonana*'s lipid content is known to rise [11.28] [11.103] [11.157]. In addition to raising the levels of certain lipids, phosphorus stress also increases the accumulation of stored lipids, particularly triacylglycerols [11.28] [11.103] [11.157]. However, because they decrease the oxidative stability of biodiesel, a large proportion of polyunsaturated fatty acids in the lipids is undesirable for the production of biodiesel [11.157].

11.4 Changes in Gene Expression in Diatoms During Stress Conditions

Diatoms are a key group of microalgae that thrives in chaotic conditions. It is still unclear whether and how diatoms might benefit from turbulence despite several experimental and computational research studies. The idea that turbulence encourages nutrition absorption is one of the most persuasive ones. Morphological traits, including the lack of flagella, the existence of a stiff exoskeleton, and the micrometer size would support the possibility that turbulence has a passive but advantageous effect on diatoms. For the assistance of fulfilling the current needs of commercial biofuel production, targeting the diatom lipids by adjusting and optimizing the growth and culture parameters, such as light, stress, and nutrients, can be an intriguing alternative. High lipid productivity has been achieved through controlling culture parameters, including temperature, salinity, and nutrient concentrations, as well as via genetic engineering techniques [11.50] [11.100]. Under unfavorable climatic conditions, a lot of microalgae tend to store triacylglycerols (TAG) as a storage lipid [11.98]. Diatoms are an underutilized resource for the biofuel sector due to their capacity to accumulate significant levels of lipids and varied compositions of fatty acids. The most common fatty acids in diatom species, both saturated and unsaturated, are 14:0, 16:0, 16:1, 16:2, 18:1, 18:2, 18:3, 20:4, and 20:5 [11.64]. The lipid production and triacylglycerol accumulation of several species of diatoms have been the subject of numerous reports in Table 11.2.

The most common dietary limiting factor to date for inducing lipid production in microalgae is nitrogen. While having a look at this, Illman *et al.* [11.105] came to the conclusion that *Chlorella emersonii*, when grown on low-N medium, may collect as much as 63% of their dry weight in lipids. The production of lipids may be boosted by additional stressors. Low light and high salinity enhanced the total fatty acid content and the polyunsaturated fatty acid (PUFA) concentration in the seaweed, namely *Grateloupia sparsa* [11.71]. Under conditions of high salinity, the marine alga *Dunaliella tertiolecta* likewise displayed a higher TAG level [11.219]. Microalgae lipid production has also been improved by the use of genetic engineering techniques, typically including the overexpression of one or more essential TAG biosynthesis pathway enzymes [11.50].

The study of an organism's metabolic and physiological response to stressors [11.63] [11.104] [11.225] is made possible by transcriptome profiling, and comparing transcriptomic responses between species might reveal aspects that are conserved and those that are divergent. According

Table 11.2 Lipid contents and productivity levels of different microalgae and diatom species.

Microalgae	Nutrient content	Lipid productivity	% of TAG	Reference
Thalassiosira weissflogii P09	Nitrogen limitation	7.27 _ 0.28	51.0 _ 3.2	[11.52]
Thalassiosira pseudonana CCMP 1335	High CO_2 20,000 ppm	--	--	[11.52]
Cyclotella cryptica CCMP 331	Nitrogen limitation	--	45 (+20%)	[11.52]
Phaeodactylum tricornutum CCMP 632	Tn19745_1 strain + nitrogen limitation	--	45-fold increase	[11.52]
	Dark	--	--	[11.16]
	H_2O	--	--	[11.110]
Navicula pelliculosa	High CO_2			[11.110]

to the findings of Dyhrman *et al., T. pseudonana* has developed a complex response to phosphorus deprivation comprising a number of biochemical tactics that are probably essential to its capacity to react to changes in environmental phosphorus availability [11.63]. By using the transcriptome as a guide, Guarnieri *et al.* developed a method to examine the triacylglycerol biosynthetic pathway in the unsequenced oleaginous microalga *Chlorella vulgaris*. The results showed an upregulation of both fatty acid and triacylglycerol biosynthetic machinery under oil-accumulating conditions and demonstrate the utility of a *de novo* assembled transgene [11.84]. In their multicomponent investigation of the N starvation response, Schmollinger *et al.* used *Chlamydomonas reinhardtii* as a reference organism and pinpointed a target to boost the nitrogen-use efficiency in the nitrogen-starved *Chlamydomonas reinhardtii* [11.197].

To better understand the fundamentals of genetic control of physiological acclimation to growth under Fe limitation, Smith *et al.* profiled the physiology, metabolite composition, and global transcripts of the pennate diatom *Phaeodactylum tricornutum* during steady-state growth at low, intermediate, and high levels of dissolved Fe over light-dark cycles [11.212]. This research may offer fresh understanding of how Fe availability affects the transcriptional regulation of diatom physiology and metabolism across light-dark cycles, elucidating the mechanisms underlying the ability of diatoms to maintain metabolic readiness for quick responses to Fe input, and exposing tactics underlying their ecological success. By allowing the study of molecular mechanism through time, such processes can shed light on the sequence of events that regulate lipid induction. Amato *et al.* investigated whether diatoms could respond to turbulence when nutrients were in excess and all other environmental factors were held constant compared to control static samples [11.7]. They used a prototypical turbulence generator to apply measured levels of turbulence [11.134] to the diatom *Chaetoceros decipiens* [11.8] in order to achieve this goal. They also conducted transcriptome studies and microscopic examination of chain lengths. The outcomes demonstrated that the therapy had an impact on the morphology and gene expression of *C. decipiens*. Because turbulence is not a stressor like nutritional deprivation [11.5] [11.14] [11.20] [11.124], high irradiance [11.165], or chemicals [11.35], the transcriptome response was moderate, as was to be expected.

The applied stress conditions, such as nutritional starvation/depletion, CO_2-enrichment, and combinations, caused early aging in *S. marinoi*, according to a study by Lauritano *et al.* [11.121]. With a few exceptions, aging did not cause changes in other gene groups. On algae cultivated in full media, the expression levels of the glutathione-related enzymes GSH-S, GR, and GST were higher in the STAT and DECL phases, but GPX was downregulated in both phases (p 0.01) compared to the EXP phase. *S. marinoi* grows in Si-depletion whereas 2b under Si-starvation exhibits altered gene expression. Gene expression levels of algae grown in full medium have been employed as the control condition for both RT-qPCR investigations. The gene with the greatest reduction—more than 4-fold—was HSP90. With an ALDH2 reduction of more than 8-fold, the downregulation was even more dramatic in the Si-depletion condition.

Through their work conducted in 2006, Davis *et al.* showed that the marine diatom *Thalassiosira pseudonana* has a fast response at the gene level to elevated concentrations of copper and that this response attenuates over 24 hr of continuous exposure. Using a PCR-based subtractive cDNA technique, 11 of the 16 copper-induced genes that they found were wholly novel; nevertheless, many of the predicted amino acid sequences showed properties that suggested they might play a role in reducing copper toxicity. The majority of the novel genes did not respond to H_2O_2 or Cd-induced stress in an identical manner, indicating that the responses varied [11.57].

In order to evaluate their potential value as species-specific genetic markers to denote Fe status in natural diatom assemblages, Whitney *et al.*

investigated the expression of a subset of genes in *Thalassiosira* genus diatoms. In this study, they compared the expression of the photosynthetic genes for ferredoxin, a protein that needs iron, and flavodoxin, a protein that doesn't need iron, in culture experiments using the isolated *Thalassiosira pseudonana* from coastal waters and *Thalassiosira weissflogii* from the open ocean that were both Fe replete and Fe stressed. Although the expression of the flavodoxin and ferredoxin genes in *T. pseudonana* showed diel periodicities, they were not responsive to the Fe status. The expression of flavodoxin in *T. weissflogii* was extremely responsive to iron levels and was only seen in cultures with low levels of Fe. According to the available genomic data, flavodoxin genes have been duplicated in the majority of diatoms, and they demonstrate that *T. pseudonana* has lost its copy in relation to the Fe-responsive copy in *T. weissflogii*. Additionally, they looked at the gene expression in *T. pseudonana* for a potential high affinity, copper (Cu)-dependent Fe uptake system. The findings showed no connection between Fe status and the genes encoding putative Cu transporters, a multi-Cu oxidase, or a Fe reductase [11.249].

Oceanic diatoms called *Chaetoceros* spp. are significant ecologically, and it is expected that their genetic responses to phosphorus (P) stress may differ from those of model species with fully sequenced genomes. In 2015, Shih *et al.* [11.208] used next-generation sequencing (NGS) to create transcriptomes for *Chaetoceros affinis* (CCMP 160) cultures grown under low phosphate and nutrient-rich conditions in order to study how phosphorus availability influences gene expression at the mRNA level. Following sequence comparisons that identified 10,972 expressed genes, of which 503 were upregulated and 553 were downregulated by P stress, this process produced 29,285 scaffolds. Quantitative reverse-transcription polymerase chain reactions were used to confirm the varied expression patterns of 41 of these genes. These findings showed that in *C. affinis* under P deprivation, genes involved in nutrition uptake are upregulated. Phosphorus deprivation caused many ribonuclease genes to be upregulated, which increased the intracellular recycling of P in *C. affinis*. Sulfolipid production-related gene expression, however, was unchanged. Due to poor maximal quantum efficiency (Fv/Fm) and high transcription levels of light-harvesting complex genes, it was assumed that extra energy was lost during P deficit in relation to photochemical activities. The NAD-malic enzyme was activated to produce pyruvate through a different pathway that requires less P. Early stationary phase cell proliferation rates were shown to be lower in low-P cultures, which is consistent with a large reduction in the transcription of the genes encoding the cell division control protein 45 and ribonucleoside-diphosphate reductase.

In 2011, Tachibana *et al.* demonstrated that the CA-1, CA-3, and CA-7 (a-CA, z-CA, and d-CA, respectively) putative CA genes in *T. pseudonana* were activated by lowering CO_2 and function in CO_2-limited conditions [11.217]. Similar to this, when CO_2 levels dropped, *C. weissflogii* increased both CO_2 and HCO_3 uptake, which was followed by an increase in both internal and external CA activity [11.31]. LCIP63, short for "low-CO_2-inducible protein of 63 kDa," was discovered in a recent proteomic investigation on *T. pseudonana* that had been acclimated to low CO_2 (50 ppm), and it was upregulated under these conditions but downregulated when nitrogen, phosphorus, or silicon were limiting factors [11.40] [11.130]. In *T. pseudonana* growing at 300 ppm vs. 1,000 ppm CO_2, LCIP63 was upregulated [11.233].

Data on the physiological and genomic responses of various diatoms to different stress conditions are available. They show how many different and intricate methods these species have developed to adapt to environmental change. In contrast to the Viridiplantae, chromalveolates which include diatoms are still poorly understood, and additional investigation is required to completely understand how this significant group of algae maintains its productivity under changing environmental conditions. Proteomics is increasingly being used in research to examine how diatoms react to stress, such as in the studies conducted by Muhseen *et al.* and Chen *et al.* [11.40] [11.149]. Candidates for thioredoxin interactions have been found in Viridiplantae using proteomics methods [11.146] [11.168]. This might also be applied to diatoms, where it can be extremely difficult to identify precise targets for each of the several thioredoxins present in diatoms [11.248]. Diatoms are beginning to be subjected to the overexpression or silencing of a gene to ascertain their metabolic function [11.66] [11.95] [11.101]. Diatoms have recently benefited from the use of customized TALEN endonucleases and the CRISPR/Cas9 system [11.53] [11.97] [11.152], enabling the creation of knockout strains with specific genetic alterations. It is obvious that the methods outlined above, along with genome sequencing, "omics," and targeted methods, will help us better understand the biology of diatoms.

11.5 Structural and Functional Aspect of Candidate Genes/Enzymes of Lipid Biosynthesis Pathway

A detailed description of the major lipid biosynthesis pathways existing in diatoms is covered in a review paper [11.15] [11.43]. The two major routes: (i) *de-novo* fatty acid biosynthesis, which occurs in plastid, and (ii) Kennedy pathway, which occurs in endoplasmic reticulum (ER), are mainly recognized in diatoms for fatty acid synthesis. Each pathway is operated and regulated through the involvement of a few genes and enzymes encoded by those genes. These genes are identified, characterized, cloned and expressed in heterologous hosts in order to understand the mechanism of lipid biosynthesis and to achieve enhanced lipid accumulation in oleaginous diatoms. Elucidation of structural and functional aspects of candidate genes/enzymes is also desirable to facilitate the process of genetic manipulation in diatoms to achieve the goal of enhanced lipid biosynthesis. Research attempts have been made for structural and functional characterization of few important regulatory enzymes like acetyl-CoA carboxylase (ACCase), diacylglycerol acetyltransferase (DGAT), and glycerol-3-phosphate acetyltransferase (GPAT), which belong to different lipid biosynthesis pathways.

A gene encoding a subunit of ACCase from the diatom *Phaeodactylum tricornutum* was cloned and expressed in *Escherichia coli* for its functional characterization. Successful expression of the gene in *E. coli* led to 1.72-fold increase in ACCase activity, hence enhanced storage lipid content up to 30% dry cell weight [11.253]. *P. tricornutum* possesses a homologous type ACCase which is encoded by a 7271 bp long gene containing five exons and four introns; hence, a tough target for genetic manipulation. Structural prediction using smart software showed that ACCase of *P. tricornutum* (PtACC) contains six domains. Out of these six, four domains (namely carbamoyl phosphate synthase l-chain, a BC domain [biotin carb c], an ATP-grasp fold, and a biotin and lipoyl attachment domain) constituted the N-terminal end of 886 amino acids which are encoded by exon-1. The ATP-grasp fold domain (located at 254-421 amino acids) was found to catalyze the ATP dependent carboxylation of biotin while the BC domain (located at 483-590 amino acids) functioned as biotin carboxylase (BC). The remaining four exons of PtACC were reported to encode only two subunits; one functions as a carboxylase catalytic subunit while the other is carboxyltransferase (CT) domain [11.252]. Acyl-ACP desaturase, a key enzyme which regulates the biosynthesis of palmitoleic acid, was characterized in *P. tricornutum*. Palmitoleic acid (16:119) is a monounsaturated fatty acid which contributes 40% of total lipid content in the diatom *P. tricornutum*. The acyl-ACP desaturase (PtAAD) gene was successfully cloned and transiently expressed in yeast and in *Nicotiana benthamiana* leaves. Three-dimensional structural modeling and protein sequence alignment of PtAAD with other known AADk demonstrated that this enzyme contains a single functional domain with eight key amino acids. These eight key amino acids determine its substrate specificity and give preference to 16:0-ACP over 18:0-ACP for the synthesis of palmitoleic acid ($16{:}1\Delta^9$) [11.132].

Triacylglycerol (TAG) is an energy-dense lipid which serves as a raw material for biofuel production in diatoms. Acyl-CoA: diacylglycerol acyltransferase (DGAT) enzyme isoforms catalyze the last and key step of TAG biosynthesis, i.e., transfer of an acyl group from acyl-CoA to a diacylglycerol molecule to make triacylglycerol (TAG). Three isoforms of DGAT enzyme, i.e., type 1, 2 and 3, were discovered in microalgae encoded by DGAT1, DGAT2 and DGAT3 genes respectively. Structural prediction studies depicted that DGAT1 and DGAT2 are transmembrane proteins containing six and two transmembrane domains respectively [11.90]. Guiheneuf *et al.* cloned and characterized a cDNA encoding Acyl-CoA: diacylglycerol acetyltransferase enzyme from the diatom *Phaeodactylum tricornutum* (PtDGAT1) which was followed by its heterologous expression in yeast. The cDNA (PtDGAT1) was reported to transcribe two types of mRNA, namely PtDGAT1short and PtDGAT1long from a single PtDGAT1 gene. Expression studies revealed that the production of active DGAT1 in *P. tricornutum* under nitrogen starvation conditions are regulated by alternative splicing of PtDGAT1short pre mRNA, which in turn regulates the TAG accumulation [11.86]. Identification and characterization of the type 2 DGAT of *P. tricornutum* (PtDGAT2) was performed by Gong *et al.* Functional analysis studies confirmed that DGAT1 prefers saturated fatty acids while DGAT2 prefers unsaturated fatty acids (C_{16} and C_{18}) as substrate for TAG biosynthesis. In contrast to the above-mentioned single-function DGAT enzymes, recently, a bifunctional DGAT isoform was discovered in *P. tricornutum*. This novel enzyme shows dual enzyme activities similar to wax ester synthase (WS) as well as acyl-CoA: diacylglycerol acetyltransferase (DGAT), and is able to catalyze the biosynthesis of wax esters as well as TAGs. Wax esters and TAGs both serve as starting material for lipid biosynthesis. Cui *et al.* characterized the dual function diacylglycerol acetyltransferase enzyme from the diatom *P. tricornutum* (PtWS/DGAT) and discovered that it has a distinct topology and phylogeny from those of DGAT1s, DGAT2s and DGAT3s having preference for saturated fatty acid substrates [11.51].

Another key regulatory enzyme for lipid biosynthesis is glycerol-3-phosphate acetyltransferase (GPAT) which catalyzes the first step of glycerol lipid (GL) as well as triacylglycerol (TAG) biosynthesis. Compared to higher plants, limited information is available about the structure and function of glycerol-3-phosphate acetyltransferase (GPAT) enzyme in microalgae. Cloning, characterization through enzymatic assay, and heterologous expression of microsomal TpGPAT from the diatom *Thalassiosira pseudonana* in yeast was achieved by Xu *et al.* Cloning and characterization of cDNAs encoding two isozymes of GPAT from a green microalga

Chlamydomonas reinhardtii confirmed the existence of ER located GPAT (CrGPATer) and chloroplast located GPAT (CrGPATcl) genes. Endoplasmic reticulum located variant of the enzyme is a transmembrane protein while the chloroplast located variant of enzyme has both soluble and membrane bound domains. Post-translational regulation of expression of CrGPATer and CrGPATcl genes through alternate splicing mechanism was reported [11.61].

11.6 Role of rDNA Technology in Improving Diatom Strains for Enhanced Lipid Production

The goal to design, manufacture, and "boot up" DNA in a selected host organism generally defines the burgeoning discipline of synthetic genomics, which involves large-scale DNA engineering [11.156]. Metabolic engineering is a potent technique for improving the biosynthetic pathways and comprehending cellular and metabolic processes in microalgae (Table 11.3 and Fig. 11.2). The research has suggested that genetic engineering can enhance TAG synthesis while changing the PUFA profile to optimize its suitability for different industrial or dietary purposes [11.3]. Contrary to the often-researched green microalgal species, diatoms are a promising candidate for biofuel production as they accumulate a large amount of lipids, the main carbon storing substances where TAGs and fatty acids typically account for 15–25% of dry biomass [11.254]. Furthermore, diatoms are regarded as a possible alternative source of essential ω3 fatty acids, EPA (eicosapentaenoic acid) and DHA (docosahexaenoic acid). *Phaeodactylum tricornutum* and *Thalassiosira pseudonana* are the two most commonly exploited model diatoms that have drawn significant interest in the field of biorefinery over the years due to their high lipid accumulation, and are model organisms due to their short genome size of about 30 Mb and rapid growth [11.262] [11.131] [11.236]. Utilizing the genetic engineering techniques on diatoms to develop more compounds or increase the output of existing ones enables the companies to maximize their usage. Diatoms serve as a reliable model for cell and genome modification. In order to further increase the economic value of diatoms, genetic engineering is a promising technique and a crucial branch to be applied in the diatoms sector [11.203]. By modifying, deleting and inserting genomic sequences, genetic engineering methods have made it possible to manipulate the genetic make-up of diatoms. Such manipulations are crucial for understanding how genes work and enhance the diatom's metabolic capability [11.101].

Table 11.3 List of diatom strains and their gene modification to enhance lipid accumulations.

S. no.	Diatom strains	Gene modified	Genome editing technique	Result	Reference
1.	*Thalassiosira pseudonana*	TpTL (TAG lipase)	Knockdown	Hyperaccumulation of TAG	[11.231]
2.	*Phaeodactylum tricornutum*	PtD5b (Delta-5-desaturase) gene	Overexpression	MUFA, PUFA and EPA levels increased by 75%, 64% and 58% than WT	[11.167]
3.	*P. tricornutum*	Nitrate reductase gene	RNAi knockdown	Increased fatty acid content	[11.123]
4.	*P. tricornutum*	PtTL (TAG lipase)	Knockdown	Hyperaccumulation of TAG	[11.17]
5.	*P. tricornutum*	PpGT (Glucose transporter) from *P. patens*	Endogenous expression	Increased accumulation of EPA and DHA	[11.89]
6.	*P. tricornutum*	OsElo5 (Delta-5-elongase) from *O. tauri*	Endogenous expression	Increased accumulation of EPA and DHA	[11.89]
7.	*Fistulifera solaris*	Glucose-6-phosphate 1-dehydrogenase, 6-phosphogluconate dehydrogenase	Overexpression	1.5-fold increase in lipid content	[11.156]
8.	*P. tricornutum*	ptTES1 (plastidial thioesterase)	TALEN	Enhanced TAG	[11.90]
9.	*P. tricornutum*	PtMCAT (Malonyl CoA-acyl carrier protein transacylase)	Overexpression	Increase in total lipids and TAG	[11.51]
10.	*P. tricornutum*	PtLPAAT1 (Lysophosphatidic acid acyltransferase)	Overexpression	Lipid content increased	[11.244]
11.	*P. tricornutum*	PtGPAT1 (Glycerol-3-phosphate acyltransferase)	Overexpression	Lipid content increased	[11.244]
12.	*P. tricornutum*	AnPhyA (Phytase) from *A. niger*	Endogenous expression	Increased EPA and DHA	[11.171]
13.	*P. tricornutum*	EcAppA (Phytase) from *E. coli*	Endogenous expression	Increased EPA and DHA	[11.171]

(*Continued*)

Table 11.3 List of diatom strains and their gene modification to enhance lipid accumulations. (*Continued*)

S. no.	Diatom strains	Gene modified	Genome editing technique	Result	Reference
14.	*P. tricornutum*	PtFAD5b (Fatty acid desaturase 5b)	Overexpression	Increased EPA	[11.246]
15.	*P. tricornutum*	PtMCAT (Malony CoA-acyl carrier protein transacylase)	Overexpression	Increased EPA	[11.246]
16.	*P. tricornutum*	PtG6PDH (Glucse-6-phosphate dehydrogenase)	Overexpression	Omega-3 accumulation and increased lipid	[11.250]
17.	*P. tricornutum*	DGAT2B (2 acyl-CoA: diacylglycerol acyltransferase)	Overexpression	Increase in DHA and TAG	[11.92]
18.	*P. tricornutum*	Lysophosphatidic acid acyltransferase gene	Overexpression	Total fatty acid concentration increased	[11.127]
19.	*P. tricornutum*	PtPTP (Plastidial pyruvate transporter)	Overexpression	Lipid content increased from 11% TO 30%	[11.200]
20.	*P. tricornutum*	PtGPAT2 (Glycerol-3-phosphate acyltransferase 2)	Overexpression	Increased TAG	[11.245]

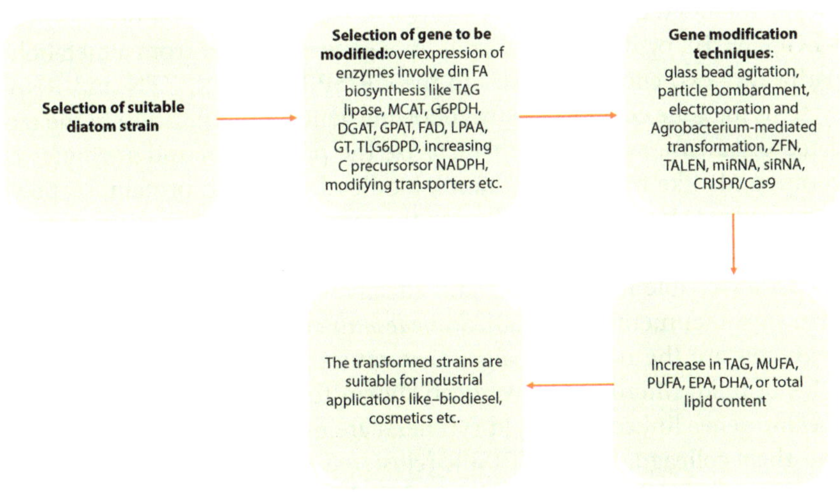

Figure 11.2 Pathway for lipid enhancement through gene modification.

To increase TAGs, recombinants can be formed to overexpress the enzymes involved in FA biosynthesis, control TAG assembly, or increase the availability of carbon precursors or NADPH needed for TAG synthesis, DAG acylation modification, and membrane lipid recycling, which have been employed to optimize metabolic flow of TAG production [11.3] [11.221]. It has been demonstrated that inhibiting ADP-glucose pyrophosphorylase increases triacylglycerol production by ten-fold, and acetyl-CoA carboxylase and type-II fatty acid synthase have been identified as the rate-limiting enzymes during fatty acid synthesis. The production of lipids can be increased by genetically altering these enzymes. *Thalassiosira pseudonana*, a diatom, has a multifunctional lipase, phospholipase, and acyltransferase that can be specifically suppressed to increase lipid production [11.221].

The four most prevalent transformation techniques for microalgae are glass bead agitation, particle bombardment, electroporation and *Agrobacterium*-mediated transformation. A new revolution in genetic engineering has been sparked by the gene editing technique, which first emerged more than 20 years ago [11.206]. To fully utilize the genomic potential, tools for genome engineering such as CRISPR (clustered regularly interspaced short palindromic repeats)/Cas (CRISPR associated system), transcription activator-like effector nucleases (TALENs), zinc-finger nucleases (ZFNs), microRNA (miRNA) and small interfering RNA (siRNA) are currently being developed at a rapid pace. For instance, the CRISPR/Cas9 system and the TALEN-mediated method were both effective in editing the genome of the diatom *P. tricornutum* [11.39]. There are four crucial steps to take into account: Choose the best microalgal strain for the intended product before using a bio-informative platform, such as KEGG, NCBI, or BioCyc, to search for a gene of interest from a metabolic pathway. Next, choose and set up the CRISPR, TALEN, ZFN and RNAi systems for gene editing. Finally, following genetic modification in the targeted organism, assess and characterize the production and attributes of compounds like biogas, bioethanol, biodiesel, pigment, protein, or polysaccharides [11.131]. This facilitates the manufacture of biofuels. Among these, CRISPR-Cas9 technologies are thought to have enormous potential since they enable more precise and effective site-specific genome editing than any other method [11.180]. *Scenedesmus obliquus* and *Neochloris oleoabundans* are the two microalgae that naturally collect significant quantities of lipid. Microalgae have been shown to produce more lipid when specific genes linked with lipid synthesis are expressed or silenced. Wang and their colleagues used CRISPR (clustered regularly interspaced short palindromic repeats)/Cas9 to successfully knock out a gene in the oleaginous model *N. oceanica* IMET1 (CRISPR associated nuclease 9) [11.27].

Several research studies have documented the application of techniques to increase acetyl-CoA/NADPH supply, improve photosynthetic efficiency or obstruct the rival routes in order to promote lipid formation [11.215].

Thus, in order to identify a potent strain of green algae with greater lipid content, genetic engineering enables us to alter genes involved in FA production. The generation of genetically enhanced microalgal strains with high lipid productivity is further facilitated by the use of the omics approach, which makes it easier to comprehend the mechanisms underlying the stress-related lipid accumulation in microalgae [11.179].

References

[11.1] Ackman, R. G., Jangaard, P. M., Hoyle, R. J., & Brockerhoff, H. (1964). Origin of marine fatty acids. I. Analyses of the fatty acids produced by the diatom *Skeletonema costatum*. *Journal of the Fisheries Board of Canada*, 21(4), 747–756.

[11.2] Adams, C., & Bugbee, B. (2014). Enhancing lipid production of the marine diatom *Chaetoceros gracilis*: synergistic interactions of sodium chloride and silicon. *Journal of Applied Phycology*, 26(3), 1351–1357.

[11.3] Ahmad, A., Banat, F., Alsafar, H., & Hasan, S. W. (2022). Recent breakthroughs in integrated biomolecular and biotechnological approaches for enhanced lipid and carotenoid production from microalgae. *Phytochemistry Reviews*, 5. https://doi.org/10.1007/s11101-022-09804-5

[11.4] Ali, M., Sultana, R., Tahir, S., Watson, I. A., & Saleem, M. (2017). Prospects of microalgal biodiesel production in Pakistan–A review. *Renewable and Sustainable Energy Reviews*, 80, 1588–1596.

[11.5] Allen, A. E., Dupont, C. L., Oborník, M., Horák, A., Nunes-Nesi, A., McCrow, J. P., Zheng, H., Johnson, D. A., Hu, H., & Fernie, A. R. (2011). Evolution and metabolic significance of the urea cycle in photosynthetic diatoms. *Nature*, 473(7346), 203–207.

[11.6] Almutairi, A. W., El-Sayed, A. E.-K. B., & Reda, M. M. (2021). Evaluation of high salinity adaptation for lipid bio-accumulation in the green microalga *Chlorella vulgaris*. *Saudi Journal of Biological Sciences*, 28(7), 3981–3988.

[11.7] Amato, A., Dell'Aquila, G., Musacchia, F., Annunziata, R., Ugarte, A., Maillet, N., Carbone, A., Ribera d'Alcalà, M., Sanges, R., & Iudicone, D. (2017). Marine diatoms change their gene expression profile when exposed to microscale turbulence under nutrient replete conditions. *Scientific Reports*, 7(1), 1–11.

[11.8] Amato, A., Fortini, S., Watteaux, R., Diano, M., Espa, S., Esposito, S., Ferrante, M. I., Peters, F., Iudicone, D., & Ribera d'Alcalà, M. (2016). TURBOGEN: Computer-controlled vertically oscillating grid system for small-scale turbulence studies on plankton. *Review of Scientific Instruments*, 87(3), 35119.

[11.9] Aoyagi, K., & Omokawa, M. (1992). Neogene diatoms as the important source of petroleum in Japan. *Journal of Petroleum Science and Engineering*, 7(3–4), 247–262.

[11.10] Appeltans, W., Ahyong, S. T., Anderson, G., Angel, M. v, Artois, T., Bailly, N., Bamber, R., Barber, A., Bartsch, I., & Berta, A. (2012). The magnitude of global marine species diversity. *Current Biology*, 22(23), 2189–2202.

[11.11] Armbrust, E. (2009). The life of diatoms in the world's oceans. *Nature*, 459(7244), 185–192.

[11.12] Armbrust, E. V., Berges, J. A., Bowler, C., Green, B. R., Martinez, D., Putnam, N. H., Zhou, S., Allen, A. E., Apt, K. E., & Bechner, M. (2004). The genome of the diatom *Thalassiosira pseudonana*: ecology, evolution, and metabolism. *Science*, 306(5693), 79–86.

[11.13] Aro, E.-M. (2016). From first generation biofuels to advanced solar biofuels. *Ambio*, 45(1), 24–31.

[11.14] Ashworth, J., Coesel, S., Lee, A., Armbrust, E. V., Orellana, M. V, & Baliga, N. S. (2013). Genome-wide diel growth state transitions in the diatom *Thalassiosira pseudonana*. *Proceedings of the National Academy of Sciences*, 110(18), 7518–7523.

[11.15] Avhad, M. R., & Marchetti, J. M. (2019). Uses of enzymes for biodiesel production. In *Advanced Bioprocessing for Alternative Fuels, Biobased Chemicals, and Bioproducts* (pp. 135–152). Elsevier.

[11.16] Bai, X., Song, H., Lavoie, M., Zhu, K., Su, Y., Ye, H., Chen, S., Fu, Z., & Qian, H. (2016). Proteomic analyses bring new insights into the effect of a dark stress on lipid biosynthesis in *Phaeodactylum tricornutum*. *Scientific Reports*, 6(1), 1–10.

[11.17] Barka, F., Angstenberger, M., Ahrendt, T., Lorenzen, W., Bode, H. B., & Büchel, C. (2016). Identification of a triacylglycerol lipase in the diatom *Phaeodactylum tricornutum*. *Biochimica et Biophysica Acta (BBA)-Molecular and Cell Biology of Lipids*, 1861(3), 239–248.

[11.18] Battarbee, R. W., Jones, V. J., Flower, R. J., Cameron, N. G., Bennion, H., Carvalho, L., & Juggins, S. (2002). Diatoms. In *Tracking Environmental Change using Lake Sediments* (pp. 155–202). Springer.

[11.19] Beattie, A., Hirst, E. L., & Percival, E. (1961). Studies on the metabolism of the Chrysophyceae. Comparative structural investigations on leucosin (Chrysolaminarin) separated from diatoms and laminarin from the brown algae. *Biochemical Journal*, 79(3), 531.

[11.20] Bender, S. J., Durkin, C. A., Berthiaume, C. T., Morales, R. L., & Armbrust, E. V. (2014). Transcriptional responses of three model diatoms to nitrate limitation of growth. *Frontiers in Marine Science*, 1, 3.

[11.21] Berges, J. A., & Falkowski, P. G. (1998). Physiological stress and cell death in marine phytoplankton: induction of proteases in response to nitrogen or light limitation. *Limnology and Oceanography*, 43(1), 129–135.

[11.22] Bhattacharjya, R., Marella, T. K., Tiwari, A., Saxena, A., Singh, P. K., & Mishra, B. (2020). Bioprospecting of marine diatoms *Thalassiosira, Skeletonema* and *Chaetoceros* for lipids and other value-added products. *Bioresource Technology*, 318, 124073.

[11.23] Bhuyar, P., Rahim, M. H. A., Yusoff, M. M., Maniam, G. P., & Govindan, N. (2019). A selective microalgae strain for biodiesel production in relation to higher lipid profile. *Maejo International Journal of Energy and Environmental Communication*, 1(1), 8–14.

[11.24] Bhuyar, P., Sundararaju, S., Rahim, M. H. A., Ramaraj, R., Maniam, G. P., & Govindan, N. (2021). Microalgae cultivation using palm oil mill effluent as growth medium for lipid production with the effect of CO_2 supply and light intensity. *Biomass Conversion and Biorefinery*, 11(5), 1555–1563.

[11.25] Boyer, J. S. (1982). Plant productivity and environment. *Science*, 218(4571), 443–448.

[11.26] Bozarth, A., Maier, U.-G., & Zauner, S. (2009). Diatoms in biotechnology: modern tools and applications. *Applied Microbiology and Biotechnology*, 82(2), 195–201.

[11.27] Brar, A., Kumar, M., Soni, T., Vivekanand, V., & Pareek, N. (2021). Insights into the genetic and metabolic engineering approaches to enhance the competence of microalgae as biofuel resource: A review. *Bioresource Technology*, 339(June), 125597. https://doi.org/10.1016/j.biortech.2021.125597

[11.28] Bromke, M. A., Giavalisco, P., Willmitzer, L., & Hesse, H. (2013). Metabolic analysis of adaptation to short-term changes in culture conditions of the marine diatom Thalassiosira pseudonana. *PLoS One*, 8(6), e67340.

[11.29] Brown, M. R., Dunstan, G. A., Norwood, S. J., & Miller, K. A. (1996). Effects of harvest stage and light on the biochemical composition of the diatom Thalassiosira pseudonana 1. *Journal of Phycology*, 32(1), 64–73.

[11.30] Burch, A. R., & Franz, A. K. (2016). Combined nitrogen limitation and hydrogen peroxide treatment enhances neutral lipid accumulation in the marine diatom Phaeodactylum tricornutum. *Bioresource Technology*, 219, 559–565.

[11.31] Burkhardt, S., Amoroso, G., Riebesell, U., & Sültemeyer, D. (2001). CO_2 and HCO_3 uptake in marine diatoms acclimated to different CO_2 concentrations. *Limnology and Oceanography*, 46(6), 1378–1391.

[11.32] Burrows, E. H., Bennette, N. B., Carrieri, D., Dixon, J. L., Brinker, A., Frada, M., Baldassano, S. N., Falkowski, P. G., & Dismukes, G. C. (2012). Dynamics of lipid biosynthesis and redistribution in the marine diatom *Phaeodactylum tricornutum* under nitrate deprivation. *BioEnergy Research*, 5(4), 876–885.

[11.33] Camacho, F. G., Grima, E. M., Mirón, A. S., Pascual, V. G., & Chisti, Y. (2001). Carboxymethyl cellulose protects algal cells against hydrodynamic stress. *Enzyme and Microbial Technology*, 29(10), 602–610.

[11.34] Cañavate, J. P., Armada, I., & Hachero-Cruzado, I. (2017). Interspecific variability in phosphorus-induced lipid remodelling among marine eukaryotic phytoplankton. *New Phytologist*, 213(2), 700–713.

[11.35] Carvalho, R. N., Bopp, S. K., & Lettieri, T. (2011). Transcriptomics responses in marine diatom *Thalassiosira pseudonana* exposed to the polycyclic aromatic hydrocarbon benzo [a] pyrene. *PloS One*, 6(11), e26985.

[11.36] Chaffin, J. D., Mishra, S., Kuhaneck, R. M., Heckathorn, S. A., & Bridgeman, T. B. (2012). Environmental controls on growth and lipid content for the freshwater diatom, *Fragilaria capucina*: a candidate for biofuel production. *Journal of Applied Phycology*, 24(5), 1045–1051.

[11.37] Cheah, W. Y., Show, P. L., Chang, J.-S., Ling, T. C., & Juan, J. C. (2015). Biosequestration of atmospheric CO_2 and flue gas-containing CO_2 by microalgae. *Bioresource Technology*, 184, 190–201.

[11.38] Chelf, P. (1990). Environmental control of lipid and biomass production in two diatom species. *Journal of Applied Phycology*, 2(2), 121–129.

[11.39] Chen, J., Huang, Y., Shu, Y., Hu, X., Wu, D., Jiang, H., Wang, K., Liu, W., & Fu, W. (2022). Recent Progress on Systems and Synthetic Biology of Diatoms for Improving Algal Productivity. *Frontiers in Bioengineering and Biotechnology*, 10(May), 1–15. https://doi.org/10.3389/fbioe.2022.908804

[11.40] Chen, X.-H., Li, Y.-Y., Zhang, H., Liu, J.-L., Xie, Z.-X., Lin, L., & Wang, D.-Z. (2018). Quantitative proteomics reveals common and specific responses of a marine diatom *Thalassiosira pseudonana* to different macronutrient deficiencies. *Frontiers in Microbiology*, 9, 2761.

[11.41] Chen, Y.-C. (2012). The biomass and total lipid content and composition of twelve species of marine diatoms cultured under various environments. *Food Chemistry*, 131(1), 211–219.

[11.42] Chiovitti, A., Molino, P., Crawford, S. A., Teng, R., Spurck, T., & Wetherbee, R. (2004). The glucans extracted with warm water from diatoms are mainly derived from intracellular chrysolaminaran and not extracellular polysaccharides. *European Journal of Phycology*, 39(2), 117–128.

[11.43] Chisti, Y. (2007). Biodiesel from microalgae. *Biotechnology Advances*, 25(3), 294–306.

[11.44] Chisti, Y. (2013). Constraints to commercialization of algal fuels. *Journal of Biotechnology*, 167(3), 201–214.

[11.45] Chisti, Y. (2016). *Algae Biotechnology; Bux, F., Chisti, Y., Eds.; Green Energy and Technology*. Springer International Publishing: Basel, Switzerland.

[11.46] Chtourou, H., Dahmen, I., Jebali, A., Karray, F., Hassairi, I., Abdelkafi, S., Ayadi, H., Sayadi, S., & Dhouib, A. (2015). Characterization of *Amphora* sp., a newly isolated diatom wild strain, potentially usable for biodiesel production. *Bioprocess and Biosystems Engineering*, 38(7), 1381–1392.

[11.47] Church, J., Hwang, J.-H., Kim, K.-T., McLean, R., Oh, Y.-K., Nam, B., Joo, J. C., & Lee, W. H. (2017). Effect of salt type and concentration on the growth and lipid content of *Chlorella vulgaris* in synthetic saline wastewater for biofuel production. *Bioresource Technology*, 243, 147–153.

[11.48] Contreras, A., García, F., Molina, E., & Merchuk, J. C. (1998). Interaction between CO_2-mass transfer, light availability, and hydrodynamic stress in the growth of *Phaeodactylum tricornutum* in a concentric tube airlift photobioreactor. *Biotechnology and Bioengineering*, 60(3), 317–325.

[11.49] Coombs, J., Darley, W. M., Holm-Hansen, O., & Volcani, B. E. (1967). Studies on the biochemistry and fine structure of silica shell formation in diatoms. Chemical composition of *Navicula pelliculosa* during silicon-starvation synchrony. *Plant Physiology*, 42(11), 1601–1606.

[11.50] Courchesne, N. M. D., Parisien, A., Wang, B., & Lan, C. Q. (2009). Enhancement of lipid production using biochemical, genetic and transcription factor engineering approaches. *Journal of Biotechnology*, 141(1–2), 31–41.

[11.51] Cui, Y., Zhao, J., Wang, Y., Qin, S., & Lu, Y. (2018). Characterization and engineering of a dual-function diacylglycerol acyltransferase in the oleaginous marine diatom *Phaeodactylum tricornutum. Biotechnology for Biofuels*, 11(1), 1–13. https://doi.org/10.1186/s13068-018-1029-8

[11.52] d'Ippolito, G., Sardo, A., Paris, D., Vella, F. M., Adelfi, M. G., Botte, P., Gallo, C., & Fontana, A. (2015). Potential of lipid metabolism in marine diatoms for biofuel production. *Biotechnology for Biofuels*, 8(1), 1–10.

[11.53] Daboussi, F., Leduc, S., Maréchal, A., Dubois, G., Guyot, V., Perez-Michaut, C., Amato, A., Falciatore, A., Juillerat, A., & Beurdeley, M. (2014). Genome engineering empowers the diatom *Phaeodactylum tricornutum* for biotechnology. *Nat Commun* 5: 3831.

[11.54] Dalvi, V., Chawla, P., & Malik, A. (2021). Year-long performance assessment of an on-site pilot scale (100 L) photobioreactor on nutrient recovery and pathogen removal from urban wastewater using native microalgal consortium. *Algal Research*, 55, 102228.

[11.55] Damsté, J. S. S., Muyzer, G., Abbas, B., Rampen, S. W., Massé, G., Allard, W. G., Belt, S. T., Robert, J.-M., Rowland, S. J., & Moldowan, J. M. (2004). The rise of the rhizosolenid diatoms. *Science*, 304(5670), 584–587.

[11.56] Darley, W. M., Sullivan, C. W., & Volcani, B. E. (1976). Studies on the biochemistry and fine structure of silica shell formation in diatoms. *Planta*, 130(2), 159–167.

[11.57] Davis, A. K., Hildebrand, M., & Palenik, B. (2006). Gene expression induced by copper stress in the diatom *Thalassiosira pseudonana. Eukaryotic Cell*, 5(7), 1157–1168.

[11.58] de Jesús-Campos, D., López-Elías, J. A., Medina-Juarez, L. Á., Carvallo-Ruiz, G., Fimbres-Olivarria, D., & Hayano-Kanashiro, C. (2020). Chemical composition, fatty acid profile and molecular changes derived from nitrogen stress in the diatom *Chaetoceros muelleri. Aquaculture Reports*, 16, 100281.

[11.59] Demirbas, A. (2008). Biofuels sources, biofuel policy, biofuel economy and global biofuel projections. *Energy Conversion and Management*, 49(8), 2106–2116.

[11.60] Doucette, G. J., Price, N. M., & Harrison, P. J. (1987). Effects of selenium deficiency on the morphology and ultrastructure of the coastal marine diatom *Thalassiosira pseudonana* (Bacillariophyceae) 1, 2. *Journal of Phycology*, 23(2), 9–17.

[11.61] Duarte-Coello, M. E., Herrera-Valencia, V. A., Echevarría-Machado, I., Casais-Molina, M. L., & Peraza-Echeverria, S. (2019). Molecular cloning and functional characterization of two glycerol-3-phosphate acyltransferases from the green microalga *Chlamydomonas reinhardtii*. *Phycological research*, 67(2), 102-111.

[11.62] Dunstan, G. A., Volkman, J. K., Barrett, S. M., Leroi, J.-M., & Jeffrey, S. W. (1993). Essential polyunsaturated fatty acids from 14 species of diatom (Bacillariophyceae). *Phytochemistry*, 35(1), 155–161.

[11.63] Dyhrman, S. T., Jenkins, B. D., Rynearson, T. A., Saito, M. A., Mercier, M. L., Alexander, H., Whitney, L. P., Drzewianowski, A., Bulygin, V. V, & Bertrand, E. M. (2012). The transcriptome and proteome of the diatom *Thalassiosira pseudonana* reveal a diverse phosphorus stress response. *PloS One*, 7(3), e33768.

[11.64] Eckart, P. (2013). *Spaceflight Life Support and Biospherics* (Vol. 5). Springer Science & Business Media.

[11.65] Efe, Ş., Ceviz, M. A., & Temur, H. (2018). Comparative engine characteristics of biodiesels from hazelnut, corn, soybean, canola and sunflower oils on DI diesel engine. *Renewable Energy*, 119, 142–151.

[11.66] Falciatore, A., Jaubert, M., Bouly, J.-P., Bailleul, B., & Mock, T. (2020). Diatom molecular research comes of age: model species for studying phytoplankton biology and diversity. *The Plant Cell*, 32(3), 547–572.

[11.67] Falkowski, P. G., & Oliver, M. J. (2007). Mix and match: how climate selects phytoplankton. *Nature Reviews Microbiology*, 5(10), 813–819.

[11.68] Falkowski, P. G., Barber, R. T., & Smetacek, V. (1998). Biogeochemical controls and feedbacks on ocean primary production. *Science*, 281(5374), 200–206.

[11.69] Field, C. B., Behrenfeld, M. J., Randerson, J. T., & Falkowski, P. (1998). Primary production of the biosphere: integrating terrestrial and oceanic components. *Science*, 281(5374), 237–240.

[11.70] Fields, F. J., & Kociolek, J. P. (2015). An evolutionary perspective on selecting high-lipid-content diatoms (Bacillariophyta). *Journal of Applied Phycology*, 27(6), 2209–2220.

[11.71] Floreto, E. A. T., & Teshima, S. (1998). *The Fatty Acid Composition of Seaweeds Exposed to Different Levels of Light Intensity and Salinity.*

[11.72] Gao, G., Wu, M., Fu, Q., Li, X., & Xu, J. (2019). A two-stage model with nitrogen and silicon limitation enhances lipid productivity and biodiesel

features of the marine bloom-forming diatom *Skeletonema costatum*. *Bioresource Technology*, 289, 121717.
[11.73] Gautam, S., Pandey, L. K., Vinayak, V., & Arya, A. (2017). Morphological and physiological alterations in the diatom *Gomphonema pseudoaugur* due to heavy metal stress. *Ecological Indicators*, 72, 67–76.
[11.74] Geider, R. J., la Roche, J., Greene, R. M., & Olaizola, M. (1993). Response of the photosynthetic apparatus of *Phaeodactylum tricornutum* (Bacillariophyceae) to nitrate, phosphate, or iron starvation 1. *Journal of Phycology*, 29(6), 755–766.
[11.75] Ginger, M. L., McFadden, G. I., & Michels, P. A. M. (2010). Rewiring and regulation of cross-compartmentalized metabolism in protists. *Philosophical Transactions of the Royal Society B: Biological Sciences*, 365(1541), 831–845.
[11.76] Giri, T., Goutam, U., Arya, A., & Gautam, S. (2022). Effect of Nutrients on Diatom Growth: A Review. *Trends in Sciences*, 19(2), 1752.
[11.77] Godbole, V., Pal, M. K., & Gautam, P. (2021). A critical perspective on the scope of interdisciplinary approaches used in fourth-generation biofuel production. *Algal Research*, 58, 102436.
[11.78] González-Fernández, C., & Ballesteros, M. (2012). Linking microalgae and cyanobacteria culture conditions and key-enzymes for carbohydrate accumulation. *Biotechnology Advances*, 30(6), 1655–1661.
[11.79] Górecka, E., Dąbek, P., Davidovich, N., Davidovich, O., Tremblay, R., Belzile, C., Gastineau, R., & Witkowski, A. (2022). Life History of the Diatom *Schizostauron trachyderma*: Cell Size and Lipid Accumulation. *Frontiers in Marine Science*.
[11.80] Govindan, N., Maniam, G. P., Ab. Rahim, M. H., Sulaiman, A. Z., Ajit, A., Chatsungnoen, T., & Chisti, Y. (2021). Production of renewable lipids by the diatom *Amphora copulata*. *Fermentation*, 7(1), 37.
[11.81] Govindan, N., Maniam, G. P., Yusoff, M. M., Chatsungnoen, T., Ramaraj, R., & Chisti, Y. (2020). Statistical optimization of lipid production by the diatom *Gyrosigma* sp. grown in industrial wastewater. *Journal of Applied Phycology*, 32(1), 375–387.
[11.82] Grantham, P. J., & Wakefield, L. L. (1988). Variations in the sterane carbon number distributions of marine source rock derived crude oils through geological time. *Organic Geochemistry*, 12(1), 61–73.
[11.83] Griffiths, M. J., & Harrison, S. T. L. (2009). Lipid productivity as a key characteristic for choosing algal species for biodiesel production. *Journal of Applied Phycology*, 21(5), 493–507.
[11.84] Guarnieri, M. T., Nag, A., Smolinski, S. L., Darzins, A., Seibert, M., & Pienkos, P. T. (2011). Examination of triacylglycerol biosynthetic pathways via de novo transcriptomic and proteomic analyses in an unsequenced microalga. *PloS One*, 6(10), e25851.
[11.85] Guerra, L. T., Levitan, O., Frada, M. J., Sun, J. S., Falkowski, P. G., & Dismukes, G. C. (2013). Regulatory branch points affecting protein and

lipid biosynthesis in the diatom *Phaeodactylum tricornutum. Biomass and Bioenergy*, 59, 306–315.

[11.86] Guihéneuf, F., Leu, S., Zarka, A., Khozin-Goldberg, I., Khalilov, I., & Boussiba, S. (2011). Cloning and molecular characterization of a novel acyl-CoA: diacylglycerol acyltransferase 1-like gene (PtDGAT1) from the diatom *Phaeodactylum tricornutum. The FEBS journal,* 278(19), 3651-3666.

[11.87] Guillou, L., Chrétiennot-Dinet, M., Medlin, L. K., Claustre, H., Goër, S. L., & Vaulot, D. (1999). Bolidomonas: a new genus with two species belonging to a new algal class, the Bolidophyceae (Heterokonta). *Journal of Phycology*, 35(2), 368–381.

[11.88] Guiry, M. D. (2012). How many species of algae are there? *Journal of Phycology*, 48(5), 1057–1063.

[11.89] Hamilton, M. L., Powers, S., Napier, J. A., & Sayanova, O. (2016). Heterotrophic production of omega-3 long-chain polyunsaturated fatty acids by trophically converted marine diatom *Phaeodactylum tricornutum. Marine Drugs*, 14(3), 53.

[11.90] Hao, X., Luo, L., Jouhet, J., Rébeillé, F., Maréchal, E., Hu, H., Pan, Y., Tan, X., Chen, Z., You, L., Chen, H., Wei, F., & Gong, Y. (2018). Enhanced triacylglycerol production in the diatom *Phaeodactylum tricornutum* by inactivation of a Hotdog-fold thioesterase gene using TALEN-based targeted mutagenesis 06 Biological Sciences 0601 Biochemistry and Cell Biology. *Biotechnology for Biofuels*, 11(1), 1–18. https://doi.org/10.1186/s13068-018-1309-3

[11.91] Harwood, J. L. (2020). Fatty acid biosynthesis. In *Plant lipids* (pp. 27–66). Blackwell.

[11.92] Haslam, R. P., Hamilton, M. L., Economou, C. K., Smith, R., Hassall, K. L., Napier, J. A., & Sayanova, O. (2020). Overexpression of an endogenous type 2 diacylglycerol acyltransferase in the marine diatom *Phaeodactylum tricornutum* enhances lipid production and omega-3 long-chain polyunsaturated fatty acid content. *Biotechnology for Biofuels*, 13(1), 1–17. https://doi.org/10.1186/s13068-020-01726-8

[11.93] Hildebrand, M., Abbriano, R. M., Polle, J. E. W., Traller, J. C., Trentacoste, E. M., Smith, S. R., & Davis, A. K. (2013). Metabolic and cellular organization in evolutionarily diverse microalgae as related to biofuels production. *Current Opinion in Chemical Biology*, 17(3), 506–514.

[11.94] Hildebrand, M., Davis, A. K., Smith, S. R., Traller, J. C., & Abbriano, R. (2012). The place of diatoms in the biofuels industry. *Biofuels*, 3(2), 221–240.

[11.95] Hildebrand, M., Manandhar-Shrestha, K., & Abbriano, R. (2017). Effects of chrysolaminarin synthase knockdown in the diatom *Thalassiosira pseudonana*: Implications of reduced carbohydrate storage relative to green algae. *Algal Research*, 23, 66–77.

[11.96] Hockin, N. L., Mock, T., Mulholland, F., Kopriva, S., & Malin, G. (2012). The response of diatom central carbon metabolism to nitrogen starvation is different from that of green algae and higher plants. *Plant Physiology*, 158(1), 299–312.

[11.97] Hopes, A., Nekrasov, V., Kamoun, S., & Mock, T. (2016). Editing of the urease gene by CRISPR-Cas in the diatom *Thalassiosira pseudonana*. *Plant Methods*, 12(1), 1–12.

[11.98] Hu, Q., Sommerfeld, M., Jarvis, E., Ghirardi, M., Posewitz, M., Seibert, M., & Darzins, A. (2008). Microalgal triacylglycerols as feedstocks for biofuel production: perspectives and advances. *The Plant Journal*, 54(4), 621–639.

[11.99] Huang, B., Marchand, J., Blanckaert, V., Lukomska, E., Ulmann, L., Wielgosz-Collin, G., Rabesaotra, V., Moreau, B., Bougaran, G., & Mimouni, V. (2019). Nitrogen and phosphorus limitations induce carbon partitioning and membrane lipid remodelling in the marine diatom *Phaeodactylum tricornutum*. *European Journal of Phycology*, 54(3), 342–358.

[11.100] Huang, G., Chen, F., Wei, D., Zhang, X., & Chen, G. (2010). Biodiesel production by microalgal biotechnology. *Applied Energy*, 87(1), 38–46.

[11.101] Huang, W., & Daboussi, F. (2017). Genetic and metabolic engineering in diatoms. *Philosophical Transactions of the Royal Society B: Biological Sciences*, 372(1728), 20160411.

[11.102] Huang, W., & Daboussi, F. (2017). Genetic and metabolic engineering in diatoms. *Philosophical Transactions of the Royal Society B: Biological Sciences*, 372(1728). https://doi.org/10.1098/rstb.2016.0411

[11.103] Hunter, J. E., Brandsma, J., Dymond, M. K., Koster, G., Moore, C. M., Postle, A. D., Mills, R. A., & Attard, G. S. (2018). Lipidomics of *Thalassiosira pseudonana* under phosphorus stress reveal underlying phospholipid substitution dynamics and novel diglycosylceramide substitutes. *Applied and Environmental Microbiology*, 84(6), e02034-17.

[11.104] Hwang, Y., Jung, G., & Jin, E. (2008). Transcriptome analysis of acclimatory responses to thermal stress in Antarctic algae. *Biochemical and Biophysical Research Communications*, 367(3), 635–641.

[11.105] Illman, A. M., Scragg, A. H., & Shales, S. W. (2000). Increase in *Chlorella* strains calorific values when grown in low nitrogen medium. *Enzyme and Microbial Technology*, 27(8), 631–635.

[11.106] Ishihara, T., Ifuku, K., Yamashita, E., Fukunaga, Y., Nishino, Y., Miyazawa, A., Kashino, Y., & Inoue-Kashino, N. (2015). Utilization of light by fucoxanthin–chlorophyll-binding protein in a marine centric diatom, *Chaetoceros gracilis*. *Photosynthesis Research*, 126(2), 437–447.

[11.107] Jacob-Lopes, E., Maroneze, M. M., Queiroz, M. I., & Zepka, L. Q. (2020). *Handbook of microalgae-based processes and products: Fundamentals and advances in energy, food, feed, fertilizer, and bioactive compounds*. Academic Press.

[11.108] Jayakumar, S., Bhuyar, P., Pugazhendhi, A., Rahim, M. H. A., Maniam, G. P., & Govindan, N. (2021). Effects of light intensity and nutrients on the lipid content of marine microalga (diatom) *Amphiprora* sp. for promising biodiesel production. *Science of the Total Environment*, 768, 145471.
[11.109] Jeffrey, S. W. (1997). Introduction to marine phytoplankton and their pigment signatures. *Phytoplankton Pigments in Oceanography*, 37–84.
[11.110] Jensen, E. L., Yangüez, K., Carrière, F., & Gontero, B. (2019). Storage compound accumulation in diatoms as response to elevated CO_2 concentration. *Biology*, 9(1), 5.
[11.111] Jiang, Y., Yoshida, T., & Quigg, A. (2012). Photosynthetic performance, lipid production and biomass composition in response to nitrogen limitation in marine microalgae. *Plant Physiology and Biochemistry*, 54, 70–77.
[11.112] Jónasdóttir, S. H. (2019). Fatty acid profiles and production in marine phytoplankton. *Marine Drugs*, 17(3), 151.
[11.113] Joux-Arab, L., Berthet, B., & Robert, J.-M. (2000). Do toxicity and accumulation of copper change during size reduction in the marine pennate diatom *Haslea ostrearia*? *Marine Biology*, 136(2), 323–330.
[11.114] Kaur, R., Kaur, L., & Gautam, S. (2021). Role of lipid droplets in diatoms for biofuel production. *Journal of Applied Biology and Biotechnology*, 9(5), 1–9.
[11.115] Khan, M. I., Shin, J. H., & Kim, J. D. (2018). The promising future of microalgae: current status, challenges, and optimization of a sustainable and renewable industry for biofuels, feed, and other products. *Microbial Cell Factories*, 17(1), 1–21.
[11.116] Knothe, G. (2008). "Designer" biodiesel: optimizing fatty ester composition to improve fuel properties. *Energy & Fuels*, 22(2), 1358–1364.
[11.117] KOOISTRA, W. H. C. F., Gersonde, R., Medlin, L. K., & Mann, D. G. (2007). The origin and evolution of the diatoms: their adaptation to a planktonic existence. *Evolution of Primary Producers in the Sea*, 207–249.
[11.118] Kroth, P. G., Chiovitti, A., Gruber, A., Martin-Jezequel, V., Mock, T., Parker, M. S., Stanley, M. S., Kaplan, A., Caron, L., & Weber, T. (2008). A model for carbohydrate metabolism in the diatom *Phaeodactylum tricornutum* deduced from comparative whole genome analysis. *PloS One*, 3(1), e1426.
[11.119] Kumar, B. R., Deviram, G., Mathimani, T., Duc, P. A., & Pugazhendhi, A. (2019). Microalgae as rich source of polyunsaturated fatty acids. *Biocatalysis and Agricultural Biotechnology*, 17, 583–588.
[11.120] Kumar, R., Strezov, V., Weldekidan, H., He, J., Singh, S., Kan, T., & Dastjerdi, B. (2020). Lignocellulose biomass pyrolysis for bio-oil production: A review of biomass pre-treatment methods for production of drop-in fuels. *Renewable and Sustainable Energy Reviews*, 123, 109763.

[11.121] Lauritano, C., Orefice, I., Procaccini, G., Romano, G., & Ianora, A. (2015). Key genes as stress indicators in the ubiquitous diatom *Skeletonema marinoi*. *BMC Genomics*, 16(1), 1–11.

[11.122] Levitan, O., Dinamarca, J., Hochman, G., & Falkowski, P. G. (2014). Diatoms: a fossil fuel of the future. *Trends in Biotechnology*, 32(3), 117–124.

[11.123] Levitan, O., Dinamarca, J., Zelzion, E., Gorbunov, M. Y., & Falkowski, P. G. (2015). An RNA interference knock-down of nitrate reductase enhances lipid biosynthesis in the diatom *Phaeodactylum tricornutum*. *Plant Journal*, 84(5), 963–973. https://doi.org/10.1111/tpj.13052

[11.124] Levitan, O., Dinamarca, J., Zelzion, E., Lun, D. S., Guerra, L. T., Kim, M. K., Kim, J., Van Mooy, B. A. S., Bhattacharya, D., & Falkowski, P. G. (2015). Remodeling of intermediate metabolism in the diatom *Phaeodactylum tricornutum* under nitrogen stress. *Proceedings of the National Academy of Sciences*, 112(2), 412–417.

[11.125] Li, D.-W., Xie, W.-H., Hao, T.-B., Cai, J.-X., Zhou, T.-B., Balamurugan, S., Yang, W.-D., Liu, J.-S., & Li, H.-Y. (2018). Constitutive and chloroplast targeted expression of acetyl-CoA carboxylase in oleaginous microalgae elevates fatty acid biosynthesis. *Marine Biotechnology*, 20(5), 566–572.

[11.126] Li, H.-Y., Lu, Y., Zheng, J.-W., Yang, W.-D., & Liu, J.-S. (2014). Biochemical and genetic engineering of diatoms for polyunsaturated fatty acid biosynthesis. *Marine Drugs*, 12(1), 153–166.

[11.127] Li, S., Chen, Y., Wang, Y., Yang, X., Wang, J., & Chen, H. (2020). Overexpression of LPAAT gene in *Phaeodactylum tricornutum* enhances fatty acid accumulation and increases fatty acid chain length. *Indian Journal of Pharmaceutical Sciences*, 82, 19–25. https://doi.org/10.36468/pharmaceutical-sciences.spl.76

[11.128] Liaud, M.-F., Lichtl, C., Apt, K., Martin, W., & Cerff, R. (2000). Compartment-specific isoforms of TPI and GAPDH are imported into diatom mitochondria as a fusion protein: evidence in favor of a mitochondrial origin of the eukaryotic glycolytic pathway. *Molecular Biology and Evolution*, 17(2), 213–223.

[11.129] Lim, C. I., & Biswas, W. (2019). Sustainability assessment for crude palm oil production in Malaysia using the palm oil sustainability assessment framework. *Sustainable Development*, 27(3), 253–269.

[11.130] Lin, Q., Liang, J.-R., Huang, Q.-Q., Luo, C.-S., Anderson, D. M., Bowler, C., Chen, C.-P., Li, X.-S., & Gao, Y.-H. (2017). Differential cellular responses associated with oxidative stress and cell fate decision under nitrate and phosphate limitations in *Thalassiosira pseudonana*: Comparative proteomics. *PLoS One*, 12(9), e0184849.

[11.131] Lin, W. R., Tan, S. I., Hsiang, C. C., Sung, P. K., & Ng, I. S. (2019). Challenges and opportunity of recent genome editing and multi-omics in cyanobacteria and microalgae for biorefinery. *Bioresource Technology*, 291(June), 121932. https://doi.org/10.1016/j.biortech.2019.121932

[11.132] Liu, B., Sun, Y., Hang, W., Wang, X., Xue, J., Ma, R., & Li, R. (2020). Characterization of a novel acyl-ACP Δ9 desaturase gene responsible for palmitoleic acid accumulation in a diatom *Phaeodactylum tricornutum*. *Frontiers in Microbiology*, 11, 584589.

[11.133] Lora-Vilchis, M. C., Huanacuni-Pilco, J. I., López-Fuerte, F. O., & Pérez-Rojas, C. A. (2018). Growth rate, lipid, fatty acids, and pigments content of *Melosira moniliformis* (Bacillariophyta) in laboratory cultures. *Revista Latinoamericana de Biotecnología Ambiental y Algal*, 9(1), 1–16.

[11.134] Lozovatsky, I., Figueroa, M., Roget, E., Fernando, H. J. S., & Shapovalov, S. (2005). Observations and scaling of the upper mixed layer in the North Atlantic. *Journal of Geophysical Research: Oceans*, 110(C5).

[11.135] Lynn, S. G., Kilham, S. S., Kreeger, D. A., & Interlandi, S. J. (2000). Effect of nutrient availability on the biochemical and elemental stoichiometry in the freshwater diatom *Stephanodiscus minutulus* (Bacillariophyceae). *Journal of Phycology*, 36(3), 510–522.

[11.136] Maeda, Y., Nojima, D., Yoshino, T., & Tanaka, T. (2017). Structure and properties of oil bodies in diatoms. *Philosophical Transactions of the Royal Society B: Biological Sciences*, 372(1728), 20160408.

[11.137] Mann, D. G., & Droop, S. J. M. (1996). Biodiversity, biogeography and conservation of diatoms. In *Biogeography of Freshwater Algae* (pp. 19–32). Springer.

[11.138] Maréchal, E., & Lupette, J. (2020). Relationship between acyl-lipid and sterol metabolisms in diatoms. *Biochimie*, 169, 3–11.

[11.139] Marella, T. K., & Tiwari, A. (2020). Marine diatom *Thalassiosira weissflogii* based biorefinery for co-production of eicosapentaenoic acid and fucoxanthin. *Bioresource Technology*, 307, 123245.

[11.140] Marella, T. K., Saxena, A., & Tiwari, A. (2020). Diatom mediated heavy metal remediation: A review. *Bioresource Technology*, 305, 123068.

[11.141] Margalef, R. (1997). *Turbulence and marine life.*

[11.142] Martín, L. A., Popovich, C. A., Martínez, A. M., Bilbao, P. G. S., Damiani, M. C., & Leonardi, P. I. (2018). Hybrid two-stage culture of *Halamphora coffeaeformis* for biodiesel production: Growth phases, nutritional stages and biorefinery approach. *Renewable Energy*, 118, 984–992.

[11.143] Mathimani, T., & Pugazhendhi, A. (2019). Utilization of algae for biofuel, bio-products and bio-remediation. *Biocatalysis and Agricultural Biotechnology*, 17, 326–330.

[11.144] McGinnis, K. M., Dempster, T. A., & Sommerfeld, M. R. (1997). Characterization of the growth and lipid content of the diatom *Chaetoceros muelleri*. *Journal of Applied Phycology*, 9(1), 19–24.

[11.145] Mishra, A., Medhi, K., Malaviya, P., & Thakur, I. S. (2019). Omics approaches for microalgal applications: Prospects and challenges. *Bioresource Technology*, 291, 121890.

[11.146] Montrichard, F., Alkhalfioui, F., Yano, H., Vensel, W. H., Hurkman, W. J., & Buchanan, B. B. (2009). Thioredoxin targets in plants: the first 30 years. *Journal of Proteomics*, 72(3), 452–474.

[11.147] Moreno-Garrido, I., Campana, O., Lubián, L. M., & Blasco, J. (2005). Calcium alginate immobilized marine microalgae: experiments on growth and short-term heavy metal accumulation. *Marine Pollution Bulletin*, 51(8–12), 823–829.

[11.148] Mourya, M., Khan, M. J., Ahirwar, A., Schoefs, B., Marchand, J., Rai, A., Varjani, S., Rajendran, K., Banu, J. R., & Vinayak, V. (2022). Latest trends and developments in microalgae as potential source for biofuels: The case of diatoms. *Fuel*, 314, 122738.

[11.149] Muhseen, Z. T., Xiong, Q., Chen, Z., & Ge, F. (2015). Proteomics studies on stress responses in diatoms. *Proteomics*, 15(23–24), 3943–3953.

[11.150] Nguyen, T.-T., Bui, X.-T., Ngo, H. H., Nguyen, K.-Q., Nguyen, H.-H., Némery, J., Fujioka, T., Duong, C. H., Dang, B.-T., & Varjani, S. (2021). Nutrient recovery and microalgae biomass production from urine by membrane photobioreactor at low biomass retention times. *Science of The Total Environment*, 785, 147423.

[11.151] Niederwieser, T., Kociolek, P., & Klaus, D. (2018). A review of algal research in space. *Acta Astronautica*, 146, 359–367.

[11.152] Nymark, M., Sharma, A. K., Sparstad, T., Bones, A. M., & Winge, P. (2016). A CRISPR/Cas9 system adapted for gene editing in marine algae. *Scientific Reports*, 6(1), 1–6.

[11.153] Olaizola, M., Bienfang, P. K., & Ziemann, D. A. (1992). Pigment analysis of phytoplankton during a subarctic spring bloom: xanthophyll cycling. *Journal of Experimental Marine Biology and Ecology*, 158(1), 59–74.

[11.154] Orcutt, D. M., & Patterson, G. W. (1974). Effect of light intensity upon lipid composition of *Nitzschia closterium* (*Cylindrotheca fusiformis*). *Lipids*, 9(12), 1000–1003.

[11.155] Orefice, I., Musella, M., Smerilli, A., Sansone, C., Chandrasekaran, R., Corato, F., & Brunet, C. (2019). Role of nutrient concentrations and water movement on diatom's productivity in culture. *Scientific Reports*, 9(1), 1–10.

[11.156] Osada, K., Maeda, Y., Yoshino, T., Nojima, D., Bowler, C., & Tanaka, T. (2017). Enhanced NADPH production in the pentose phosphate pathway accelerates lipid accumulation in the oleaginous diatom *Fistulifera solaris*. *Algal Research*, 23, 126–134. https://doi.org/10.1016/j.algal.2017.01.015

[11.157] Palanisamy, K. M., Maniam, G. P., Sulaiman, A. Z., Ab. Rahim, M. H., Govindan, N., & Chisti, Y. (2022). Palm Oil Mill Effluent for Lipid Production by the Diatom *Thalassiosira pseudonana*. *Fermentation*, 8(1), 23.

[11.158] Palanisamy, K. M., Paramasivam, P., Jayakumar, S., Maniam, G. P., Rahim, M. H. A., & Govindan, N. (2021). Economical cultivation system

of microalgae *Spirulina platensis* for lipid production. *IOP Conference Series: Earth and Environmental Science*, 641(1), 012022.
[11.159] Pampuch, M., Walker, E. J. L., & Karas, B. J. (2022). Towards synthetic diatoms: The *Phaeodactylum tricornutum* Pt-syn 1.0 project. *Current Opinion in Green and Sustainable Chemistry*, 35, 100611. https://doi.org/10.1016/j.cogsc.2022.100611
[11.160] Pancha, I., Chokshi, K., Maurya, R., Trivedi, K., Patidar, S. K., Ghosh, A., & Mishra, S. (2015). Salinity induced oxidative stress enhanced biofuel production potential of microalgae *Scenedesmus* sp. CCNM 1077. *Bioresource Technology*, 189, 341–348.
[11.161] Pandey, A., Lee, D. J., Chang, J.-S., Chisti, Y., & Soccol, C. R. (2018). *Biomass, Biofuels, Biochemicals: Biofuels from Algae*. Elsevier.
[11.162] Pandey, A., Tyagi, R. D., & Varjani, S. (2021). *Biomass, Bofuels, Biochemicals: Circular Bioeconomy—Current Developments and Future Outlook.* Elsevier.
[11.163] Pandey, L. K., & Bergey, E. A. (2016). Exploring the status of motility, lipid bodies, deformities and size reduction in periphytic diatom community from chronically metal (Cu, Zn) polluted waterbodies as a biomonitoring tool. *Science of the Total Environment*, 550, 372–381.
[11.164] Pandit, P. R., Fulekar, M. H., & Karuna, M. S. L. (2017). Effect of salinity stress on growth, lipid productivity, fatty acid composition, and biodiesel properties in *Acutodesmus obliquus* and *Chlorella vulgaris. Environmental Science and Pollution Research*, 24(15), 13437–13451.
[11.165] Park, S., Jung, G., Hwang, Y., & Jin, E. (2010). Dynamic response of the transcriptome of a psychrophilic diatom, *Chaetoceros neogracile*, to high irradiance. *Planta*, 231(2), 349–360.
[11.166] Peltier, G., & Schmidt, G. W. (1991). Chlororespiration: an adaptation to nitrogen deficiency in *Chlamydomonas reinhardtii. Proceedings of the National Academy of Sciences*, 88(11), 4791–4795.
[11.167] Peng, K. T., Zheng, C. N., Xue, J., Chen, X. Y., Yang, W. D., Liu, J. S., Bai, W., & Li, H. Y. (2014). Delta 5 fatty acid desaturase upregulates the synthesis of polyunsaturated fatty acids in the marine diatom *Phaeodactylum tricornutum. Journal of Agricultural and Food Chemistry*, 62(35), 8773–8776. https://doi.org/10.1021/jf5031086
[11.168] Pérez-Pérez, M. E., Mauriès, A., Maes, A., Tourasse, N. J., Hamon, M., Lemaire, S. D., & Marchand, C. H. (2017). The deep thioredoxome in *Chlamydomonas reinhardtii*: new insights into redox regulation. *Molecular Plant*, 10(8), 1107–1125.
[11.169] Popovich, C. A., Pistonesi, M., Hegel, P., Constenla, D., Bielsa, G. B., Martin, L. A., Damiani, M. C., & Leonardi, P. I. (2019). Unconventional alternative biofuels: Quality assessment of biodiesel and its blends from marine diatom *Navicula cincta*. *Algal Research*, 39, 101438.

[11.170] Price, N. M., Thompson, P. A., & Harrison, P. J. (1987). Selenium: An essential element for growth of the coastal marine diatom *Thalassiosira pseudonana* (Bacillariophyceae) 1, 2. *Journal of Phycology*, 23(1), 1–9.
[11.171] Pudney, A., Gandini, C., Economou, C. K., Smith, R., Goddard, P., Napier, J. A., Spicer, A., & Sayanova, O. (2019). Multifunctionalizing the marine diatom *Phaeodactylum tricornutum* for sustainable co-production of omega-3 long chain polyunsaturated fatty acids and recombinant phytase. *Scientific Reports*, 9(1), 1–10. https://doi.org/10.1038/s41598-019-47875-1
[11.172] Pye, S., Bradley, S., Hughes, N., Price, J., Welsby, D., & Ekins, P. (2020). An equitable redistribution of unburnable carbon. *Nature Communications*, 11(1), 1–9.
[11.173] Radakovits, R., Eduafo, P. M., & Posewitz, M. C. (2011). Genetic engineering of fatty acid chain length in *Phaeodactylum tricornutum. Metabolic Engineering*, 13(1), 89–95.
[11.174] Rajaram, M. G., Nagaraj, S., Manjunath, M., Boopathy, A. B., Kurinjimalar, C., Rengasamy, R., Jayakumar, T., Sheu, J. R., & Li, J. Y. (2018). Biofuel and Biochemical Analysis of *Amphora coffeaeformis* RR03, a Novel Marine Diatom. *Cultivated in an Open Raceway Pond Energies*, 11, 1341–1352.
[11.175] Ramachandra, T. v, & Saranya, G. (2022). Sustainable Bioeconomy prospects of diatom biorefineries in the Indian west coast. *Renewable and Sustainable Energy Reviews*, 161, 112399.
[11.176] Ramachandra, T. v, Mahapatra, D. M., & Gordon, R. (2009). Milking diatoms for sustainable energy: biochemical engineering versus gasoline-secreting diatom solar panels. *Industrial & Engineering Chemistry Research*, 48(19), 8769–8788.
[11.177] Rathour, R., Gupta, J., Mishra, A., Rajeev, A. C., Dupont, C. L., & Thakur, I. S. (2020). A comparative metagenomic study reveals microbial diversity and their role in the biogeochemical cycling of Pangong lake. *Science of the Total Environment*, 731, 139074.
[11.178] RAVEN, J. A. (1983). The transport and function of silicon in plants. *Biological Reviews*, 58(2), 179–207.
[11.179] Rawat, J., Gupta, P. K., Pandit, S., Prasad, R., & Pande, V. (2021). Current perspectives on integrated approaches to enhance lipid accumulation in microalgae. *3 Biotech*, 11(6), 1–27. https://doi.org/10.1007/s13205-021-02851-3
[11.180] Reduction, D., Of, A., Reducing, D., In, T., & Animals, O. F. D. (2021). *Plant Archives*. 21(table 1), 2184–2185.
[11.181] Reitan, K. I., Rainuzzo, J. R., & Olsen, Y. (1994). Effect of nutrient limitation on fatty acid and lipid content of marine microalgae 1. *Journal of Phycology*, 30(6), 972–979.

[11.182] Renaud, S. M., Thinh, L.-V., & Parry, D. L. (1999). The gross chemical composition and fatty acid composition of 18 species of tropical Australian microalgae for possible use in mariculture. *Aquaculture*, 170(2), 147–159.

[11.183] Renaud, S. M., Thinh, L.-V., Lambrinidis, G., & Parry, D. L. (2002). Effect of temperature on growth, chemical composition and fatty acid composition of tropical Australian microalgae grown in batch cultures. *Aquaculture*, 211(1–4), 195–214.

[11.184] Renuka, N., Guldhe, A., Prasanna, R., Singh, P., & Bux, F. (2018). Microalgae as multi-functional options in modern agriculture: current trends, prospects and challenges. *Biotechnology Advances*, 36(4), 1255–1273.

[11.185] Roessler, P. G. (1988a). Changes in the activities of various lipid and carbohydrate biosynthetic enzymes in the diatom *Cyclotella cryptica* in response to silicon deficiency. *Archives of Biochemistry and Biophysics*, 267(2), 521–528.

[11.186] Roessler, P. G. (1988b). Effects of silicon deficiency on lipid composition and metabolism in the diatom *Cyclotella cryptica*. *Journal of Phycology*, 24(3), 394–400.

[11.187] Round, F. E., Crawford, R. M., & Mann, D. G. (1990). *Diatoms: Biology and Morphology of the Genera*. Cambridge University Press.

[11.188] Rukminasari, N., Omar, S. A., & Lukman, M. (2021). Temperature and nitrate concentration effect on the abundance and growth rate of *Melosira* sp. *BIOLINK (Jurnal Biologi Lingkungan Industri Kesehatan)*, 7(2), 185–194.

[11.189] Sabu, S., Singh, I. S. B., & Joseph, V. (2019). Improved lipid production in oleaginous brackish diatom *Navicula phyllept*a MACC8 using two-stage cultivation approach. *3 Biotech*, 9(12), 1–15.

[11.190] Saha, S., Enugutti, B., Rajakumari, S., & Rajasekharan, R. (2006). Cytosolic triacylglycerol biosynthetic pathway in oilseeds. Molecular cloning and expression of peanut cytosolic diacylglycerol acyltransferase. *Plant Physiology*, 141(4), 1533-1543.

[11.191] Saranya, G., & Ramachandra, T. v. (2020). Novel biocatalyst for optimal biodiesel production from diatoms. *Renewable Energy*, 153, 919–934.

[11.192] Savio, S., Farrotti, S., Paris, D., Arnaìz, E., Díaz, I., Bolado, S., Muñoz, R., Rodolfo, C., & Congestri, R. (2020). Value-added co-products from biomass of the diatoms *Staurosirella pinnata* and *Phaeodactylum tricornutum*. *Algal Research*, 47, 101830.

[11.193] Saxena, A., Marella, T. K., Singh, P. K., & Tiwari, A. (2021). Indoor mass cultivation of marine diatoms for biodiesel production using induction plasma synthesized nanosilica. *Bioresource Technology*, 332, 125098.

[11.194] Saxena, A., Mishra, B., & Tiwari, A. (2022). Mass cultivation of marine diatoms using local salts and its impact on growth and productivity. *Bioresource Technology*, 352, 127128.

[11.195] Saxena, A., Tiwari, A., Kaushik, R., Iqbal, H. M. N., & Parra-Saldívar, R. (2021). Diatoms recovery from wastewater: Overview from an ecological and economic perspective. *Journal of Water Process Engineering*, 39, 101705.

[11.196] Sayanova, O., Mimouni, V., Ulmann, L., Morant-Manceau, A., Pasquet, V., Schoefs, B., & Napier, J. A. (2017). Modulation of lipid biosynthesis by stress in diatoms. *Philosophical Transactions of the Royal Society B: Biological Sciences*, 372(1728), 20160407.

[11.197] Schmollinger, S., Mühlhaus, T., Boyle, N. R., Blaby, I. K., Casero, D., Mettler, T., Moseley, J. L., Kropat, J., Sommer, F., & Strenkert, D. (2014). Nitrogen-sparing mechanisms in *Chlamydomonas affect* the transcriptome, the proteome, and photosynthetic metabolism. *The Plant Cell*, 26(4), 1410–1435.

[11.198] Schreiber, V., Dersch, J., Puzik, K., Bäcker, O., Liu, X., Stork, S., Schulz, J., Heimerl, T., Klingl, A., & Zauner, S. (2017). The central vacuole of the diatom *Phaeodactylum tricornutum*: identification of new vacuolar membrane proteins and of a functional di-leucine-based targeting motif. *Protist*, 168(3), 271–282.

[11.199] Semenenko, V., & Vladimirova, M. (1961). Effect of cosmic flight conditions in the Sputnik-ship on the viability of *Chlorella*. *Physiol. Plants*, 8, 743–749.

[11.200] Seo, S., Kim, J., Lee, J. W., Nam, O., Chang, K. S., & Jin, E. (2020). Enhanced pyruvate metabolism in plastids by overexpression of putative plastidial pyruvate transporter in *Phaeodactylum tricornutum*. *Biotechnology for Biofuels*, 13(1), 1–11. https://doi.org/10.1186/s13068-020-01760-6

[11.201] Shahid, A., Rehman, A. U., Usman, M., Ashraf, M. U. F., Javed, M. R., Khan, A. Z., Gill, S. S., & Mehmood, M. A. (2020). Engineering the metabolic pathways of lipid biosynthesis to develop robust microalgal strains for biodiesel production. *Biotechnology and Applied Biochemistry*, 67(1), 41–51.

[11.202] Shahsavarani, H., Sugiyama, M., Kaneko, Y., Chuenchit, B., & Harashima, S. (2012). Superior thermotolerance of *Saccharomyces cerevisiae* for efficient bioethanol fermentation can be achieved by overexpression of RSP5 ubiquitin ligase. *Biotechnology Advances*, 30(6), 1289–1300.

[11.203] Sharma, N., Simon, D. P., Diaz-Garza, A. M., Fantino, E., Messaabi, A., Meddeb-Mouelhi, F., Germain, H., & Desgagné-Penix, I. (2021). Diatoms Biotechnology: Various Industrial Applications for a Greener Tomorrow. *Frontiers in Marine Science*, 8(February). https://doi.org/10.3389/fmars.2021.636613

[11.204] Sheehan, J., Dunahay, T., Benemann, J., & Roessler, P. (1998). A look back at the US Department of Energy's aquatic species program: biodiesel from algae. *National Renewable Energy Laboratory*, 328, 1–294.

[11.205] Shen, Q.-H., Gong, Y.-P., Fang, W.-Z., Bi, Z.-C., Cheng, L.-H., Xu, X.-H., & Chen, H.-L. (2015). Saline wastewater treatment by *Chlorella vulgaris* with simultaneous algal lipid accumulation triggered by nitrate deficiency. *Bioresource Technology*, 193, 68–75.

[11.206] Shi, Q., Chen, C., Zhang, W., Wu, P., Sun, M., Wu, H., Wu, H., Fu, P., & Fan, J. (2021). Transgenic eukaryotic microalgae as green factories: providing new ideas for the production of biologically active substances. *Journal of Applied Phycology*, 33(2), 705–728. https://doi.org/10.1007/s10811-020-02350-7

[11.207] Shifrin, N. S., & Chisholm, S. W. (1981). Phytoplankton lipids: Interspecific differences and effects of nitrate, silicate and light-dark cycles 1. *Journal of Phycology*, 17(4), 374–384.

[11.208] Shih, C. Y., Kang, L. K., & Chang, J. (2015). Transcriptional responses to phosphorus stress in the marine diatom, *Chaetoceros affinis*, reveal characteristic genes and expression patterns in phosphorus uptake and intracellular recycling. *Journal of Experimental Marine Biology and Ecology*, 470, 43–54. https://doi.org/10.1016/j.jembe.2015.05.001

[11.209] Sicko-Goad, L. M., Schelske, C. L., & Stoermer, E. F. (1984). Estimation of intracellular carbon and silica content of diatoms from natural assemblages using morphometric techniques 1. *Limnology and Oceanography*, 29(6), 1170–1178.

[11.210] Sims, P. A., Mann, D. G., & Medlin, L. K. (2006). Evolution of the diatoms: insights from fossil, biological and molecular data. *Phycologia*, 45(4), 361–402.

[11.211] Smith, S. R., Abbriano, R. M., & Hildebrand, M. (2012). Comparative analysis of diatom genomes reveals substantial differences in the organization of carbon partitioning pathways. *Algal Research*, 1(1), 2–16.

[11.212] Smith, S. R., Gillard, J. T. F., Kustka, A. B., McCrow, J. P., Badger, J. H., Zheng, H., New, A. M., Dupont, C. L., Obata, T., & Fernie, A. R. (2016). Transcriptional orchestration of the global cellular response of a model pennate diatom to diel light cycling under iron limitation. *PLoS Genetics*, 12(12), e1006490.

[11.213] Smol, J. P., & Stoermer, E. F. (2010). *The Diatoms: Applications for the Environmental and Earth Sciences*. Cambridge University Press.

[11.214] Stonik, V., & Stonik, I. (2015). Low-molecular-weight metabolites from diatoms: structures, biological roles and biosynthesis. *Marine Drugs*, 13(6), 3672–3709.

[11.215] Sun, X. M., Ren, L. J., Zhao, Q. Y., Ji, X. J., & Huang, H. (2019). Enhancement of lipid accumulation in microalgae by metabolic engineering. *Biochimica et Biophysica Acta - Molecular and Cell Biology of Lipids*, 1864(4), 552–566. https://doi.org/10.1016/j.bbalip.2018.10.004

[11.216] Suryaningtyas, I. T., Permadi, S., Sapulete, S., & Widyartini, D. S. (2022). Growth and Lipid Profiles of *Melosira* sp. in response to Different Salinity Levels. *Journal of Aquaculture and Fish Health*, 11(2), 216–226.

[11.217] Tachibana, M., Allen, A. E., Kikutani, S., Endo, Y., Bowler, C., & Matsuda, Y. (2011). Localization of putative carbonic anhydrases in two marine diatoms, *Phaeodactylum tricornutum* and *Thalassiosira pseudonana*. *Photosynthesis Research*, 109(1), 205–221.

[11.218] Taguchi, S., Hirata, J. A., & Laws, E. A. (1987). Silicate deficiency and lipid synthesis of marine diatoms 1, 2. *Journal of Phycology*, 23, 260–267.

[11.219] Takagi, M., & Yoshida, T. (2006). Effect of salt concentration on intracellular accumulation of lipids and triacylglyceride in marine microalgae *Dunaliella* cells. *Journal of Bioscience and Bioengineering*, 101(3), 223–226.

[11.220] Tanaka, T., Maeda, Y., Veluchamy, A., Tanaka, M., Abida, H., Maréchal, E., Bowler, C., Muto, M., Sunaga, Y., & Tanaka, M. (2015). Oil accumulation by the oleaginous diatom *Fistulifera solaris* as revealed by the genome and transcriptome. *The Plant Cell*, 27(1), 162–176.

[11.221] Tanvir, R. U., Zhang, J., Canter, T., Chen, D., Lu, J., & Hu, Z. (2021). Harnessing solar energy using phototrophic microorganisms: A sustainable pathway to bioenergy, biomaterials, and environmental solutions. *Renewable and Sustainable Energy Reviews*, 146(August 2020), 111181. https://doi.org/10.1016/j.rser.2021.111181

[11.222] Terashima, M., Specht, M., & Hippler, M. (2011). The chloroplast proteome: a survey from the *Chlamydomonas reinhardtii* perspective with a focus on distinctive features. *Current Genetics*, 57(3), 151–168.

[11.223] Thomas Kiran, M., Archana, T., & Bhaskar, M. v. (2015). A new novel solution to grow diatom algae in large natural water bodies and its impact on CO_2 capture and nutrient removal. *Journal of Algal Biomass Utilization*, 6(2), 22–27.

[11.224] Thomas, W. H., Dodson, A. N., & Reid, F. M. H. (1978). Diatom productivity compared to other algae in natural marine phytoplankton assemblages. *Journal of Phycology*, 14(3), 250–253.

[11.225] Thompson, A. W., Huang, K., Saito, M. A., & Chisholm, S. W. (2011). Transcriptome response of high-and low-light-adapted *Prochlorococcus* strains to changing iron availability. *The ISME Journal*, 5(10), 1580–1594.

[11.226] Tiwari, A., & Marella, T. K. (2019). Potential and application of diatoms for industry-specific wastewater treatment. *Application of Microalgae in Wastewater Treatment*, 321–339.

[11.227] Tiwari, A., Marella, T. K., & Saxena, A. (2022). Diatom biorefinery: From carbon mitigation to high-value products. In *Biomass, Biofuels, Biochemicals* (pp. 401–420). Elsevier.

[11.228] Tiwari, A., Melchor-Martínez, E. M., Saxena, A., Kapoor, N., Singh, K. J., Saldarriaga-Hernández, S., Parra-Saldívar, R., & Iqbal, H. M. N. (2021). Therapeutic attributes and applied aspects of biological macromolecules (polypeptides, fucoxanthin, sterols, fatty acids, polysaccharides, and polyphenols) from diatoms—a review. *International Journal of Biological Macromolecules*, 171, 398–413.

[11.229] Tokushima, H., Inoue-Kashino, N., Nakazato, Y., Masuda, A., Ifuku, K., & Kashino, Y. (2016). Advantageous characteristics of the diatom *Chaetoceros gracilis* as a sustainable biofuel producer. *Biotechnology for Biofuels*, 9(1), 1–19.

[11.230] Treguer, P., Nelson, D. M., van Bennekom, A. J., DeMaster, D. J., Leynaert, A., & Quéguiner, B. (1995). The silica balance in the world ocean: a reestimate. *Science*, 268(5209), 375–379.

[11.231] Trentacoste, E. M., Shrestha, R. P., Smith, S. R., Glé, C., Hartmann, A. C., Hildebrand, M., & Gerwick, W. H. (2013). Metabolic engineering of lipid catabolism increases microalgal lipid accumulation without compromising growth. *Proceedings of the National Academy of Sciences*, 110(49), 19748–19753.

[11.232] Upadhyay, S., Lawrence, K., Shekhar, S. K., & Anantharaman, P. (2017). Lipid screening, extraction and production of biodiesel from marine diatom *Odontella aurita*. *Int. J. Curr. Microbiol. App. Sci*, 6(3), 1259–1269.

[11.233] Valenzuela, J. J., López García de Lomana, A., Lee, A., Armbrust, E. V, Orellana, M. V, & Baliga, N. S. (2018). Ocean acidification conditions increase resilience of marine diatoms. *Nature Communications*, 9(1), 1–10.

[11.234] Valenzuela, J., Mazurie, A., Carlson, R. P., Gerlach, R., Cooksey, K. E., Peyton, B. M., & Fields, M. W. (2012). Potential role of multiple carbon fixation pathways during lipid accumulation in *Phaeodactylum tricornutum*. *Biotechnology for Biofuels*, 5(1), 1–17.

[11.235] Varjani, S., Shah, A. v, Vyas, S., & Srivastava, V. K. (2021). Processes and prospects on valorizing solid waste for the production of valuable products employing bio-routes: a systematic review. *Chemosphere*, 282, 130954.

[11.236] Vavitsas, K., Kugler, A., Satta, A., Hatzinikolaou, D. G., Lindblad, P., Fewer, D. P., Lindberg, P., Toivari, M., & Stensjö, K. (2021). Doing synthetic biology with photosynthetic microorganisms. *Physiologia Plantarum*, 173(2), 624–638. https://doi.org/10.1111/ppl.13455

[11.237] Vella, F. M., Sardo, A., Gallo, C., Landi, S., Fontana, A., & d'Ippolito, G. (2019). Annual outdoor cultivation of the diatom *Thalassiosira weissflogii*: productivity, limits and perspectives. *Algal Research*, 42, 101553.

[11.238] Villanova, V., & Spetea, C. (2021). Mixotrophy in diatoms: Molecular mechanism and industrial potential. *Physiologia Plantarum*, 173(2), 603–611.

[11.239] Vinayak, V., Gordon, R., Gautam, S., & Rai, A. (2014). Discovery of a diatom that oozes oil. *Advanced Science Letters*, 20(7–8), 1256–1267.

[11.240] Vinayak, V., Manoylov, K. M., Gateau, H., Blanckaert, V., Hérault, J., Pencréac'h, G., Marchand, J., Gordon, R., & Schoefs, B. (2015). Correction: Vinayak, V., et al. Diatom Milking: A Review and New Approaches. Marine Drugs 2015, 13, 2629–2665. *Marine Drugs*, 13(12), 7301.

[11.241] Wagner, H., Jakob, T., & Wilhelm, C. (2006). Balancing the energy flow from captured light to biomass under fluctuating light conditions. *The New Phytologist*, 169(1), 95–108.

[11.242] Wang, J.-K., & Seibert, M. (2017). Prospects for commercial production of diatoms. *Biotechnology for Biofuels*, 10(1), 1–13.

[11.243] Wang, Q., Lu, Y., Xin, Y., Wei, L., Huang, S., & Xu, J. (2016). Genome editing of model oleaginous microalgae *Nannochloropsis* spp. by CRISPR/Cas9. *The Plant Journal*, 88(6), 1071–1081.

[11.244] Wang, X., Dong, H.-P., Wei, W., Balamurugan, S., Yang, W.-D., Liu, J.-S., & Li, H.-Y. (2018). Dual expression of plastidial GPAT1 and LPAT1 regulates triacylglycerol production and the fatty acid profile in *Phaeodactylum tricornutum*. *Biotechnology for Biofuels*, 11(1), 1–14.

[11.245] Wang, X., Liu, S.-F., Li, R.-Y., Yang, W.-D., Liu, J.-S., Lin, C. S. K., Balamurugan, S., & Li, H.-Y. (2020). TAG pathway engineering via GPAT2 concurrently potentiates abiotic stress tolerance and oleaginicity in *Phaeodactylum tricornutum*.

[11.246] Wang, X., Luo, S.-W., Luo, W., Yang, W.-D., Liu, J.-S., & Li, H.-Y. (2019). Adaptive evolution of microalgal strains empowered by fulvic acid for enhanced polyunsaturated fatty acid production. *Bioresource Technology*, 277, 204–210.

[11.247] Wang, Y., Fang, J., Leonard, S. S., & Rao, K. M. K. (2004). Cadmium inhibits the electron transfer chain and induces reactive oxygen species. *Free Radical Biology and Medicine*, 36(11), 1434–1443.

[11.248] Weber, T., Gruber, A., & Kroth, P. G. (2009). The presence and localization of thioredoxins in diatoms, unicellular algae of secondary endosymbiotic origin. *Molecular Plant*, 2(3), 468–477.

[11.249] Whitney, L. P., Lins, J. J., Hughes, M. P., Wells, M. L., Chappell, P. D., & Jenkins, B. D. (2011). Characterization of putative iron responsive genes as species-specific indicators of iron stress in thalassiosiroid diatoms. *Frontiers in Microbiology*, 2, 234.

[11.250] Wu, S., Gu, W., Huang, A., Li, Y., Kumar, M., Lim, P. E., Huan, L., Gao, S., & Wang, G. (2019). Elevated CO_2 improves both lipid accumulation and growth rate in the glucose-6-phosphate dehydrogenase engineered *Phaeodactylum tricornutum*. *Microbial Cell Factories*, 18(1), 1–16.

[11.251] Xiaodong, D., Xiaowen, F., & Yajun, L. (2011). The effects of nutritional restriction on neutral lipid accumulation in *Chlamydomonas* and *Chlorella*. *African Journal of Microbiology Research*, 5(3), 260–270.

[11.252] Xie, W., & Wang, J. (2012). Immobilized lipase on magnetic chitosan microspheres for transesterification of soybean oil. *Biomass and Bioenergy*, 36, 373-380.

[11.253] Xie, W. H., Pang, F., Niu, Y. F., Zhang, M. H., Yang, W. D., Liu, J. S., & Li, H. Y. (2013). Functional characterization of an ACCase subunit from the diatom *Phaeodactylum tricornutum* expressed in *Escherichia coli*. Biotechnology and applied biochemistry, 60(3), 330-335.

[11.254] Xin, L., Hong-Ying, H., & Yu-Ping, Z. (2011). Growth and lipid accumulation properties of a freshwater microalga *Scenedesmus* sp. under different cultivation temperature. *Bioresource Technology*, 102(3), 3098–3102.

[11.255] Xuan, R. N., Safitri, I., Mouget, J. L., Pruvost, J., Turpin, V., & Jaouen, P. (2020). Design of an artificial culture medium to optimize *Haslea ostrearia* biomass and marennine production. *Algal Research*, 45, 101653.

[11.256] Yang, Z.-K., Niu, Y.-F., Ma, Y.-H., Xue, J., Zhang, M.-H., Yang, W.-D., Liu, J.-S., Lu, S.-H., Guan, Y., & Li, H.-Y. (2013). Molecular and cellular mechanisms of neutral lipid accumulation in diatom following nitrogen deprivation. *Biotechnology for Biofuels*, 6(1), 1–14.

[11.257] Yi, Z., Xu, M., Di, X., Brynjolfsson, S., & Fu, W. (2017). Exploring valuable lipids in diatoms. *Frontiers in Marine Science*, 4(JAN), 1–10. https://doi.org/10.3389/fmars.2017.00017

[11.258] Yodsuwan, N., Sawayama, S., & Sirisansaneeyakul, S. (2017). Effect of nitrogen concentration on growth, lipid production and fatty acid profiles of the marine diatom *Phaeodactylum tricornutum*. *Agriculture and Natural Resources*, 51(3), 190–197.

[11.259] Yu, E. T., Zendejas, F. J., Lane, P. D., Gaucher, S., Simmons, B. A., & Lane, T. W. (2009). Triacylglycerol accumulation and profiling in the model diatoms *Thalassiosira pseudonana* and *Phaeodactylum tricornutum* (Baccilariophyceae) during starvation. *Journal of Applied Phycology*, 21(6), 669–681.

[11.260] Zhang, M., Xing, Y., Wang, F., Mi, T., & Zhen, Y. (2020). Responses of triacylglycerol synthesis in *Skeletonema marinoi* to nitrogen and phosphate starvations. *Journal of Phycology*, 56(6), 1505–1520.

[11.261] Zhu, X.-G., Long, S. P., & Ort, D. R. (2008). What is the maximum efficiency with which photosynthesis can convert solar energy into biomass? *Current Opinion in Biotechnology*, 19(2), 153–159.

[11.262] Zulu, N. N., Zienkiewicz, K., Vollheyde, K., & Feussner, I. (2018). Current trends to comprehend lipid metabolism in diatoms. *Progress in Lipid Research*, 70, 1–16. https://doi.org/10.1016/j.plipres.2018.03.001

12

Implications of Diatoms for Heavy Metal Bioremediation

Varad Nagar[1], Vinay Aseri[2], Rushikesh Chopade[3], Pritam P. Pandit[4], Badal Mavry[5], Apoorva Singh[3], Garima Awasthi[6], Kumud Kant Awasthi[6] and Mahipal Singh Sankhla[7,8]*

[1]*Department of Forensic Science, Banaras Hindu University, Varanasi, Uttar Pradesh, India*
[2]*School of Forensic Science and Risk Management, Rashtriya Raksha University, Gandhinagar, Gujrat, India*
[3]*Department of Forensic Science, Vivekananda Global University, Jaipur, Rajasthan, India*
[4]*Department of Forensic Science, School of Science, Sandip University, Nashik, Maharashtra, India*
[5]*Department of Forensic Science (SBMS), Galgotias University, Greater Noida, Uttar Pradesh, India*
[6]*Department of Life Sciences, Vivekananda Global University, Jaipur, Rajastha, India*
[7]*Department of Forensic Science, (UIAHS), Chandigarh University, Mohali, Punjab, India*
[8]*University Centre for Research and Development (UCRD), Chandigarh University, Mohali, Punjab, India*

Abstract

Alarming levels of heavy metals are hazardous to the aquatic ecology, resulting in a significant loss of species variety. Pollutant biosorption employing naturally inspired sources such as microalgae provides a number of advantages. Diatoms, a varied species of phytoplankton that makes up approximately 45% of marine primary production, have a remarkable capacity for survival in contaminated aquatic environments. They are important for metal biogeochemistry in settings with fresh and salt water. Chemical pollutants and dangerous metals from contaminated locations are degraded, speciated, and detoxified in large parts. Diatoms have several advantages, including plentiful availability, low cost, high

**Corresponding author*: mahi4n6@gmail.com

Vandana Vinayak and Richard Gordon (eds.) Diatom Cultivation for Biofuel, Food and High-Value Products, (323–340) © 2025 Scrivener Publishing LLC

metal removal efficiency, and environmental friendliness. This chapter discusses new developments and methods in diatom-based bioremediation and biosorption of hazardous heavy metals, with the intention of providing relevant information to aid the development of efficient and economically viable heavy metal bioremediation technology by implication of diatoms.

Keywords: Diatoms, heavy metals, toxicity, remediation

12.1 Introduction

Heavy metals are an extreme danger to sea-going organic entities and a worldwide issue for the climate, biosphere, hydrosphere, and lithosphere—the four major circles of the Earth. Clean water is a fundamental need for all living things in the world. Water bodies are being undermined by developing mechanical leaps forwards and contemporary anthropogenic activities [12.1]–[12.3]. Plenty of new toxins are being created consistently, with the latest items originating from river beds, industrialization, fuel ignition, vehicles, and purifying tasks [12.4] [12.5]. Disturbing degrees of water contamination are perilous to oceanic nature, bringing about a huge loss of different animal species. In view of their ability to continue, gather, and bio-enhance in the sea-going pecking order, heavy metals can hurt living things [12.6] [12.7]. Intracellular and extracellular absorption frameworks in phytoplankton come directly from metal particles in water. Phytoplankton, being an essential creator, is one of the first to be presented to and impacted by these contaminants. Diatom green growth is the most widely recognized phytoplankton in the amphibian biological system, and they assume a significant part in eliminating heavy metals from the contaminated climate and making them innocuous, permitting the "dirty" climate to revert to its generally expected condition. This is a savvy and ecologically harmless technique for taking out heavy metal from the climate [12.7].

Diatoms are intriguing bioindicators for really monitoring changes in environmental setups due to their remarkable capacity to survive in polluted water bodies. Diatoms are the most remarkable arrangement of photosynthetic eukaryotic microorganisms today, and they can be tracked down in basically any sort of oceanic biological system. Dinoflagellates can have them as endosymbionts [12.8] [12.9]. A long time ago, diatoms emerged because of an endosymbiotic connection between red green growth and heterotrophic bacteria [12.10]. In spite of the fact that water contains an assortment of microalgae, diatoms are the most significant and are unmistakable since they are the main species having silica in their cells [12.10] [12.11]. Frustules are a kind of silica cell wall that safeguards plants from photoinhibition, gives legitimate supplement take-up, controls the sinking rate, and fills in as a mechanical barrier against slow eaters [12.12]. Their silica focus differs by species and is impacted by elements such as size, development design, and ecological factors like light, temperature, saltiness, supplements, and development stage [12.13]. There are north of 200,000 living diatom species, every one of which is found in an alternate climate [12.14]. They can be tracked down in every single oceanic territory, including lentic, lotic, and sand living spaces. Over two-thirds of species are tracked down in freshwater, with 83% being benthic [12.15]. Diatom green growth has an assortment of abilities to survive that permit them to flourish in various oceanic conditions. Subsequently, it fills in as a critical essential creator on an overall scale, representing 20% of carbon obsession and, generally speaking, 40% of marine creation [12.16]. Because of their developmental heredity, they contrast extraordinarily from other green growth species as far as cell and metabolic pathways [12.17] [12.18]. Diatoms can be tracked down in every photic zone and are very sensitive to even small changes in the physicochemical elements of their environmental factors, making them the ideal possibility for biological observation [12.19] [12.20]. Their broad range of natural versatility [12.21] legitimizes their future applications in various fields [12.22].

The continuing entry of partially treated and untreated wastewater could jeopardize the sea-going pecking order. Heavy metals are incredibly unsafe because of their dependable nature [12.23] [12.24]. Thus, there is a pressing need to foster innovation that can lessen their utilization while likewise taking into consideration their proficient removal from the climate. Adsorption utilizing enacted carbon, film filtration, particle trade, and substance precipitation are existing innovations for heavy metal removal from wastewater, yet they are all energy intensive, costly, and require ongoing upkeep [12.25]. Heavy metals, herbicides, and other side-effects can be eliminated by means of green growth-based wastewater treatment, known as phytoremediation [12.25] [12.26]. Since the rapid growth of diatoms makes them pervasive, they are profoundly versatile in climate. Their non-poisonous, eco-accommodating, financially savvy, supportable framework is reusable, recyclable, and equipped for remediating heavy metal with low energy input due to autotrophy, which has energized the interest of researchers everywhere. Diatoms have developed to take advantage of their profoundly productive cell hardware to fight pollutants and to involve wastewater as a development substrate [12.27] [12.28]. Their three-layered frustule nanoarchitecture can possibly eliminate heavy metal that taint wastewater [12.29]. To battle impurities, diatoms utilize biotic and abiotic instruments that are very versatile in amassing poisons under various ecological circumstances [12.30] [12.31]. Subsequently, utilization of diatoms is a financially savvy inventive methodology which purifies harmful substances as well as fills in as a suitable long-haul system for reusing heavy metals. Recent years have seen a major increase in interest in phytoremediation, including the use of microalgae or macroalgae for removal of heavy metal particles detected in water testing [12.32].

12.2 Mechanism for Heavy Metal Removal by Diatoms

The use of miniature or full-scale green growth in the treatment of waste and wastewaters is known as phycoremediation [12.33]. Due to their viable cell components and versatile methods, diatom green growth has the ability to phycoremediate various wastewaters. They might integrate an assortment of large-scale and micro-nutrients from wastewater into their biomass. At the point when diatoms are filled in wastewater, their biomass can be utilized to make an extensive variety of high-esteem compounds with applications in bioenergy, restorative science, and food and nourishment [12.34]. Diatoms control their enzymatic action, permitting them to detox heavy metals by dehalogenation and denitrification [12.28]. Diatoms enact a metal exclusion component because of heavy metal openness to keep up with the balance of the metal fixation in the climate. Heavy metal-tolerant diatoms can produce phytochelatins, macromolecules, and exocellular polysaccharides to act as a barrier to extracellular adsorption in order to reduce their toxicity [12.35]. Phytochelatins and metallothioneins, which make metal chelated structures in diatoms, help in the amassing and stockpiling of heavy metals away from vacuoles and, furthermore, as insoluble salts in the cytoplasm [12.36]. Cadmium (Cd) and other heavy metals are moved intracellularly through the manganese (Mn) and zinc (Zn) retention frameworks [12.37]. *Skeletonema costatum* showed a heavy metal chelated complex vehicle and capacity component because of cadmium (Cd) and copper (Cu) stress, with upgraded capacity in the vacuole.

12.3 Bioremediation and Biosorption of Heavy Metals

Heavy metal waste is treated in various ways. Bioremediation is a procedure where a living being goes about as an organic debasing specialist. The development of ecological biotechnology in the field of contaminant control through the use of organic cycles has led to the development of bioremediation, which is both practical and engaging. As a rule, two bioremediation strategies have been created: *in-situ* bioremediation and *ex-situ* bioremediation [12.38].

The heavy metal bioremediation process mainly takes place through two mechanisms: dynamic uptake and latent uptake. Heavy metal particles bind to the cell wall in two distinct ways, the first of which is the exchange of monovalent and divalent particles, such as the replacement of Na, Mg, and Ca in the cell walls by heavy metal particles; the next step is the formation of a compound between the heavy metal particle and functional groups like carbonyl, amino, thiol, hydroxy, and carboxyl hydroxy located on the contact wall. While dynamic uptake may occur in a variety of live cells, this cycle rotates quickly. At the same time, this component happens in accordance with the utilization of the metal particle in the development of microorganisms and intracellular collection of the metal particle. The heavy metals can likewise be stored in the digestion and discharge process on the subsequent level. The pH and the presence of various particles are two factors that have an influence on the uptake unrelated interaction. pH and the presence of various particles have an influence on the detached uptake process, whereas temperature, the strength of the ionic bonds, light, and other factors have an impact on the dynamic uptake process [12.39]–[12.43].

One form of remediation that uses elements of science as a remediator is bioremediation. This approach doesn't affect the environment or spread tainted biota. It has been determined from a few studies that green growth is capable of acting as bioremediation. *Porphyra leucosticta*, a red-green growth, is 90% capable of removing Pb(II) [12.39]. *Skeletonema* sp. has the ability to retain heavy metals in two ways: retention and adsorption. Given that *Skeletonema* sp. has cellulose in its cell walls, adsorption occurs. Cellulose in the cell walls has functional groups, such as hydroxyl, which can bind with heavy metals [12.44]. Heavy metal biosorption utilizing organic material has arisen as a possible option rather than the current traditional physicochemical strategies [12.45]–[12.47]. Among the organic material, green growth has ended up being worthwhile on the grounds that it provides a few benefits, for example, low cost, possible metal recovery, lesser volume of synthetic and/or natural sludge, high proficiency in weakening effluents, and enormous surface region to volume proportion [12.48].

The ability of natural materials to absorb heavy metals from wastewater via metabolically mediated or physicochemical pathways of uptake is known as biosorption [12.49]. Biosorption, which is the capacity of certain microbial biomaterials to accumulate heavy metals from even the weakest fluid arrangements, offers a possible and monetarily appealing other option. The biosorption interaction includes a solid phase (sorbent or biosorbent; biological material) and a fluid phase (solvent, normally water) containing dissolved species to be adsorbed (adsorbate, metal). Because of the preference of the adsorbent for the adsorbate species, the last option is to accumulate various components, thereby binding them. The cycle goes on till harmony is achieved between how much the strongly bound adsorbate species and its part stay in the arrangement. The level of adsorbent partiality for the adsorbate decides its dissemination between the solid and fluid phases [12.50].

12.4 Challenges

Notwithstanding the numerous appealing benefits of microalgae in HM bioremediation, their broad application actually faces a few difficulties, like pollution by different microorganisms, popularity for certain supplements, trouble in gathering, and early downstream processing. Specifically, the plan of action of utilizing microalgae to settle HM contamination has not yet been shaped. It's somewhat intricate innovation and costly expense make it difficult to advance microalgae for wider use. Microalgae are small and carry a negative surface charge that keeps a steady dispersion state, which is why harvesting microalgae from water or culture media is a difficult task [12.51]. Due to their somewhat faster rate of development, bacteria, parasites, and other microorganisms found in heavy metal effluent may be seeking micronutrients and perhaps posing a threat to the environment. Wastewater could contain an enormous amount of total suspended solids (TSS) and has high turbidity that diminishes light entrance and blocks microalgae development. The presence of heavy metal left over in a microalgal cell could obstruct downstream handling of an important element, which is part of the difficulty in bioremediation of weighty metals [12.52].

A few elements influence adsorption of heavy metals by diatoms, for example, concentration of heavy metals, size of biomass, temperature, pH, and so on. The underlying centralization of metals assumes a huge part in heavy metal removal. Adsorption increases and eventually gets submerged when wastewater's metal fixation increases. Diatoms have several functional groups on their surface, and low-priority functional groups adsorb metal at high concentrations, whereas high-priority functional groups adsorb metal at low concentrations. Subsequently, the expulsion of the heavy metal declines with expansions in fixation [12.53].

Essentially, pH is likewise an element of heavy metal adsorption from wastewater. Be that as it may, the ideal pH to improve most extreme metal removal by diatoms is as yet not completely examined. According to studies, diatoms' metal absorption increases as the pH rises [12.54].

12.5 Advantage of Diatoms Over Other Techniques and Algae

Diatoms are unique microalgae that may be categorized as araphid pennants, raphid pennants, spiral, bipolar, and multipolar centric based on their cell walls. These five significant categories were started and differentiated consecutively during the Mesozoic time under diminishing carbon dioxide (CO_2) levels forming a high-level carbon concentrating instrument [12.55]. Throughout the Eocene epoch, this led to an increase in the variety of diatom species, producing global cooling with a sharp decline in barometer CO_2 levels. Because of their hard silica cell wall, diatoms display higher sinking rates which help in quicker remote ocean entombment of nutrients, and they are known to be essential supporters of present-day oil holds because of their high lipid content. Their biosilica, or silica cell wall, is a permeable, three-layered, three-dimensional structure known as a frustule. It serves as a possible pH cushion framework, improving carbonic anhydrase activity close to the cell surface, which improves the conversion of bicarbonate to carbon dioxide (CO_2) [12.56].

Diatoms have far higher photosynthetic activity than plants do in low light conditions. For example, the marine diatom *Phaeodactylum tricornutum* converts light energy into biomass much more efficiently than *Chlorella vulgaris* does under low irradiance [12.57]. Diatoms are useful for examining the water composition of lakes, streams, and oceans. They ultimately reduce dangerous metal fixation in water resources and can serve as an alternative to treating wastewater since they are more resistant to the damaging metals that are diverted from other microbial groups. Since diatoms are bountiful photoautotrophs in new and marine waters, they take on various procedures for endurance and development when presented with high metal focus. Some of the diatoms used in remediation of heavy metals and their mechanisms are discussed in Table 12.1. Their atomic system is very extraordinary, which empowers them to separate essential and non-essential heavy metals for their development [12.58].

12.6 Production of Diatoms on a Commercial Scale and Its Application

Most green *Thalassiosira* growth, including diatoms, typically combine excellent nutrients, vital unsaturated fats, amino acids, and dietary supplements, and is comparable to conventional vegetable sources. Earlier

Table 12.1 Species of diatoms used in the remediation of heavy metals.

Diatom species	Removed heavy metal	Mechanism involved	Reference
Planothidium lanceolatum	Cd; Cu; Zn	Surface ligand modification	[12.59]
Achnanthidium minutissimum	Cd; Pb	Ionic strength, carboxyl and silanol groups, and adsorption	[12.60]
Nitzschia closterium	Cu; Zn; Co; Mn	Adsorption	[12.61]
Thalassiosira oceanica	Cd	Absorption (Fe deficiency; increased Cd intake)	[12.60]
Nitzschia palea	Cd	Absorption (Phytochelation)	[12.60]
Phaeodactylum tricornutum	Cd; Pb; Cu	(Phytochelatins and antioxidant enzymes) Absorption	[12.62]
Thalassiosira weissflogii	Cd	(Metallothionein Transport System) Absorption	[12.60]
Nitzschia closterium	Zn, Cu, Co, and Mn	Adsorption, primarily van der Waals contacts and hydrogen bonds	[12.61]
Phaeodactylum tricornutum	Cu, Cd, and Pb	Phytochelatins and antioxidant enzymes absorb nutrients	[12.63]
Thalassiosira pseudonana	Zn, Cd, and Cu	Chemical adsorption	[12.64]
Chaetoceros costatum	Hg	Living and non-living diatoms play a role in the absorption mechanism, although they are all dependent on the separation of Hg via hydrogen bonds	[12.65]
Thalassiosira oceanica	Cd	Mostly through chemical bonding, chemical absorption	[12.60]

impediments forced upon large-scale manufacturing of such items from diatoms were basically restricted by creation expenses and agreeability, eventually resulting in such algal items being restricted to such wholesome applications as health food, beauty products, and animal feed [12.66].

Normally, the physicochemical synthesis of NPs for commercial applications is time-consuming, energy-intensive, and necessitates hazardous/harmful synthetic ingredients, high temperatures, and tensions—all of which are climate-disrupting factors. As it was demonstrated, control over the temperature and hydrolysis time greatly affected the normal molecule size and shape. Consequently, there is increasing interest in applying more secure, practical, quick, and greener techniques for creating nanostructures. Natural, inexhaustible, and reasonable assets can be used by means of financially savvy and eco-accommodating methodologies to produce these NPs, subsequently wiping out or diminishing the difficult and muddled procedures [12.67] [12.68]. Permeable silica micro shells acquired from diatoms have shown unique possibilities for medication and quality drug delivery applications. The interesting models of diatoms, with high temperature tolerance, high surface region (up to 200 m^2 g^{-1}), basic substance functionalization/adjustment methods, adequate biocompatibility, simplicity of hereditary controls, optical/photonic attributes, mechanical opposition, and eco-friendliness, lift them as brilliant medication/quality drug delivery systems. For working on the adequacy of anticancer medication delivery and diminishing/forestalling their related secondary effects, core-shell biosilica-based frameworks were used for the chemotherapeutic medications' exemplification with controlled and synergistic conveyance highlights [12.68]–[12.70]. Diatoms are considered significant wellsprings of anticancer agents that can possibly be applied as diatom-based anticancer drug-delivery frameworks. With certain nutrients and antibodies added, diatom-based NPs were modified and functionalized so they could be utilized against cancerous cells. These NPs also had a longer drug release time, more particularity/selectivity, and fewer unintended side effects. Diatom-based anticancer combinations and drug delivery systems present interesting avenues for further preclinical and clinical research in drug development and clinical trials. Different species of diatoms were used to create a concentrate, which was then tested for potential cell reinforcement, alleviating anticancer, antidiabetic, antibacterial, and antibiofilm actions [12.69] [12.71]. For biofuel creation, diatoms are promising up-and-comers because of their bountiful nature, practical handling conventions, and quick and controllable development. It was shown that growing *Cyclotella cryptica* and *Thalassiosira weissflogii* diatoms in a two-stage process under nitrogen restriction might surprisingly increase the amounts of lipids. Strikingly, nitrogen limit didn't influence the pace of development; be that as it may, it impacted the rebuilding of lipid and also creation of triacylglycerols in *T. weissflogii* and *C. cryptica*. These diatoms are a compelling candidate for the up-versatile production of biofuels since *T. weissflogii* might account for up to 88% and 82% of all glycerolipids, respectively. Additionally, the oil recognized from *Nitzschia punctata* diatom was employed as a biocatalyst for the enzymatic transesterification of the lipase protein with high strength obtained from *Cladosporium tenuissimum* growths. This biocatalyst demonstrated excellent efficiency of biodiesel synthesis [12.70]. Diatoms are extensively used for the phytoremediation of heavy metal contamination in wastewaters due to their exceptional cell, metabolic, and photosynthetic capabilities. In one study, the diatoms *Navicula pelliculosa* strain CCMP543 and *P. tricornutum* strain CCY0033 were used to bioremediate dichromate and produce lipids that might be used as a biofuel. To accomplish critical yields of biomass, amassing of lipids, and dichromate disposal, different development conditions were assessed. Thus, the ideal end of dichromate and creation of lipids could be achieved for the two diatoms. Essentially, the diatom *Planothidium lanceolatum* was used to remove Cd, Cu, and Zn by biosorption from aqueous environments; the best pH values for the maximum Cu removal were 6.0 and 8.0, respectively. The diatom's capacity for biosorption may also be improved by raising the temperature (from 15 to 25 °C) and metal particle fixation. Apparently, further insightful examinations are still expected to track down the specific components and involved functional groups for the biosorption of tainted metal by diatoms. Moreover, it was shown that a few proteins of the frustule layer, for example, frustules, could be related to ecological pressure [12.71] [12.72]. The financial practicality of a novel creation of biomolecules by designed creatures is reliant upon modest refining. Like other photosynthetic organic entities, diatoms require simply carbon dioxide, water, inorganic salts, and light to develop, so refined media for up-and-comers of interest would be quickly accessible in ocean or hard water, in this manner making medium improvement with natural accumulation superfluous [12.73]. Among the misleadingly combined substances for use in medication and drug storage are antibodies, immunizations, chemicals, and proteins. These are at present created in microorganisms, yeast, and mammalian cell culture, which have the downside of being poisonous, infectious, and costly Plants have been utilized in the past to avoid these negative effects; however, they have the additional weakness of transgenic dust, which perhaps prompts undesired engendering of hereditary modifications [12.74]. There are a few strategies for the remediation of wastewater like existing techniques, compound strategies, natural strategies, and aggregate techniques. The existing strategies, which include the utilization of initiated carbon, graphene oxide, filtration and flocculation for the adsorption of heavy metal pollutants though synthetic techniques, suggests ozonation, oxidation-decrease responses, ionizing radiations, and UV therapies, yet these strategies involve costly hardware, harmful synthetic substances, skilled labor, and is tedious, not very practical, and may only somewhat eliminate heavy metals from wastewater. However, these methods are not a manageable answer for the long-term objective. Consequently, to keep up with and support our sea-going biological system from contamination, a delayed mechanical idea which remediates wastewater in addition to being nature-accommodating is high on the request list. Algal-based remediation is a natural solution that skillfully addresses the problem by degrading the pollutants in the wastewater. The major practical method for phytoremediation of heavy metals from wastewater is the diatom algal intervention invention, which is extremely promising in biodegradation and photodegradation of heavy metal pollution in wastewater [12.75].

12.7 Future Aspects

The need to improve data in the largest number of diatoms is the focus of research involving various light systems for the development of marine diatoms. Light control in the development framework has been demonstrated to be a critical method for supporting microalgal efficiency and quality including marine diatoms, albeit the response to light isn't just summed up among photosynthetic microorganisms. Finding the best light conditions for developing marine diatoms could help their worth as biotechnology organisms [12.76]. Different types of life are impacted by trophic status and natural micropollutants, with tube-forming provincial diatoms more sensitive to these pollutants and the following diatoms experiencing increased levels of trophic and ecological pollution. Natural societies are associated with actual aggravation because of flows, in addition to supplement focuses. High-profile cells are prone to low streams and unpleasant situations, low-profile cells are prone to high streams, and motile cells are common in surroundings with abundant supplements [12.77]. The blend of scientific categorization-free datasets and AI strategies to evaluate water quality in view of diatom arrays appears to be encouraging; nonetheless, more examinations need to be finished to approve steady outcomes to be utilized for customary biomonitoring. As opposed to other microalgae, diatoms so far have been utilized exclusively for a couple of biotechnological applications. By and large, the fundamental justification behind the low interest in algae products is the expense of photoautotrophic development contrasted with different sources and the unfortunate capacities to hereditarily alter them [12.78]. Experiences from concentrating on disintegration might distinguish species- or condition-explicit changes in the silicification cycle intended to shield the silica from the climate. The examination of diatom cell silicon digestion may likewise have biomechanical applications in envisioning biomimetic ways to deal with amalgamation of materials [12.79]. To accomplish expanded financial advantages, reproducing, refinement, gathering, and coextraction ought to be carefully thought of. It is significant that the adaptable inward lighting framework outfitted with a maturation framework addresses clever biotechnology for the profoundly effective and financial creation of diatoms in the future [12.80].

12.8 Conclusion

Diatoms are a significant taxon, not only as a result of their assortment and nature, but also their use in ecological observation, nanotechnology, biofuel, medication, horticulture, and creation of food products. Their siliceous cell wall identifies them. Notwithstanding their small size, they have a significant environmental impact because of their staggering surplus as well as proficiency. Their unmistakable actual qualities and physiological cycle, custom fitted to the unbending cell wall and the range of territories they cover, have always been interesting. Investigating diatom variety from neglected locales won't just assist in fostering a superior comprehension of their conveyance, but will also result in getting novel mixtures for different business purposes. The investigation of diatoms alluded to as Phycology has progressed alongside improvements in science and includes a huge range of subjects, including subatomic science, phylogeny, environment, geography, biophysics, hereditary qualities, and transformative science. Included in various levels of novel permeable designs with a high surface region, tailorable surface, along with high biocompatibility, compound strength, and explicit optical and photoluminescence properties, diatom frustules are being taken advantage of as a new and minimal expense platform for an expansive scope of biomedical applications.

References

[12.1] Martín, J. R., De Arana, C., Ramos-Miras, J. J., Gil, C., & Boluda, R. (2015). Impact of 70 years urban growth associated with heavy metal pollution. *Environmental Pollution*, 196, 156-163.

[12.2] Ganeshkumar, V., Subashchandrabose, S. R., Dharmarajan, R., Venkateswarlu, K., Naidu, R., & Megharaj, M. (2018). Use of mixed wastewaters from piggery and winery for nutrient removal and lipid production by Chlorella sp. MM3. *Bioresource Technology*, 256, 254-258.

[12.3] Marella, T. K., Saxena, A., & Tiwari, A. (2020). Diatom mediated heavy metal remediation: A review. *Bioresource Technology*, 305, 123068.

[12.4] Geissen, V., Mol, H., Klumpp, E., Umlauf, G., Nadal, M., van der Ploeg, M., ... & Ritsema, C. J. (2015). Emerging pollutants in the environment: a challenge for water resource management. *International Soil and Water Conservation Research*, 3(1), 57-65.

[12.5] John, Y., David Jr, V. E., & Mmereki, D. (2018). A comparative study on removal of hazardous anions from water by adsorption: a review. *International Journal of Chemical Engineering*, 2018(1), 3975948.

[12.6] Ouyang, W., Wang, Y., Lin, C., He, M., Hao, F., Liu, H., & Zhu, W. (2018). Heavy metal loss from agricultural watershed to aquatic system: A scientometrics review. *Science of the Total Environment*, 637, 208-220.

[12.7] Hayashi, T., Krebs, W. N., Saito-Kato, M., & Tanimura, Y. (2018). The turnover of continental planktonic diatoms near the middle/late Miocene boundary and their Cenozoic evolution. *PloS One*, 13(6), e0198003.

[12.8] Marella, T. K., & Tiwari, A. (2020). Marine diatom *Thalassiosira weissflogii* based biorefinery for co-production of eicosapentaenoic acid and fucoxanthin. *Bioresource Technology*, 307, 123245.

[12.9] Hehenberger, E., Burki, F., Kolisko, M., & Keeling, P. J. (2016). Functional relationship between a dinoflagellate host and its diatom endosymbiont. *Molecular Biology and Evolution*, 33(9), 2376-2390.

[12.10] Karaman, E. S., Wang, Z., Di Benedetto, G., Zunino III, J. L., Meng, X., & Mitra, S. (2019). Fabrication of supercapacitors and flexible electrodes using biosilica from cultured diatoms. *Materials Today Energy*, 11, 166-173.

[12.11] Pawolski, D., Heintze, C., Mey, I., Steinem, C., & Kröger, N. (2018). Reconstituting the formation of hierarchically porous silica patterns using diatom biomolecules. *Journal of Structural Biology*, 204(1), 64-74.

[12. 12] Zhang, S., Liu, H., Ke, Y., & Li, B. (2017). Effect of the silica content of diatoms on protozoan grazing. *Frontiers in Marine Science*, 4, 202.

[12.13] Saxena, A., Prakash, K., Phogat, S., Singh, P. K., & Tiwari, A. (2020). Inductively coupled plasma nanosilica based growth method for enhanced biomass production in marine diatom algae. *Bioresource Technology*, 314, 123747.

[12.14] Sánchez, C., Cristóbal, G., & Bueno, G. (2019). Diatom identification including life cycle stages through morphological and texture descriptors. *PeerJ*, 7, e6770.

[12.15] Pla-Rabés, S., Hamilton, P. B., Ballesteros, E., Gavrilo, M., Friedlander, A. M., & Sala, E. (2016). The structure and diversity of freshwater diatom assemblages from Franz Josef Land Archipelago: a northern outpost for freshwater diatoms. *PeerJ*, 4, e1705.

[12.16] McNair, H. M., Brzezinski, M. A., & Krause, J. W. (2018). Diatom populations in an upwelling environment decrease silica content to avoid growth limitation. *Environmental Microbiology*, 20(11), 4184-4193.

[12.17] Sims, P. A., Mann, D. G., & Medlin, L. K. (2006). Evolution of the diatoms: insights from fossil, biological and molecular data. *Phycologia*, 45(4), 361-402.

[12.18] Marella, T. K., López-Pacheco, I. Y., Parra-Saldívar, R., Dixit, S., & Tiwari, A. (2020). Wealth from waste: Diatoms as tools for phycoremediation of wastewater and for obtaining value from the biomass. *Science of the Total Environment*, 724, 137960.

[12.19] Wasmund, N. (2017). The diatom/dinoflagellate index as an indicator of ecosystem changes in the Baltic Sea. 2. Historical data for use in determination of good environmental status. *Frontiers in Marine Science*, 4, 153.

[12.20] Dalu, T., Wasserman, R. J., Wu, Q., Froneman, W. P., & Weyl, O. L. (2018). River sediment metal and nutrient variations along an urban–agriculture gradient in an arid austral landscape: implications for environmental health. *Environmental Science and Pollution Research*, 25(3), 2842-2852.

[12.21] Bozarth, A., Maier, U. G., & Zauner, S. (2009). Diatoms in biotechnology: modern tools and applications. *Applied Microbiology and Biotechnology*, 82, 195-201.

[12.22] Kumar, K. S., Dahms, H. U., Won, E. J., Lee, J. S., & Shin, K. H. (2015). Microalgae–a promising tool for heavy metal remediation. *Ecotoxicology and Environmental Safety*, 113, 329-352.

[12.23] Awasthi, G., Nagar, V., Mandzhieva, S., Minkina, T., Sankhla, M. S., Pandit, P. P., ... & Srivastava, S. (2022). Sustainable amelioration of heavy metals in soil ecosystem: Existing developments to emerging trends. *Minerals*, 12(1), 85.

[12.24] Singh Sankhla, M., Kumar, R., & Prasad, L. (2021). Seasonal variations of lead and chromium concentrations in the water samples from Yamuna River in Delhi, India. *Iranian Journal of Toxicology*, 15(2), 109-114.

[12.25] Kiran, M. T., Bhaskar, M. V., & Tiwari, A. (2016). Phycoremediation of eutrophic lakes using diatom algae. *Lake Sciences and Climate Change*, 103-115.

[12.26] Tiwari, A., & Marella, T. K. (2019). Potential and application of diatoms for industry-specific wastewater treatment. *Application of Microalgae in Wastewater Treatment: Volume 1: Domestic and Industrial Wastewater Treatment*, 321-339.

[12.27] Marella, T. K., Datta, A., Patil, M. D., Dixit, S., & Tiwari, A. (2019). Biodiesel production through algal cultivation in urban wastewater using algal floway. *Bioresource Technology*, 280, 222-228.

[12.28] Xiong, J. Q., Kurade, M. B., & Jeon, B. H. (2018). Can microalgae remove pharmaceutical contaminants from water?. *Trends in Biotechnology*, 36(1), 30-44.

[12.29] Jamali, A. A., Akbari, F., Ghorakhlu, M. M., de la Guardia, M., & Khosroushahi, A. Y. (2012). Applications of diatoms as potential microalgae in nanobiotechnology. *BioImpacts: BI*, 2(2), 83.

[12.30] Congestri, R., Cox, E. J., Cavacini, P., & Albertano, P. (2005). Diatoms (Bacillariophyta) in phototrophic biofilms colonising an Italian wastewater treatment plant. *Diatom Research*, 20(2), 241-255.

[12.31] Guzzon, A., Bohn, A., Diociaiuti, M., & Albertano, P. (2008). Cultured phototrophic biofilms for phosphorus removal in wastewater treatment. *Water Research*, 42(16), 4357-4367.

[12.32] Azimi, A., Azari, A., Rezakazemi, M., & Ansarpour, M. (2017). Removal of heavy metals from industrial wastewaters: a review. *ChemBioEng Reviews*, 4(1), 37-59.

[12.33] Phang, S. M., Chu, W. L., & Rabiei, R. (2015). Phycoremediation. *The Algae World*, 357-389.

[12.34] Olguín, E. J. (2012). Dual purpose microalgae–bacteria-based systems that treat wastewater and produce biodiesel and chemical products within a Biorefinery. *Biotechnology Advances*, 30(5), 1031-1046.

[12.35] Maršálek, B., & Rojíčková, R. (1996). Stress factors enhancing production of algal exudates: a potential self-protective mechanism?. *Zeitschrift für Naturforschung C*, 51(9-10), 646-650.

[12.36] Perrein-Ettajani, H., Amiard, J. C., Haure, J., & Renaud, C. (1999). Effets des métaux (Ag, Cd, Cu) sur la composition biochimique et compartimentation de ces métaux chez deux microalgues *Skeletonema costatum* et *Tetraselmis suecica*. *Canadian Journal of Fisheries and Aquatic Sciences*, 56(10), 1757-1765.

[12.37] Sunda, W. G., & Huntsman, S. A. (2000). Effect of Zn, Mn, and Fe on Cd accumulation in phytoplankton: Implications for oceanic Cd cycling. *Limnology and Oceanography*, 45(7), 1501-1516.

[12.38] Priadie, B. (2012). Teknik bioremediasi sebagai alternatif dalam upaya pengendalian pencemaran air. *Jurnal ilmu lingkungan*, 10(1), 38-48.

[12.39] Knauer, K., Behra, R., & Sigg, L. (1997). Adsorption and uptake of copper by the green alga *Scenedesmus subspicatus* (Chlorophyta) 1. *Journal of Phycology*, 33(4), 596-601.

[12.40] Sankhla, M. S., Kumar, R., & Prasad, L. (2022). Impact of Variation in Climatic Changes in Concentration of Lead & Nickel in Yamuna River Water, Delhi, India. *Materials Today: Proceedings*, 69, 1540-1547.

[12.41] Manwani, S., Vanisree, C. R., Jaiman, V., Awasthi, K. K., Yadav, C. S., Sankhla, M. S., ... & Awasthi, G. (2022). Heavy metal contamination in vegetables and their toxic effects on human health. *Sustainable Crop Production: Recent Advances*, 181.

[12.42] Vanisree, C. R., Sankhla, M. S., Singh, P., Jadhav, E. B., Verma, R. K., Awasthi, K. K., ... & Nagar, V. (2022). Heavy metal contamination of food crops: Transportation via food chain, human consumption, toxicity and management strategies. In *Environmental Impact and Remediation of Heavy Metals*. IntechOpen.

[12.43] Ye, J., Xiao, H., Xiao, B., Xu, W., Gao, L., & Lin, G. (2015). Bioremediation of heavy metal contaminated aqueous solution by using red algae *Porphyra leucosticta. Water Science and Technology*, 72(9), 1662-1666.

[12.44] Wisudyawati, D. (2015). *Studi perbandingan kemampuan Skeletonema sp. Dan Chaetoceros sp. Sebagai agen bioremediasi (fito-akumulasi) terhadap logam berat timbal (Pb)* (Doctoral dissertation, UNIVERSITAS AIRLANGGA).

[12.45] Gupta, R., Ahuja, P., Khan, S., Saxena, R. K., & Mohapatra, H. (2000). Microbial biosorbents: meeting challenges of heavy metal pollution in aqueous solutions. *Current Science*, 967-973.

[12.46] Aksu, Z. (1998). Biosorption of heavy metals by microalgae in batch and continuous systems. In *Wastewater treatment with algae* (pp. 37-53). Berlin, Heidelberg: Springer Berlin Heidelberg.

[12.47] Torres, E., Cid, A., Herrero, C., & Abalde, J. (1998). Removal of cadmium ions by the marine diatom *Phaeodactylum tricornutum* Bohlin accumulation and long-term kinetics of uptake. *Bioresource Technology*, 63(3), 213-220.

[12.48] Gupta, V. K., Shrivastava, A. K., & Jain, N. (2001). Biosorption of chromium (VI) from aqueous solutions by green algae *Spirogyra* species. *Water Research*, 35(17), 4079-4085.

[12.49] Gupta, V. K., & Rastogi, A. (2008). Biosorption of lead (II) from aqueous solutions by non-living algal biomass *Oedogonium* sp. and *Nostoc* sp.—a comparative study. *Colloids and Surfaces B: Biointerfaces*, 64(2), 170-178.

[12.50] Fourest, E., & Roux, J. C. (1992). Heavy metal biosorption by fungal mycelial by-products: mechanisms and influence of pH. *Applied Microbiology and Biotechnology*, 37, 399-403.

[12.51] Karnwal, A. (2024). Unveiling the Promise of Biosorption for Heavy Metal Removal from Water Sources. *Desalination and Water Treatment*, 100523.

[12.52] Yan, C., Qu, Z., Wang, J., Cao, L., & Han, Q. (2022). Microalgal bioremediation of heavy metal pollution in water: Recent advances, challenges, and prospects. *Chemosphere*, 286, 131870.

[12.53] Leong, Y. K., & Chang, J. S. (2020). Bioremediation of heavy metals using microalgae: recent advances and mechanisms. *Bioresource Technology*, 303, 122886.

[12.54] Mehta, S. K., & Gaur, J. P. (2005). Use of algae for removing heavy metal ions from wastewater: progress and prospects. *Critical Reviews in Biotechnology*, 25(3), 113-152.

[12.55] Zhang JianYing, Z. J., Ding TengDa, D. T., Zhang ZhiJian, Z. Z., Xu LiPing, X. L., & Zhang ChunLong, Z. C. (2015). Enhanced adsorption of trivalent arsenic from water by functionalized diatom silica shells.

[12.56] Armbrust, E. V. (2009). The life of diatoms in the world's oceans. *Nature*, 459(7244), 185-192.

[12.57] Milligan, A. J., & Morel, F. M. (2002). A proton buffering role for silica in diatoms. *Science*, 297(5588), 1848-1850.

[12.58] Wagner, H., Jakob, T., & Wilhelm, C. (2006). Balancing the energy flow from captured light to biomass under fluctuating light conditions. *New Phytologist*, 169(1).

[12.59] Sbihi, K., Cherifi, O., El Gharmali, A., Oudra, B., & Aziz, F. (2012). Accumulation and toxicological effects of cadmium, copper and zinc on the growth and photosynthesis of the freshwater diatom *Planothidium lanceolatum* (Brébisson) Lange-Bertalot: a laboratory study. *J Mater Environ Sci*, 3(3), 497-506.

[12.60] Loix, C., Huybrechts, M., Vangronsveld, J., Gielen, M., Keunen, E., & Cuypers, A. (2017). Reciprocal interactions between cadmium-induced cell wall responses and oxidative stress in plants. *Frontiers in Plant Science*, 8, 1867.

[12.61] Jadoon, S., & Malik, A. D. N. A. (2017). DNA damage by heavy metals in animals and human beings: an overview. *Biochem Pharmacol*, 6(3), 1-8.

[12.62] Morelli, E., & Scarano, G. (2004). Copper-induced changes of non-protein thiols and antioxidant enzymes in the marine microalga *Phaeodactylum tricornutum*. *Plant Science*, 167(2), 289-296.

[12.63] Hernández-Ávila, J., Salinas-Rodríguez, E., Cerecedo-Sáenz, E., Reyes-Valderrama, M. I., Arenas-Flores, A., Román-Gutiérrez, A. D., & Rodríguez-Lugo, V. (2017). Diatoms and their capability for heavy metal removal by cationic exchange. *Metals*, 7(5), 169.

[12.64] Łukowski, A., & Dec, D. (2018). Influence of Zn, Cd, and Cu fractions on enzymatic activity of arable soils. *Environmental Monitoring and Assessment*, 190, 1-12.

[12.65] Volland, S., Bayer, E., Baumgartner, V., Andosch, A., Lütz, C., Sima, E., & Lütz-Meindl, U. (2014). Rescue of heavy metal effects on cell physiology of the algal model system Micrasterias by divalent ions. *Journal of Plant Physiology*, 171(2), 154-163.

[12.66] Becker, E. W. (2007). Micro-algae as a source of protein. *Biotechnology Advances*, 25(2), 207-210.

[12.67] Bondareva, J. V., Aslyamov, T. F., Kvashnin, A. G., Dyakonov, P. V., Kuzminova, Y. O., Mankelevich, Y. A., ... & Evlashin, S. A. (2020). Environmentally Friendly method of silicon recycling: synthesis of silica nanoparticles in an aqueous solution. *ACS Sustainable Chemistry & Engineering*, 8(37), 14006-14012.

[12.68] Aw, M. S., Simovic, S., Yu, Y., Addai-Mensah, J., & Losic, D. (2012). Porous silica microshells from diatoms as biocarrier for drug delivery applications. *Powder Technology*, 223, 52-58.

[12.69] Terracciano, M., De Stefano, L., & Rea, I. (2018). Diatoms green nanotechnology for biosilica-based drug delivery systems. *Pharmaceutics*, 10(4), 242.

[12.70] Kabir, A., Nazeer, N., Bissessur, R., & Ahmed, M. (2020). Diatoms embedded, self-assembled carriers for dual delivery of chemotherapeutics in cancer cell lines. *International Journal of Pharmaceutics*, 573, 118887.

[12.71] Hussein, H. A., & Abdullah, M. A. (2020). Anticancer compounds derived from marine diatoms. *Marine Drugs*, 18(7), 356.

[12.72] Lauritano, C., Andersen, J. H., Hansen, E., Albrigtsen, M., Escalera, L., Esposito, F., ... & Ianora, A. (2016). Bioactivity screening of microalgae for antioxidant, anti-inflammatory, anticancer, anti-diabetes, and antibacterial activities. *Frontiers in Marine Science*, 3, 68.

[12.73] Saranya, G., & Ramachandra, T. V. (2020). Novel biocatalyst for optimal biodiesel production from diatoms. *Renewable Energy*, 153, 919-934.

[12.74] Santos, J., Almeida, S. F., & Figueira, E. (2013). Cadmium chelation by frustulins: a novel metal tolerance mechanism in *Nitzschia palea* (Kützing) W. Smith. *Ecotoxicology*, 22, 166-173.

[12.75] Sbihi, K., Cherifi, O., Bertrand, M., & El Gharmali, A. (2014). Biosorption of metals (Cd, Cu and Zn) by the freshwater diatom *Planothidium lanceolatum*: a laboratory study. *Diatom Research*, 29(1), 55-63.

[12.76] Nigh, R., Benbrook, C., Brush, S., Garcia-Barrios, L., Ortega-Paczka, R., & Perales, H. R. (2000). Transgenic crops: A cautionary tale. *Science*, 287(5460), 1927-1927.

[12.77] Ghosh, A., Khanra, S., Mondal, M., Halder, G., Tiwari, O. N., Saini, S., ... & Gayen, K. (2016). Progress toward isolation of strains and genetically engineered strains of microalgae for production of biofuel and other value added chemicals: A review. *Energy Conversion and Management*, 113, 104-118.

[12.78] Larras, F., Coulaud, R., Gautreau, E., Billoir, E., Rosebery, J., & Usseglio-Polatera, P. (2017). Assessing anthropogenic pressures on streams: A random forest approach based on benthic diatom communities. *Science of the Total Environment*, 586, 1101-1112.

[12.79] Kroger, N., Deutzmann, R., & Sumper, M. (1999). Polycationic peptides from diatom biosilica that direct silica nanosphere formation. *Science*, 286(5442), 1129-1132.

[12.80] Zhao, P., Gu, W., Wu, S., Huang, A., He, L., Xie, X., ... & Wang, G. (2014). Silicon enhances the growth of *Phaeodactylum tricornutum* Bohlin under green light and low temperature. *Scientific Reports*, 4(1), 3958.

13

Optimizing Bioenergy from Diatoms through Biofilms

G. Saranya[1,2] and T.V. Ramachandra[1,2,3]*

[1]*Energy & Wetlands Research Group, Centre for Ecological Sciences (CES), Indian Institute of Science (IISc), Bangalore, India*

[2]*Centre for Sustainable Technologies (ASTRA), Indian Institute of Science, Bangalore, Karnataka, India*

[3]*Centre for Infrastructure, Sustainable Transportation and Urban Planning (CiSTUP), Indian Institute of Science, Bangalore, Karnataka, India*

Abstract

The impetus for the exploration of environmentally friendly fuels during the past decade has increased with the concerns about escalating greenhouse gas (GHG) emissions due to the burning of fossil fuels and its associated climate change. In this regard, biofuel from algae, especially diatoms, is emerging as a viable liquid fuel alternative in transportation. Diatoms are metabolically versatile microorganisms with broad applicability as feedstock for aquaculture feed, biofuel, nutraceutical, and pharmaceutical/cosmetic industries. The economic feasibility of diatom biofuel depends on minimizing the capital and operational costs involved in upstream and downstream bioprocessing of diatoms. Exploiting the biofilm-forming potential of benthic diatoms for growing them in attached forms is emerging as a prudent cultivation technology for sustainable diatom biomass production. This chapter discusses various design considerations and optimization techniques useful for diatom cultivation using biofilm. The key areas to improve the performance of such engineered biosystems include the design, construction, and field validations of cultivation technologies. Various factors influencing diatom productivity, such as species selection, algal-bacterial interaction, volumetric mass transfer coefficient, and the oxygen transfer rate, are presented in this chapter. The concluding part discusses the scope for the valorization of diatom biomass for bioenergy and value-added bioproducts.

**Corresponding author*: tvr@iisc.ac.in; energy.ces@iisc.ac.in; ORCID: 0000-0001-5528-1565

Vandana Vinayak and Richard Gordon (eds.) Diatom Cultivation for Biofuel, Food and High-Value Products, (341–376) © 2025 Scrivener Publishing LLC

Keywords: Bioenergy, biomass, diatom biofilm technology, diatom cultivation, design consideration, valorization, value-added products

13.1 Introduction

Algal cultivation for fuels, platform chemicals, and nutraceutical products dates back several decades. Microalgae with higher lipid productivity and the ability to grow in barren lands are emerging as promising feedstocks [13.1]. Most microalgae cultivation ventures are based on cells grown in suspension forms using open/raceway ponds and photobioreactors. These suspended cell cultures require a liquid medium of 12–2000 kg more than per unit of algal cell dry weight (kg) [13.2] and the reported algal cell concentrations were minimal, with 0.5 g L^{-1} for open ponds and 2–6 g L^{-1} for photobioreactors [13.3]. Harvesting such dilute suspensions (~1% dry weight) through centrifugation to achieve a dry weight of 20% accounts for 20–30% of the total production costs [13.4] and has posed challenges to commercializing microalgal biofuel/bioproducts. This necessitates exploring an optimal biofuel process for economical microalgae production with technically feasible downstream processing.

Ecological engineering based on ecosystem-based approaches provides scope for economical solutions to optimize environmental and energy issues due to unplanned human-induced activities [13.5]. It involves utilizing free energies from nature (solar insolation) and wastewater nutrients with nominal interferences of man-made technologies. Bioremediation of wastewater through constructed wetlands coupled with bioenergy generation [13.6] [13.7] has been a successful and widely used engineering technology in line with ecology. Constructed wetlands are man-made sewage treatment options that take up organic and inorganic nutrients and bioremediate wastewater in addition to biomass production through diverse microbial and biochemical processes [13.8]. Ecologically engineered systems were implemented along the coastlines of India, especially on the flood plains (Gazani lands), by providing suitable low-cost substrates like gravel stones as a source of attachment for the benthic diatoms that are predominant in estuaries and harvesting of microalgal biomass through algal turf scrubbers (ATS) [13.9]. This would fetch dual benefits for the farmers that ensure bioenergy feedstock and livestock fodder ranging between 0.012–0.045 kg m^{-2} d^{-1} [13.5] [13.9] [13.10]. Conventional microalgal cultivation (open pond/raceway pond) systems showed much less algal productivity (~ five times less) than such engineered biofilm cultivation systems.

Engineered microalgal cultivation systems using biofilms [13.11] has gained significance during the past decade. Biofilm [13.12] refers to the complex structures formed by diverse microbial colonies on any solid surface exposed to favorable conditions for them to thrive, such as a moist environment or submerged inside a water column for a prolonged period. The consortium of microorganisms is usually embedded in extracellular polymeric substances (EPS) [13.13] consisting of high molecular weight carbon-rich molecules [13.14] such as nucleic acids, proteins, polysaccharides, and phospholipids [13.15]. These are usually secreted external to the plasmalemma [13.16], and form an essential component in forming biofilms. EPS, otherwise known as jelly stalks, serve multiple benefits, including physical resistance to turbulent/flowing waters and mechanical resistance by preventing desiccation during sudden exposure to sunlight or extreme environmental conditions such as dehydration, harmful chemicals, pH, and temperature fluctuations [13.2] [13.17].

Biofilm is composed of diverse microbes such as microalgae, microzooplankton, protozoa, larvae, bacteria, archaea and macroalgae [13.18]. Adsorption is the foremost reversible process in biofilm formation that is established through cell adherence to a solid surface, followed by an irreversible cell fixation to the substratum through EPS secretion [13.19]. This is succeeded by the rapid growth of biofilms owing to the cell division of adhered microorganisms leading to the development of mature biofilms with raised surfaces exhibiting a three-dimensional structure with multiple layers through clustering of cells and its subsequent immobilization [13.20]. The maturation of biofilm influences the variability and relative abundances of the participating microorganisms as there is a decline in the biofilm thickness after a threshold, leaving the microorganisms exposed to grazing pressures by parasites with eventual cell death [13.21]. Biofilm grown microalgae show enhanced resistance and resilience to hostile environmental conditions [13.22] as the EPS matrix present in biofilms aids in avoiding water desiccation and provides enough room for nutrient storage and water, thus helps to defend the cells from adverse environmental conditions [13.2].

Photosynthetic biofilms are a consortium of microorganisms consisting of diverse green algae, diatoms, bacteria, cyanobacteria, protozoa, and other multicellular organisms [13.23]. Nevertheless, diatoms, green algae, and filamentous algae form significant portions of photosynthetic biofilm biomass in the case of substrate-based biofilm cultivation [13.24]. Early-stage biofilm development contains EPS secreted by diatoms and provides a favorable environment for bacterial communities. The early phase is considered as "conditioning" of the growth surface. The conditioning phase is followed by the early colonization phase, where the algal cells begin to proliferate in the upper layers [13.25]. After early colonization, the rapid proliferation of species occurs due to the abundance of EPS and organic carbon sources that are conducive to microbes' community establishment, leading to species succession (in some cases by filamentous microalgae and cyanobacteria). The biofilm matrix with such a dense and diverse algal population facilitates retention and sharing of nutrients, enabling a symbiotic relationship between autotrophic algae and heterotrophic bacteria [13.26].

There has been considerable attention and a shift in research focus from suspended algal cultivation to biofilm-based cultivation techniques, with the preliminary reports of a substantial reduction in harvesting costs proving economic viability [13.27]. A considerable reduction (of ~12 times) is evident from a water requirement of 17 tons for producing one ton of microalgal biomass. A higher quantity (12–2000 kg) of water is required for the algae cultivated under suspended cultivation using open ponds/photobioreactors [13.18] [13.28] or a large-scale outdoor production system per kg of dry biomass [13.2].

13.2 Different Configurations of Biofilm Cultivation Systems

The biofilm bioreactor systems exposed to the ambient atmosphere are submerged continuously or intermittently, and moisture is provided by a porous substrate [13.2]. Constantly submerged systems use mechanized pumping systems for water and media recycling with a slightly inclined flow ways to enable gravity-driven flow [13.20]. The constantly submerged bioreactor systems designed earlier were made of diverse construction materials such as polycarbonates [13.29], PVC (polyvinyl chloride) [13.30], concrete [13.31] and GFRP (glass-fiber-reinforced plastic) [13.32]. The liquid cultivation medium explored so far includes synthetic nutrient recipes like BG-11, synthetic wastewater, and artificial seawater. The algal biomass productivity in such submerged cultivation systems varied between 0.71 g m^{-2} d^{-1} to 30 g m^{-2} d^{-1} [13.2]. Compared to these, intermittently submerged systems are of two different configurations: (i) Algal Turf Systems (ATS) [13.5], based on pulsed periodic submersion of microalgal cells in order to mimic wave action experienced by algal turfs along shorelines, and reefs on broader flow ways that are constructed identical to those of constantly submerged systems, and (ii) intermittently submerged cultivation systems with the diverse design configurations of different surface materials that include algal turf scrubbers - plastic screens [13.9], rocking attached cultivation system - polystyrene, bench-scale rotating algal biofilm reactor - cotton rope [13.33], rocking cultivation chamber - glass-reinforced plastic [13.34], photo-rotating biological contactor - polyvinyl chloride [13.35] and alga disk (rotating biological contactor) - stainless steel mesh, and cotton duct [13.27] [13.36]. Biomass productivity in such intermittently submerged systems using a natural consortium of species prevalent in wastewaters and single-cell cultures using agricultural runoffs, manures, and wastewaters of municipal and industrial origins varied between 0.04–54 g m^{-2} d^{-1} [13.2]. Perfused systems, also known as "twin-layer systems" or "porous substrate bioreactors" [13.37], utilize a semi-permeable membrane to support adherence to microalgal growth and a barricade to restrict microalgae entering the porous substratum. The liquid conducting most, the permeable substratum is a source of nutrients and moisture supply [13.38]. A minimal liquid medium with direct exposure of microalgal cells to the ambient conditions facilitates improved gas exchange and reduces the constraints of light attenuation and self-shading effects of the cells [13.39]. However, solid-state cultivation systems pose the risk of desiccation of the microalgal biofilms when the setup is under less humid conditions. The substrata investigated so far for such perfused systems include unprinted newspapers [13.39], cellulose acetate/nitrate filters [13.40–13.42], polyethylene oven geotextile [13.43], and chromatography filter [13.44]. The estimated biomass productivity in such ventures ranged between 0.42–25 g m^{-2} d^{-1}. Earlier work provides different design configurations of biofilm cultivation systems with the details of microalgae [13.2].

13.3 Surface Materials for Biofilm Cultivation

Different solid surface materials (substrates) used so far include polystyrene foam, polyurethane, cotton duct/rope [13.45], concrete, stainless steel [13.32], nylon sponge, loofah sponge, nylon mesh, polyethylene fiber, polycarbonate [13.46], glass fibre reinforced plastic (GFRP) [13.19], nylon nets and plastic scrubber screens [13.9], Perspex, titanium, stainless steel 316-L, glass, copper, aluminum brass and admiralty brass [13.47], polypropylene, high-density polyethylene, polyester, aluminum and stainless steel [13.48], cellulose acetate [13.49], flexible fiber bundles [13.50], cotton, linen, mohair [13.51], glass [13.52] and printing paper [13.39]. Diverse microalgae (green algae and diatoms) were grown as biofilms using such supporting materials. *Scenedesmus dimorphous* grown on a stainless-steel sheet treated with a microscale laser had higher attachment due to increased surface area than in control [13.53]. *Chlorella* sp. cultivated using polystyrene foam in dairy wastewater showed a high biomass yield of 25.65 g (dry weight) m^{-2} [13.54]. The diatom *Cylindrotheca fusiformis* grown using glass plate and filter paper as supporting materials showed biomass productivity of 50–80 g $m^{-2}d^{-1}$, which is 400–700% higher with a potential saving in water consumption than traditional open ponds [13.41].

13.3.1 Biofilm Bioreactor Design

Biofilm bioreactors were designed considering the following operational factors: i) inert properties of the substrate; ii) effective light distribution across the bioreactor; iii) enhanced CO_2 mass transfer; iv) liquid mixing in the reservoir; v) biofilm thickness; and vi) shear stress applied to the biofilm.

13.3.2 Lab-Scale Biofilm Bioreactor

The biofilm bioreactor given in Figure 13.1 was designed using a non-reactive fiber-reinforced plastic tray, having a dimension of (0.51 × 0.27) m with a footprint area of 0.1377 m^2 illuminated by two white fluorescent tube lights. For diatom attachment, gravel stones were evenly distributed across the tray and used as substrate. The culture media (f/2 seawater media) was recycled using a peristaltic pump throughout the growth period using a sprinkler system.

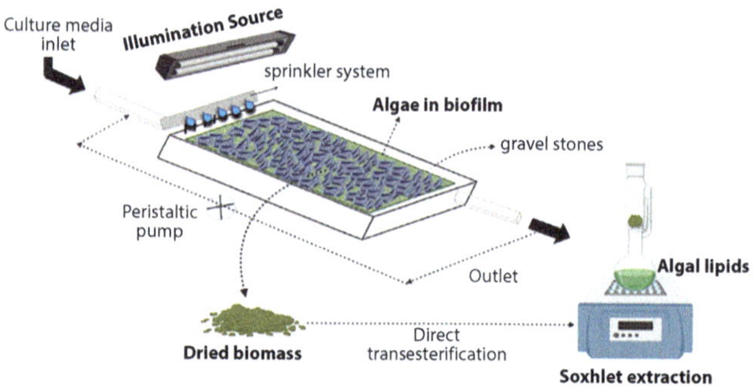

Figure 13.1 Laboratory-scale biofilm reactor setup for diatom cultivation.

13.3.3 Operating Conditions for Diatom Cultivation

The diatom *Amphora* sp. (isolated from the flood plains of the Aghanashini estuary) was cultured using f/2 media [13.55] under laboratory conditions of 25 ± 2 °C, salinity: 35 ppt and illumination intensity was set at ~360 μmol m^{-2} s^{-1} with 14 h light:10 h dark cycle. The cell viability was maintained by sequential subculturing at periodic intervals. The variation in biomass productivity in the biofilm reactor was compared with a control maintained in an Erlenmeyer culture flask in suspension form with 3.5 L of media as control volume for the batch reaction. A pure culture of *Amphora* sp. having an initial biomass concentration of 187 mg L^{-1} was used as a seed inoculum for culture experimentation.

13.3.4 Field Biofilm Bioreactor

Figure 13.2 gives the location details of the field bioreactors installed in salinity gradients of the Aghanashini estuary's Gazani lands (colloquial term for flood plains).

The sites chosen were:

(i) Bargi (BA): This is an enclosed water body with prolonged water retention and is not affected by water mixing due to diurnal tidal undulations as the sluice gates are kept closed. This location is being used for a crab cultivation site at Kirmani village of Kumta Taluk in Uttara Kannada district.

(ii) Kagal (KA): This location is leased for shrimps and crustacean catching based on tidal undulations and is situated

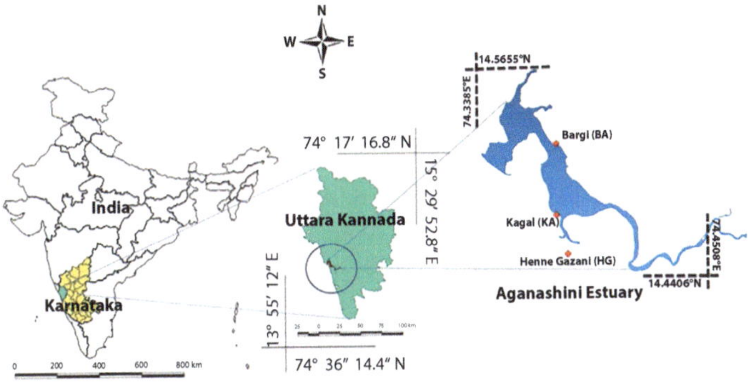

Figure 13.2 Study sites of bioreactor installation.

approximately 13 km from the river mouth possessing multiple pockets of natural Gazani.

(iii) Henne Gazani (HG): This is an abandoned paddy field located in Hegde village and is now being used for traditional shrimp farming.

13.3.5 Evaluation of Algal Growth in Biofilms Grown under Field Conditions

Each bioreactor consisting of multiple rows of gravel stones (used as the substrate for diatom cultivation) was monitored for bacterial growth and influencing water parameters on alternate days. Diatoms were sampled to determine biomass productivity by scraping algae attached to three stones in an orderly fashion and replaced back at the original location. Water quality parameters measured include air temperature (AT), water temperature (WT), pH, salinity, dissolved oxygen (DO), total dissolved solids (TDS), electrical conductivity (EC), and solar insolation. The temperature was measured using a mercury thermometer, while all other onsite parameters, including DO, were measured using electronic probes. Solar insolation was measured using a Beetech lux meter. Water sampling was carried out at each site to determine nutrient concentrations by APHA 2005 standard protocols for the examination of water and wastewater.

Air temperature varied between 31.5–35 °C with a mean temperature of 33.3 ± 0.22 °C across stations. Water temperature varied between 31–35 °C with a mean surface water temperature of 31.5 ± 0.22 °C. The estuary exhibited warm waters, and the temperature was high with minimal variation recorded between air and surface water temperatures (~1 to 3 °C) as the experimentations were carried out during the post-monsoon (summer) season. Variations in temperatures observed in air and surface waters were comparable to the values recorded in the tropical southwest coastline estuaries of India [13.56] [13.57]. The pH of water in all the stations was alkaline ranging between 8.1–9.7, and a mean pH of 8.9 ± 0.20 with pH variations observed to be identical to mesotrophic water conditions having lesser nutrients. The estuarine salinity across stations varied between 8–29 ppt with mean salinity of 17.5 ± 8.08 ppt. Wider fluctuations in salinity were observed as the stations chosen spanned between upstream and downstream of the estuary that experienced different salinity levels (with the mixing of freshwater).

The DO values varied between 5.6 and 9.1 mg L^{-1} across stations. The mean DO value across sites was found to be 7.24 ± 0.50 mg/L. Oligotrophic lakes with tidal perturbations that are turbulent exhibit higher DO levels. A similar DO trend was observed in the chosen study sites. The solar insolation showed a mixed trend across stations, with the highest mean solar insolation of 63 ± 19.08 Kilolux at Bargi (BA) station with the minimum and maximum value of 8.1 K Lux 88.9 Kilolux, respectively. A predictable trend in nutrient variation across seasons among the different study stations was not found; most nutrient variations were subject to non-point source pollutions. TDS and EC values were highest for Bargi (BA) site, with mean TDS and EC values being 25 ± 3.73 ppt. Higher TDS and EC at BA station are due to the prolonged water stagnation with fewer tidal undulations. TDS of BA varied between 15.8 to 28.2 ppt, whereas for KA and HG, the minimum and maximum values ranged between 10.65 to 22 ppt, respectively. EC varied between 1 mS to 53.5 mS with a mean EC across stations of 36.3 ± 12.2 mS.

Nitrates and phosphate levels showed only marginal variations of 0.03 ± 0.01 mg L^{-1} and 0.06 ± 0.03 mg L^{-1}, respectively, across stations. Silicate levels showed wider variability of 0.12 mg L^{-1} to 0.36 mg L^{-1} across stations.

13.3.6 Growth Dynamics of Lab-Cultivated Diatom and In-Field Bioreactor

Chlorophyll-a concentration of the *Amphora* sp. was used as a surrogate variable for growth, and it was measured at three-day intervals during the complete growth cycle. Chlorophyll-a and biomass productivity on per day basis was estimated following Strickland and Parson's protocol for sea water [13.58]. The diatom cell viability assessment was also carried out through visual microscopic examination of the cells under a high-resolution fluorescent microscope model: Olympus BX51. Initial lag phase was observed during the first three days of the growth cycle followed by the log phase until the 10th day. Diatoms grown in biofilms were found to be performing better than suspended cultivation grown diatoms. The chlorophyll content of the biofilm cultivation diatoms reached around 160 mg/μL on the 12th day, while the chlorophyll content of suspended mode cultivated diatoms reached 120 mg/μL around the 10th day and decreased afterward. The growth dynamics in the field bioreactor were different from the lab-scale bioreactor, and Figure 13.3 illustrates mean areal cell densities at study stations.

It was observed that the species composition and cell density showed a steady increase of diatoms after 0–5 days of the initial conditioning stage, then early colonization followed by a mature phase till 17 days for stations

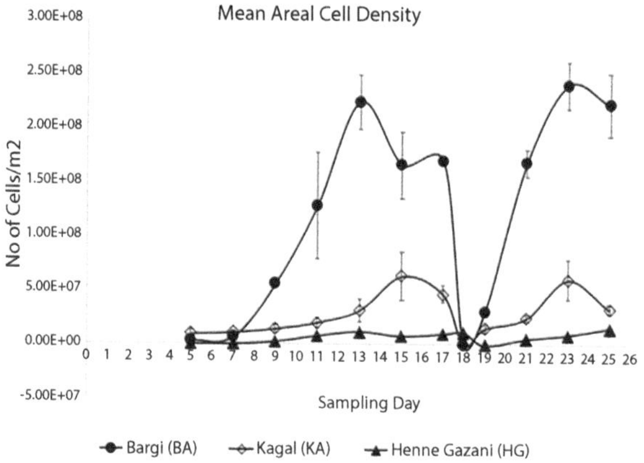

Figure 13.3 Mean areal cell density during the growth period.

BA and KA, and then species succession by cyanobacteria (*Oscillatoria* sp.). In comparison, the cell maturation stage was attained only on the 19th day after introducing the substrate (stones) at station HG. Diatom cell density was maximum at station BA, with a steady increase in the number of cells after the initial five days of incubation. Diatoms at KA stations showed intermediate cell density with a slow and steady increase in cell numbers which reached saturation on the 17th day. The least cell density was recorded at station HG. Harvesting of diatom was carried out at stations BA and KA on the 18th day, and reestablishment studies were initiated from the 19th day to understand the dynamics of biomass production in stones that are acclimatized to already established biofilm in substrates. During this phase, the biomass reached a maximum on the 5th day after the initial harvest. Thus, if large-scale production of such biofilm-based bioreactors is practiced, the initial harvest immediately after the introduction of substrata takes two weeks and two days, followed by a subsequent harvest every five days. Earlier studies have also reported that diatoms have accounted for a significant constituent of the early 15–20 days of the biofilm formation while filamentous cyanobacteria and Chlorophyceae members dominate later [13.59].

13.3.7 Isolation and Identification of Bacteria in Biofilm

Bacteria that are part of the microbial consortium in the biofilm were isolated through serial dilution with agar plating in nutrient agar. Multiple bacterial colonies obtained in initial spread plating were monocultures developed using quadrant streaking onto agar plates. The typical bacterial colonies among the pure bacterial isolates at three different study sites were subjected to gram staining and 16SrRNA-based molecular identification. Genomic DNA was extracted, and the concentration of extracted DNA was determined using a NanoDrop spectrophotometer. The extracted DNA was subjected to 16SrRNA amplification using 16SrRNAF and 16SrRNAR forward and reverse primers using PCR. The amplified DNA exhibited a single discrete PCR amplicon of 1500 bp when resolved on Agarose gel. The DNA sequencing of PCR amplicon was carried out using BDT v3.1 cycle sequencing kit on ABI 3730xl genetic analyzer. The forward and reverse sequence obtained was used to generate a consensus sequence in MEGA software version 7.0.

13.3.8 Bacterial Morphology Studies using DAPI

The morphology of the bacteria was studied through a confocal microscope (Zeiss LSM 880 Leica confocal microscope with Airyscan). Pure bacterial colonies were inoculated in nutrient broth and incubated for 24 h at 35 °C. The grown bacterial culture was centrifuged at 5000 rpm for 5 min, and the cell pellets obtained were resuspended in phosphate-buffered saline (PBS). Bacterial cell fixation was carried out by treating the cells with 2.5% glutaraldehyde for 2 h. The fixed bacterial cells were treated with DAPI (4′,6-diamidino-2-phenylindole), a fluorescent stain that binds with the DNA. The excitation and emission wavelengths were fixed at 358 nm and 461 nm, respectively.

The embedded bright field and fluorescent images of the bacteria are given in Figure 13.4, and the morphology of bacteria assessed under the confocal microscope revealed significant variations in the structure of the most abundant and common bacteria associated with biofilm formation. Microscopic observations of commonly found bacteria at three sites revealed Gram-positive short rods resembling coccobacillus, exhibiting opaque, yellow-colored colony-forming bacteria on agar plates, and Gram-negative long rod-shaped cream-colored colonies.

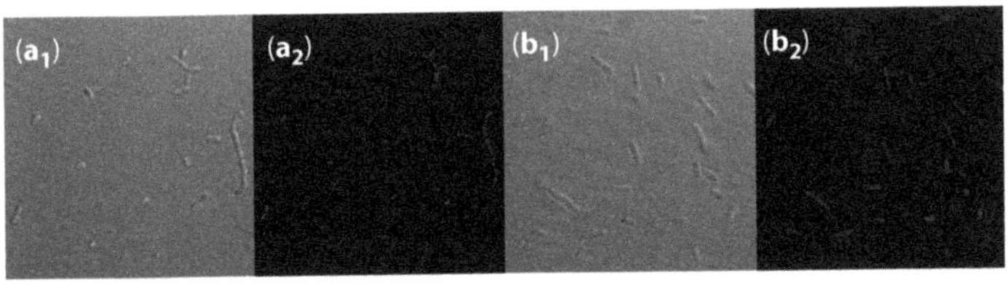

Figure 13.4 Embedded bright field and fluorescent images of bacterial strains: (a_1-a_2) *Exiguobacterium aurantiacum*, (b_1-b_2) *Aeromonas veronii*.

13.3.9 Species Interaction during Biofilm Cultivation

Attached cultivation technology is appropriate for diatoms [13.21] as it facilitates biofilm formation with the symbiotic interactions between bacteria and algae. Algae acts as a secondary habitat for bacteria by providing the conducive oxygen supply much required by heterotrophic bacteria in mineralizing organic substances [13.60]. Algae are facilitated by bacteria through the provision of inorganic nutrients for cell growth through the decomposition of organic fractions. The bacterial communities often associated with algal biofilms are Proteobacteria (alpha and beta) and *Bacteroides* [13.61]. The abundance of *Bacteroides* in biofilms could be attributed to their capability to glide along with the chemoorganotrophic potential to degrade high molecular weight algal exudates like EPS [13.62]. Hence, bacterial samples were collected and analyzed for their morphology and phylogeny to understand the bacterial community interaction between microalgae in the biofilm bioreactor.

13.3.10 Biofilm Bacteria Identification through Molecular Sequencing

The 16SrRNA gene sequences of commonly associated biofilm facilitating bacteria among three sites were used for the generation of consensus sequence in MEGA version 7.0. This sequence was used to carry out BLAST in the NCBI GenBank database repository. The sequences showed significant alignments in BLAST and exhibited 100% homology to alkaliphilic, thermotolerant Gram-positive strain *Exiguobacterium profundum* and a Gram-negative strain *Aeromonas veronii. Exiguobacterium profundum* is an alkaliphilic, halotolerant, non-spore-forming Gram-positive bacillus with small irregular rods, otherwise known as coccobacilli. The colonies exhibited yellow-orange pigmented colonies similar to the strain isolated from a microbial mat from a western lake [13.63]. *Exiguobacterium* colonies showed a yellow, shiny, round umbonate shape with a smooth, entire margin [13.64]. The first ten sequences were selected based on the maximum identity score and aligned using multiple sequence alignment software programs of ClustalW and distance matrix was generated. The phylogenetic tree was constructed using MEGA 7.0. Figures 13.5 and 13.6 depict the phylogenetic tree representing the species homology with respective bacterial strains.

The bacteria belonging to the genus *Aeromonas* are Gram-negative bacilli that are pathogenic to humans and animals. They form some of the important pathogens of reptiles and fishes in aquatic environments and in

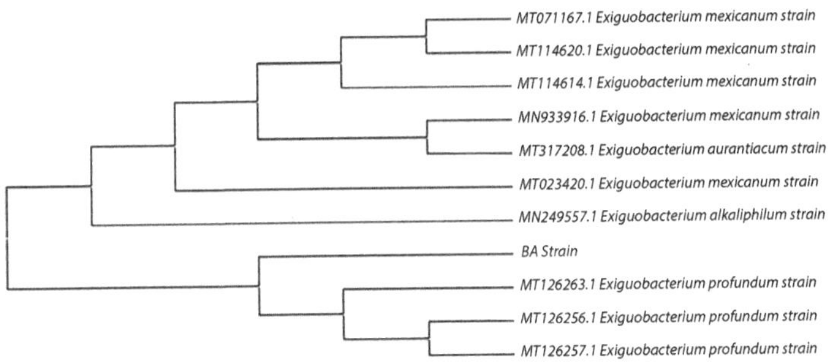

Figure 13.5 Phylogenetic tree revealing species homology of the bacterial strain (BA) with *Exiguobacterium profundum*.

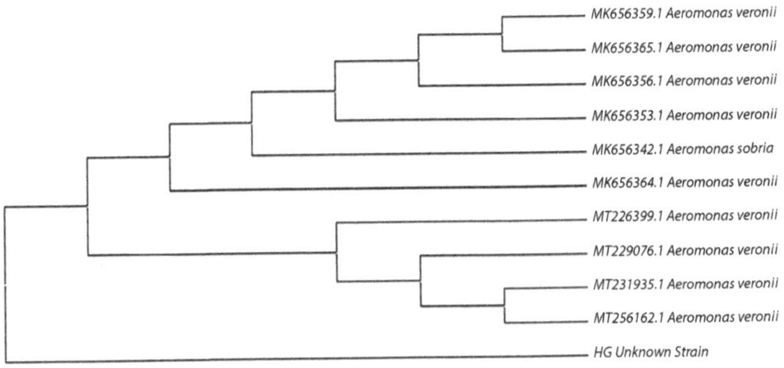

Figure 13.6 Phylogenetic tree revealing species homology of the bacterial strain (HG) with *Aeromonas veronii*.

immunocompromised human hosts [13.65]. *Aeromonas* sp. has characteristics of adhering to biotic and abiotic surfaces, forming biofilms [13.66]. Bacteria secreted a matrix of extracellular polymeric substances, which become structured clusters of cells that provide increased resistance to antibiotics [13.67]. Earlier studies on the Foglia river connecting Italian estuaries with livestock farming practices have recorded *Aeromonas* genera [13.68]. *Aeromonas veronii* was reported in Lake Erie in the United States, with evidence of biofilm formation as a critical characteristic of antibiotic resistance [13.69].

13.3.11 Diatom Sampling and Analysis

Diatoms adhered to the substrate (stones) of the bioreactor were scrubbed using a scrubber and collected in an acid-cleaned sampling bottle. Three stones from each bioreactor (of the study station) were collected, processed and enumerated by counting 400 valves per view. The spatial extent of substrates was determined by measuring the stones' dimensions (length, width, and diagonal), which were validated by mapping on the graph sheet and measuring the area using a planimeter for determination of volumetric and areal productivity.

The harvesting period was finalized by assessing the saturation cell density over the stones during the sampling days. Harvesting of the biomass was done when the diatom cell saturation was reached on the stones. On the day of harvest, the attached biomass was removed from all stones (90 stones) in each bioreactor. The harvested biomass (diatoms) was suspended in 50 mL distilled water and fixed with 2% Lugol's iodine. The volumetric productivity of the harvested biomass was calculated by counting the number of cells in a defined volume (10 μL) which was then normalized to cell density per mL. The areal productivity of the biomass that was adhered to stones was determined after normalizing the area of the stone to m^2. The areal productivity is computed as per Equation 13.1 and expressed as cell density per m^2 as well as ash-free dry weight (AFDW) per m^2. The AFDW of the harvested biomass was determined using a muffle furnace.

$$Areal\ productivity = \frac{No\ of\ cells}{Area\ of\ the\ stone\ (m^2)} \tag{13.1}$$

The AFDW of dried algal samples yielded 59% for substrate-based cultivation, whereas for suspended cultivation, AFDW was found to be 70%. Ash content is a critical parameter that determines the quality of biodiesel feedstocks. Lower ash content means higher utilizable biomass for biofuel production. Earlier investigations indicated higher ash content in the substrate-based cultivated biomass as a disadvantage over algae grown in suspended cultivation mode [13.70] [13.71]. A recent study reveals that microalgae growing in a mixotrophic mode (using readily available sugars) could solve the problem of higher ash content as the sugar source provided help in enhanced lipid accumulation through the supply of extra organic carbon [13.70].

13.3.12 Biomass Yield and Productivity

As the cell density reached saturation, biomass was harvested from the stones deployed by scrubbing diatom cells and determining the area of the stones. Harvested biomass was centrifuged at 2500 rpm for 5 mins, and the pellets obtained were washed twice with distilled water and dried in an oven at 60 °C overnight until they reached constant weight. Biomass yield and productivity were calculated as per Equations 13.2 and 13.3. All calculations were made on an AFDW (ash-free dry weight) basis.

$$Biomass\ yield = \frac{Biomass\ dry\ weight\ (g)}{Cumulative\ area\ of\ the\ stones\ (m^2)} \tag{13.2}$$

$$Biomass\ productivity = \frac{Biomass\ dry\ weight\ (g)}{cummulative\ area\ (m^2) \times No\ of\ days\ (d)} \tag{13.3}$$

The biomass concentration of diatoms harvested at 12 days growth cycle from the biofilm bioreactor and the culture flasks (suspension cultivation) was found to be 0.57 g L^{-1} and 0.045 g L^{-1}, respectively, resulting in biomass productivity of 0.048 g $L^{-1}d^{-1}$ for biofilm cultivation and 3.78 mg $L^{-1}d^{-1}$ for suspended cultivation. Biomass concentration estimated for biofilm and suspension cultivation of *Amphora* sp. was found to be ~12 times higher for biofilm cultivation when compared to flask cultures. The biomass concentration and productivity on an aerial scale in the biofilm bioreactor was found to be 4190 mg m^{-2} and 349 mg m^{-2} d^{-1}, respectively. The aerial productivity of *Amphora* sp. was comparatively less than the reported productivity in an earlier study on the diatom *Nitzschia palea* [13.21]. However, it is to be noted that the media used for the growth of *Nitzschia palea* was synthetic wastewater, while the present study was carried out using f/2 media. This experimental result proves the viability of biofilm cultivation of diatoms over conventional shake-flask culturing.

The biofilm bioreactors installed under field conditions exhibited an aerial productivity of 0.02, 0.012, and 0.002 g m^{-2} at the study stations BA, KA and HG, respectively, after an eighteen day growth period. The biomass productivity and yield of diatoms were found to be significantly influenced by local environmental conditions prevailing at the three different study stations. The average biomass productivity of diatoms across stations were 123 mg m^{-2}. The biomass productivity values were calculated on an ash-free dry weight (AFDW) basis and showed aerial biomass productivity of 1240, 710, and 110 mg m^{-2} d^{-1}. An earlier study that used aluminium-based algal turf scrubbing (ATS) system showed 0.014 kg m^{-2} of average aerial biomass productivity, and algae grown in biofilm cultivation system made of wood in another study showed a productivity of 0.012 kg m^{-2} [13.72]. Both the reported productivities were comparable to the present study.

13.3.13 Statistical Analysis

A two-way ANOVA was carried out on the field collected data to understand the variations in physicochemical parameters across stations and the significance of each of the parameters in biomass and lipid productivity. The principal component analysis was carried out to visualize the effect of physicochemical parameters on diatom productivity at each station. Data analysis was carried out using Excel 2016 (ANOVA) and PAST statistical software ver. 4.02. ANOVA performed using physicochemical parameters at the respective study stations (BA, KA, and HG) demonstrated significant variations in its environmental parameters and diatom productivity among stations (listed as columns) for analysis as against physicochemical parameters (listed as rows) evident from the p-value (< 0.05). Thus, the statistical analysis reveals the role of environmental conditions at each station in biofilm formation on the introduced substrate (stones) with positive correlations towards biomass adherence and hence the diatom productivity. Table 13.1 lists the summary of ANOVA.

The principal component analysis (PCA) (Figure 13.7) carried out considering daily physicochemical parameters on biomass and lipid productivity of diatoms during harvest at the end of the growth cycle represents a strong correlation of biomass to solar insolation of the study stations (evident from the PCA loading). The environmental conditions at BA (1st quadrant in Figure 13.7) favored higher biomass due to the prevailing solar insolation and TDS. Stations HG and KA in the 2nd, 3rd, and 4th quadrants show low levels of biomass productivity due to low solar insolation and TDS.

Table 13.1 Analysis of variance table of data collected from different stations.

Source of variation	SS	df	MS	F	P-value	F crit
Rows	1.60E+16	32	5.03E+14	1.0	0.471	1.479
Columns	9.22E+16	10	9.22E+15	18.32	**1.82E-26**	1.860
Error	1.60E+17	320	5.03E+14			
Total	2.69E+17	362				

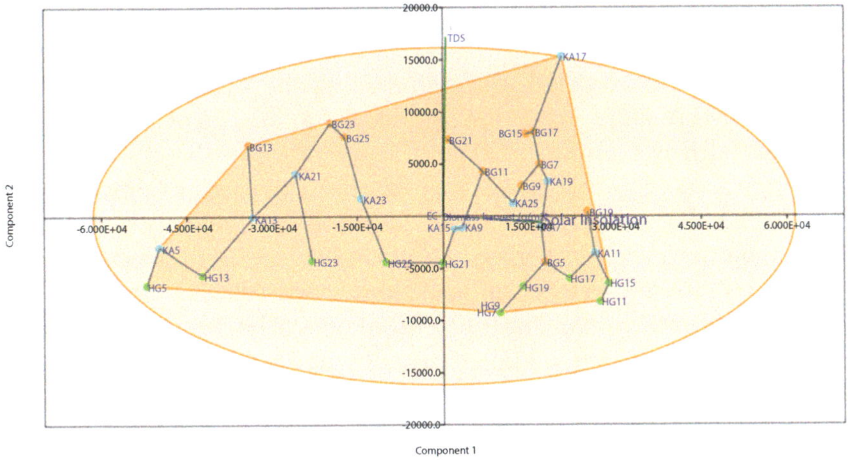

Figure 13.7 PCA plot showing correlation among station's physicochemical parameters and diatom productivity.

Table 13.2 Eigenvalues of principal components.

PC	Eigenvalue	% Variance
1	5.53E08	93.53
2	3.82E07	6.4704
3	104.07	1.75E-05
4	20.55	3.47E-06
5	13.31	2.25E-06
6	2.32	3.92E-07
7	0.64	1.09E-07
8	0.34	5.76E-08
9	0.24	4.21E-08

Table 13.2 lists the eigenvalue and % variance for principal components. Principal components 1 and 2 explain 100% variability of environmental parameters depicting variations in the biomass productivities with higher correlations across stations.

13.3.14 Optimization of Reaction Parameters for Direct Transesterification

Direct transesterification is the direct conversion of algal biomass into biodiesel involving the esterification of free fatty acids (FFAs) and transesterification of triglycerides present in the microalgal biomass [13.73]. This process enables the direct conversion of microalgal cell-bound lipids into biodiesel, thus eliminating the lipid extraction step, which otherwise incurs excess energy and hazardous halogenated chemical input such as chloroform. Reports suggest that the direct transesterification process improves biodiesel yield when compared to conventional extraction-transesterification, as the later process might incur lipid loss. Thus, to optimize the reaction conditions and volume of reagents required for direct transesterification, the algal biomass was subjected to a range of operating conditions involving different volumes of hexane, methanol, and H_2SO_4 under different reaction temperatures and reaction times. The various factors considered with their input levels are given in Table 13.3. The FAME conversion efficiency of direct transesterification was determined using Equation 13.4. The results obtained in optimization were used to generate response surface graphs using Minitab 14 statistical software.

$$FAME\ conversion\ efficiency\ (\%) = \frac{weight\ of\ FAME\ in\ DT\ \left({}^{mg}/_{mg}\right)}{weight\ of\ FAME\ in\ LEAT\ \left({}^{mg}/_{mg}\right)} \times 100 \tag{13.4}$$

The optimal concentrations of hexane, methanol, and sulfuric acid were determined for direct transesterification and computed biodiesel conversion efficiency, which was compared with the conventional two-step lipid extraction and transesterification process. The maximum concentration of fatty acid methyl esters achievable in the conventional lipid extraction-transesterification process was considered as the reference for calculating the biodiesel conversion yield using Equation 13.4.

Hexane of varying volumes ranging from 0.5–2.5 mL was used per gram of dried algal biomass, and at 2 mL, hexane volume showed a higher biodiesel conversion efficiency of 83.33%. This is contrary to the earlier reports of a reduced conversion yield with increasing volumes of n-hexane due to the reduction in the interactions, and hence mass and heat transfer between methanol and the lipids [13.74].

Methanol optimization showed a biodiesel conversion efficiency (96.49%) at a volume of 3 mL. This highlights that any increase in the methanol to co-solvent ratio significantly improves the biodiesel yield from 41.67% to 96.49% (Figure 13.8).

Table 13.3 Variation in operating parameters considered for the study.

Varying parameter	Volume of hexane (mL)	Volume of methanol (mL)	Volume of catalyst (mL)	Reaction temperature (°C)	Reaction time (min)	Biodiesel conversion efficiency (%)
Hexane	0.5	3	0.5	90	40	31.25
	1	3	0.5	90	40	31.25
	1.5	3	0.5	90	40	72.92
	2	3	0.5	90	40	83.33
	2.5	3	0.5	90	40	41.67
Methanol	2	1	0.5	90	40	20.83
	2	2	0.5	90	40	41.67
	2	3	0.5	90	40	96.49
	2	4	0.5	90	40	93.75
	2	5	0.5	90	40	83.33

(Continued)

Table 13.3 Variation in operating parameters considered for the study. (*Continued*)

Varying parameter	Volume of hexane (mL)	Volume of methanol (mL)	Volume of catalyst (mL)	Reaction temperature (°C)	Reaction time (min)	Biodiesel conversion efficiency (%)
H_2SO_4	2	3	0.2	90	40	26.32
	2	3	0.4	90	40	41.67
	2	3	0.6	90	40	87.72
	2	3	0.8	90	40	20.83
	2	3	1	90	40	31.25
Temperature	2	3	0.6	60	40	37.09
	2	3	0.6	70	40	72.92
	2	3	0.6	80	40	97.22
	2	3	0.6	90	40	23.81
	2	3	0.6	100	40	36.04
Time	2	3	0.6	60	20	35.09
	2	3	0.6	60	40	26.32
	2	3	0.6	60	60	96.49
	2	3	0.6	60	80	81.25
	2	3	0.6	60	100	41.67

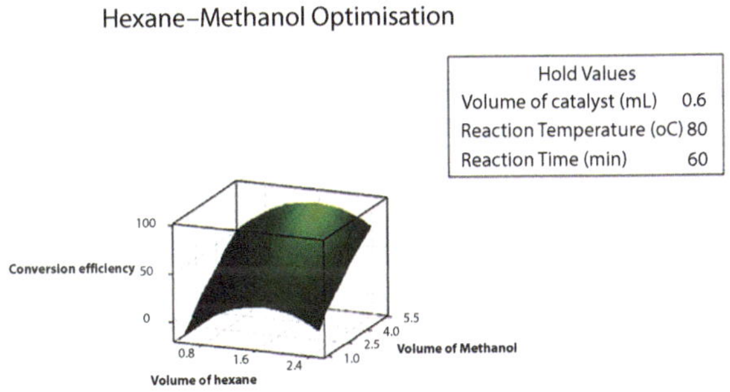

Figure 13.8 Response surface plots of hexane-methanol optimization.

The catalyst plays a vital role in the biodiesel conversion process. Using appropriate concentrations of the acid catalyst during transesterification reduces the efforts, energy, and materials spent on neutralizing acid and acid trace removal from the biodiesel produced [13.75]. The optimal concentration of sulfuric acid was 0.6 mL for the effective conversion of oil into fatty acid methyl esters (FAMEs) (Figure 13.9).

The reaction temperature at 80°C and a reaction time of 60 min yielded a higher biodiesel conversion efficiency as FAME yield increased from 37.09% (at 60°C) to 97.22% (at 80°C) (Figure 13.10). This could be attributed to the disruption of algal cell walls with higher temperatures, accelerating the transesterification reaction. Similar results were reported earlier [13.74] [13.76] [13.77].

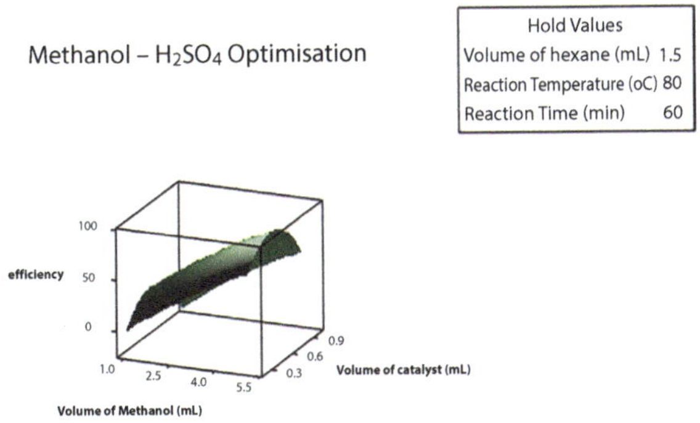

Figure 13.9 Response surface plots of methanol-H_2SO_4 optimization.

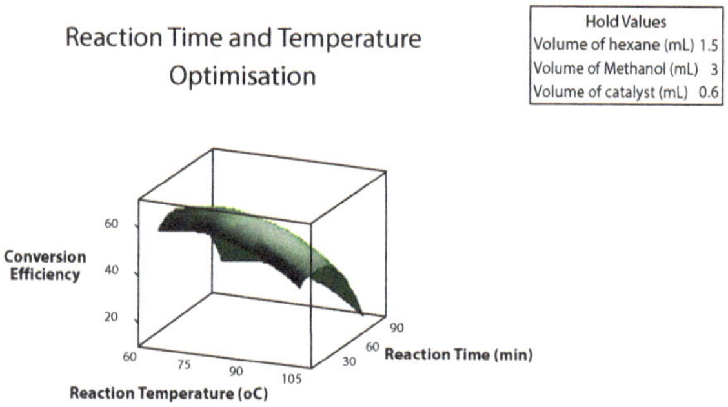

Figure 13.10 Response surface plots of reaction time and temperature optimization.

13.3.15 Direct Transesterification of the Field Harvested Biomass

Dried algal biomass (~1 g) was treated with optimized volumes of reactants and catalyst (3 mL methanol, 0.6 mL conc. H_2SO_4, and 2.5 mL hexane) in a screw-capped vial and subjected to transesterification at 80°C for 60 min (based on the optimization results) in a Hach DRB200 heat block with intermittent mixing. Then, vials were allowed to cool at room temperature and 2 mL of distilled water was added and thoroughly mixed to enable phase separation. The organic hexane layer consisting of biodiesel (FAME) was recovered, and FAME quantity was determined using GC-MS analysis by using (~2 mg) capric acid methyl ester (C10:0) as an internal standard.

FAME composition analysis of the field harvested biomass showed different fatty acid profiles with reference to stations. Figure 13.11 highlights GC peaks of fatty acids derived from algal biomass harvested from different stations. Palmitic acid (C16:0) was found to be the dominant fatty acid methyl ester contributing up to 42.034 ± 1.56% in diatoms of station BA, 61.58 ± 3.12% of station KA, and 73.73 ± 0.52% of HG station, respectively. Short-chain myristic acid (C14:0) methyl ester was highest for biomass harvested from KA with 12.91 ± 1.14%, 11.76 ± 0.55% in station HG, and 5.02 ± 0.61% in BA. Table 13.4 shows the variation in fatty acid methyl esters profile of algal biomass harvested from different stations of the Aghanashini estuary. Behenic acid methyl ester (C22:0) is present only in station BA extracted lipids with 13.50 ± 1.36%. As station BA was observed to have the highest biomass with lentic waters, variations in fatty acid profiles were noted with other stations.

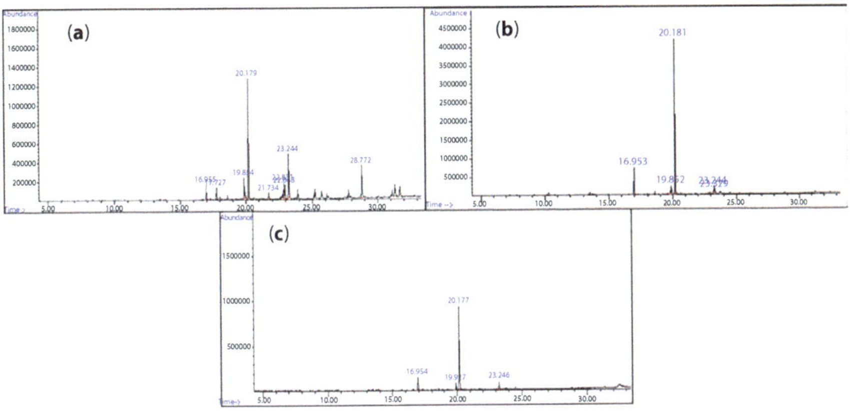

Figure 13.11 FAME composition analysis of field harvested biofilm biomass at different stations - (a) Kagal (KA), (b) Henne Gazani (HG), and (c) Bargi (BA).

13.3.16 Biodiesel Extraction and Determination of Its Quality

Increasing global energy demand and diminishing petroleum resources have led to a focus shift towards alternative renewable and sustainable fuels. Biodiesel is a realistic fuel of the future [13.78] owing to its environmentally friendly characteristics [13.79]. It is known for its good quality exhaust, sustainability, and biodegradability. Transesterification of oils, especially triacylglycerols (TAGs), with monohydric alcohols in the presence of catalysts usually yields biodiesel. The transesterification product is usually mono-alkyl esters of its corresponding fatty acids [13.80]. Biodiesel is considered a better substitute for petro-diesel due to its biodegradability, nontoxicity, and renewability with reduced CO_2, SO_2, and particulate matter emissions. Biodiesel as drop-in fuel could be used in compression ignition (CI) engines without any modifications, provided the physical and chemical properties conform with the international standard specifications of biodiesel (ASTM 6751, EN 14214) [13.81]. The unit operations involved in refining the feedstocks and the method used for the conversion of fatty acids into fatty acid alkyl esters act as critical factors in determining the fuel's quality to meet the designed specification standards.

Biodiesel fuel is generally characterized by its density, viscosity, cetane number, flash point, acid value, higher heating value (HHV), cloud, and pour points [13.82]. Biodiesel quality estimation of lipids extracted from field harvested algal biomass met the specifications set by ASTM and EN

Table 13.4 FAME composition of biomass harvested from different stations.

Peak no.	RT (min.)	FAMEs	Carbon chain length	BA % out of total	KA % out of total	HG % out of total
1	13.73	Undecanoic acid, 10-methyl-, methyl ester	C12:0	–	2.54 ± 0.52	-
2	16.954	Myristic acid, methyl ester	C14:0	5.207 ± 0.61	12.91 ± 1.14	11.76 ± 0.55
3	17.727	Tridecanoic acid 4,8,12-trimethyl-, methyl ester	C13:0	3.976 ± 0.02	–	–
4	18.577	Pentadecanoic acid methyl ester	C15:0	–	6.49 ± 1.27	-
5	19.853	Palmitoleic acid, methyl ester	C16:1	7.88 ± 0.04	6.618 ± 0.54	6.26 ± 0.23
6	20.178	Palmitic acid, methyl ester	C16:0	42.034 ± 1.56	61.58 ± 3.12	73.73 ± 0.52
7	21.733	Margaric acid, methyl ester	C17:0	1.742 ± 0.41	–	–
8	22.86	Oleic acid, methyl ester	C18:1n-9	6.008 ± 0.20	4.16 ± 1.10	-
9	23.243	Stearic acid, methyl ester	C18:0	14.34 ± 0.27	7.25 ± 0.72	8.23 ± 0.30
10	28.772	Behenic acid, methyl ester	C22:0	13.50 ± 1.36	–	–

quality standards. Cetane number is a critical factor in determining the delay in time between fuel injection and ignition onset. The cetane number of BA, HG, and KA station-derived biodiesel was ~ CN 62, which falls well ahead of the minimum requirement of CN 51. Cloud point (CP) is the temperature at which the first visible solid begins to appear upon cooling, and hence it determines the cold filter plugging property of biodiesel [13.83]. The CP value of biodiesel derived from the diatoms of biofilm bioreactor ranged between 18.6 and 19.2 °C. The cloud point standard specifications are usually location-dependent due to substantial variations in weather conditions across geographical locations. The cloud point values are higher, yet it is suitable for tropical countries like India.

Iodine value is the measure of the degree of unsaturation and is usually determined by the amount of I_2 that usually reacts with C=C bonds. Earlier studies on biodiesel quality determination of several microalgae have reported biodiesel quality satisfying ASTM specifications of biodiesel [13.84] [13.85]. The HHV of the biodiesel in the present study was 38.7 MJ/kg, which was comparable with other microalgal-derived biodiesel [13.86], which is usually 10–12% less than what is achievable in petroleum-derived biodiesel (46 MJ/kg) [13.87].

13.4 Microalgal Biorefinery

Insights obtained from lab-scale and field prototype biofilm bioreactor, aided in designing a microalgal bioreactor using granite stones as substrates for microalgal cultivation in an area of one-hectare in the Gazani lands situated along the coastlines of Karnataka. Bio-products possible from microalgal biorefinery include biodiesel, glycerol, biogas, and algal meal fertilizer. Acid/biocatalyst-based transesterification of algal biomass results in biodiesel as the main product and crude glycerol as the byproduct. The crude glycerol, after refining, has diverse applications in the pharmaceutical and cosmetic industries. The spent algal biomass yields biogas when subjected to anaerobic digestion. The solid residue left over after biogas production was considered a source of biofertilizer as a source of nitrogen in agricultural fields.

13.4.1 Material Balance

A mass balance of algal refinery considering 100 kg dry algal biomass byproducts other than biodiesel (Figure 13.12) was performed. Biodiesel yield of 15–23 kg is possible through transesterification of lipid content of 18–26% on a dry weight basis. Crude glycerol (a transesterification byproduct) of 1.5–2.3 kg, 10% of the biodiesel produced [13.88]. A biogas potential of 0.272 m^3/kg of spent algal biomass after lipid extraction is possible through anaerobic digestion, which would provide a biogas yield rate of 24.48 m^3 per kg of biomass [13.89] [13.90]. The raw biogas thus obtained can either be directly fed into cooking pipelines or can be upgraded using appropriate absorption columns and then used for domestic cooking/heating applications. With an assumption of 10% of biomass loss during the direct transesterification process, the solid digestate in a slurry form ranging between 55 and 70 kg that is rich in nitrogen is useful as an organic biofertilizer.

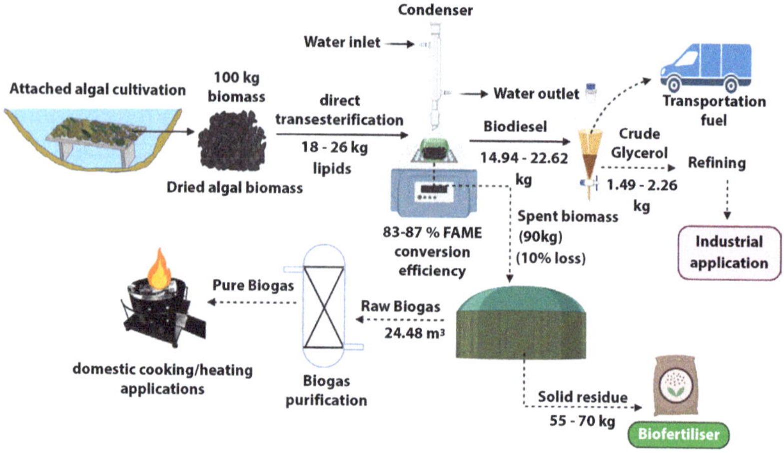

Figure 13.12 Material balance of microalgal biorefinery.

Thus, microalgal biorefinery utilizing microalgal biomass provides bioenergy (biodiesel and biogas) with value-added products such as glycerol and biofertilizer. Deployment of such substrate-based microalgal bioreactor in the brackish water flood plains (that are left abandoned) along the coastal regions of Karnataka would enhance job opportunities for a better livelihood for the coastal population at a decentralized level with the scope for CO_2 sequestration to mitigate GHG footprint.

13.5 Conclusion and Future Perspectives

Biofilm bioreactor designed using gravel stones as substrates for diatom attachment at a laboratory scale and in the field provided insights into the areal productivity of diatoms and necessary environmental conditions for optimal growth. Lab-scale prototype on single-species biofilm studies using the diatom *Amphora* sp. yielded aerial biomass of 4.19 g m^{-2} with an aerial productivity rate of 0.35 g m^{-2} d^{-1} in 12 days of cultivation. Bioreactor setup deployed in the field at different locations along an estuary yielded higher biomass of 0.022 kg m^{-2} (AFDW) with biomass productivity comparable to productivities of similar attached cultivation systems of varying reactor configurations. Two bacterial isolates, *Exiguobacterium aurantiacum* and *Aeromonas veronii,* were found to play a key role in a biofilm by establishing symbiotic community associations of bacteria and diatoms. Optimization of reaction parameters and solvents used for direct transesterification of microalgal biomass resulted in 1.5 mL hexane, 3 mL methanol, and 0.6 mL sulfuric acid as optimized concentrations for higher biodiesel yield when the reaction time and temperature were maintained at 80°C for 60 min. Lipid and biodiesel characterization yielded distinct fatty acid profiles ranging between C12 to C22 carbon chain length fatty acids, primarily influenced by diatom communities and their prevalent composition at each site. The quality of lipids extracted from the field bioreactor proved its compatibility with diesel engines with properties well within the standard specification limits. Microalgal biomass production designed based on the insights obtained from lab and field bioreactor prototypes demonstrates the prospects of valorization of biomass for biodiesel extraction and commercialization.

Acknowledgments

The authors are grateful to the EIACP division of the Ministry of Environment, Forests and Climate Change, Government of India [SP/DEOO-22-001]; and the Indian Institute of Science [R1011] for the infrastructure support for ecological research. We thank Mr. Sreekanth Naik, Mr. Vishnu D. Mukri, and Ms. Gayatri Naik for their assistance during the field data collection. We thank all stakeholders of the study region for actively participating in the scientific discussions and cooperation during field data compilation.

Funding

The research was financially supported by ISRO–IISc Space Technology Cell (ISTC/BES/TVR/389) and Gajanana and Madhavi Hegde (CES/TVR/GJH001) endowment for biofuel research.

Research Ethics

The publication is based on the original research and has not been submitted elsewhere for publication or web hosting.

Animal Ethics

The research does not involve either humans, animals, or tissues.

References

[13.1] Kang, S., Heo, S. and Lee, J.H. (2019) Techno-economic Analysis of Microalgae-Based Lipid Production: Considering Influences of Microalgal Species. *Industrial and Engineering Chemistry Research*, 58, (2). https://doi.org/10.1021/acs.iecr.8b03999

[13.2] Berner, F., Heimann, K. and Sheehan, M. (2015) Microalgal biofilms for biomass production. *Journal of Applied Phycology*, Springer Netherlands. 27, (5). https://doi.org/10.1007/s10811-014-0489-x

[13.3] Davis, R., Aden, A. and Pienkos, P.T. (2011) Techno-economic analysis of autotrophic microalgae for fuel production. *Applied Energy*, Elsevier. 88, (10). https://doi.org/10.1016/J.APENERGY.2011.04.018

[13.4] Fasaei, F., Bitter, J.H., Slegers, P.M. and van Boxtel, A.J.B. (2018) Techno-economic evaluation of microalgae harvesting and dewatering systems. *Algal Research*, Elsevier. 31. https://doi.org/10.1016/J.ALGAL.2017.11.038

[13.5] Adey, W.H., Kangas, P.C. and Mulbry, W. (2011) Algal Turf Scrubbing: Cleaning Surface Waters with Solar Energy while Producing a Biofuel. *BioScience*, 61, (6). https://doi.org/10.1525/bio.2011.61.6.5

[13.6] Ramachandra, T. V., Mahapatra, D.M., Bhat, S.P. and Joshi, N. V. (2015) Biofuel Production Along with Remediation of Sewage Water Through Algae. *Algae and Environmental Sustainability*, Springer India, New Delhi. p. 33–51. https://doi.org/10.1007/978-81-322-2641-3_4

[13.7] Irwin, N.B., Irwin, E.G., Martin, J.F. and Aracena, P. (2018) Constructed wetlands for water quality improvements: Benefit transfer analysis from Ohio. *Journal of Environmental Management*, Academic Press. 206. https://doi.org/10.1016/J.JENVMAN.2017.10.050

[13.8] Ramachandra, T. V., Sudarshan, P.B., Mahesh, M.K. and Vinay, S. (2018) Spatial patterns of heavy metal accumulation in sediments and macrophytes of Bellandur wetland, Bangalore. *Journal of Environmental Management*, Academic Press. 206. https://doi.org/10.1016/j.jenvman.2017.10.014

[13.9] Craggs, R.J., Adey, W.H., Jenson, K.R., St. John, M.S., Green, F.B. and Oswald, W.J. (1996) Phosphorus removal from wastewater using an algal turf scrubber. *Water Science and Technology*, International Association on Water Quality. 33, (7). https://doi.org/10.1016/0273-1223(96)00354-X

[13.10] Adey, W.H., Laughinghouse, H.D., Miller, J.B., Hayek, L.A.C., Thompson, J.G., Bertman, S. *et al.* (2013) Algal turf scrubber (ATS) floways on the Great Wicomico River, Chesapeake Bay: Productivity, algal community structure, substrate and chemistry. *Journal of Phycology*, 49, (3). https://doi.org/10.1111/jpy.12056

[13.11] Mantzorou, A. and Ververidis, F. (2019) Microalgal biofilms: A further step over current microalgal cultivation techniques. *Science of The Total Environment*, Elsevier. 651. https://doi.org/10.1016/J.SCITOTENV.2018.09.355

[13.12] Costerton, J.W., Geesey, G.G. and Cheng, K.-J. (1978) How Bacteria Stick. *Scientific American*, 238, (1). https://doi.org/10.1038/scientificamerican0178-86

[13.13] Mantzorou, A., Navakoudis, E., Paschalidis, K. and Ververidis, F. (2018) Microalgae: a potential tool for remediating aquatic environments from toxic metals. *International Journal of Environmental Science and Technology*, Springer Berlin Heidelberg. 15, (8). https://doi.org/10.1007/s13762-018-1783-y

[13.14] Gerbersdorf, S.U., Westrich, B. and Paterson, D.M. (2009) Microbial Extracellular Polymeric Substances (EPS) in Fresh Water Sediments. *Microbial Ecology*, Springer-Verlag. 58, (2). https://doi.org/10.1007/s00248-009-9498-8

[13.15] Christenson, L. and Sims, R. (2011) Production and harvesting of microalgae for wastewater treatment, biofuels, and bioproducts. *Biotechnology Advances*, Elsevier. 29, (6). https://doi.org/10.1016/j.biotechadv.2011.05.015

[13.16] Hoagland, K.D., Rosowski, J.R., Gretz, M.R. and Roemer, S.C. (1993) Diatom Extracellular Polymeric Substances: Function, Fine Structure, Chemistry, and Physiology. *Journal of Phycology*, John Wiley & Sons, Ltd (10.1111). 29, (5). https://doi.org/10.1111/j.0022-3646.1993.00537.x

[13.17] Galié, S., García-Gutiérrez, C., Miguélez, E.M., Villar, C.J. and Lombó, F. (2018) Biofilms in the Food Industry: Health Aspects and Control Methods. *Frontiers in Microbiology*, Frontiers Media SA. 9. https://doi.org/10.3389/fmicb.2018.00898

[13.18] Katarzyna, L., Sai, G. and Singh, O.A. (2015) Non-enclosure methods for non-suspended microalgae cultivation: literature review and research needs. *Renewable and Sustainable Energy Reviews*, Pergamon. 42. https://doi.org/10.1016/J.RSER.2014.11.029

[13.19] Shen, Y., Xu, X., Zhao, Y. and Lin, X. (2014) Influence of algae species, substrata and culture conditions on attached microalgal culture. *Bioprocess and Biosystems Engineering*, Springer Berlin Heidelberg. 37, (3). https://doi.org/10.1007/s00449-013-1011-6

[13.20] Zippel, B. and Neu, T.R. (2005) Growth and structure of phototrophic biofilms under controlled light conditions. *Water Science and Technology*, IWA Publishing. 52, (7). https://doi.org/10.2166/wst.2005.0202

[13.21] Schnurr, P.J., Espie, G.S. and Allen, D.G. (2013) Algae biofilm growth and the potential to stimulate lipid accumulation through nutrient starvation. *Bioresource Technology*,. https://doi.org/10.1016/j.biortech.2013.03.036

[13.22] Palma, H., Killoran, E., Sheehan, M., Berner, F. and Heimann, K. (2017) Assessment of microalga biofilms for simultaneous remediation and biofuel generation in mine tailings water. *Bioresource Technology*, Elsevier. 234. https://doi.org/10.1016/J.BIORTECH.2017.03.063
[13.23] Ogbonna, J.C. and Tanaka, H. (1998) Cyclic autotrophic/heterotrophic cultivation of photosynthetic cells: A method of achieving continuous cell growth under light/dark cycles. *Bioresource Technology*, Elsevier Sci Ltd. 65, (1–2). https://doi.org/10.1016/S0960-8524(98)00018-2
[13.24] Avendaño-Herrera, R.E. and Riquelme, C.E. (2007) Production of a diatom-bacteria biofilm in a photobioreactor for aquaculture applications. *Aquacultural Engineering*, Elsevier. 36, (2). https://doi.org/10.1016/j.aquaeng.2006.08.001
[13.25] Mack, W.N., Mack, J.P. and Ackerson, A.O. (1975) Microbial film development in a trickling filter. *Microbial Ecology*, Springer-Verlag. 2, (3). https://doi.org/10.1007/BF02010441
[13.26] Barranguet, C., van Beusekom, S., Veuger, B., Neu, T., Manders, E., Sinke, J. *et al.* (2004) Studying undisturbed autotrophic biofilms: still a technical challenge. *Aquatic Microbial Ecology*, Inter-Research. 34, (1). https://doi.org/10.3354/ame034001
[13.27] Gross, M., Henry, W., Michael, C. and Wen, Z. (2013) Development of a rotating algal biofilm growth system for attached microalgae growth with *in situ* biomass harvest. *Bioresource Technology*, Elsevier. 150. https://doi.org/10.1016/J.BIORTECH.2013.10.016
[13.28] Podola, B., Li, T. and Melkonian, M. (2017) Porous Substrate Bioreactors: A Paradigm Shift in Microalgal Biotechnology? *Trends in Biotechnology*, Elsevier. 35, (2). https://doi.org/10.1016/j.tibtech.2016.06.004
[13.29] Zippel, B., Rijstenbil, J. and Neu, T.R. (2007) A flow-lane incubator for studying freshwater and marine phototrophic biofilms. *Journal of Microbiological Methods*, Elsevier. 70, (2). https://doi.org/10.1016/J.MIMET.2007.05.013
[13.30] Boelee, N.C., Temmink, H., Janssen, M., Buisman, C.J.N. and Wijffels, R.H. (2011) Nitrogen and phosphorus removal from municipal wastewater effluent using microalgal biofilms. *Water Research*, 45, (18). https://doi.org/10.1016/j.watres.2011.08.044
[13.31] Ozkan, A., Kinney, K., Katz, L. and Berberoglu, H. (2012) Reduction of water and energy requirement of algae cultivation using an algae biofilm photobioreactor. *Bioresource Technology*, Elsevier. 114. https://doi.org/10.1016/J.BIORTECH.2012.03.055
[13.32] Shen, Y., Chen, C., Chen, W. and Xu, X. (2014) Attached culture of *Nannochloropsis oculata* for lipid production. *Bioprocess and Biosystems Engineering*, 37, (9). https://doi.org/10.1007/s00449-014-1147-z
[13.33] Johnson, M.B. and Wen, Z. (2010) Development of an attached microalgal growth system for biofuel production. *Applied Microbiology and Biotechnology*, Springer-Verlag. 85, (3). https://doi.org/10.1007/s00253-009-2133-2

[13.34] Christenson, L.B. and Sims, R.C. (2012) Rotating algal biofilm reactor and spool harvester for wastewater treatment with biofuels by-products. *Biotechnology and Bioengineering*, Wiley-Blackwell. 109, (7). https://doi.org/10.1002/bit.24451

[13.35] Orandi, S., Lewis, D.M. and Moheimani, N.R. (2012) Biofilm establishment and heavy metal removal capacity of an indigenous mining algal-microbial consortium in a photo-rotating biological contactor. *Journal of Industrial Microbiology & Biotechnology*, 39, (9). https://doi.org/10.1007/s10295-012-1142-9

[13.36] Blanken, W., Janssen, M., Cuaresma, M., Libor, Z., Bhaiji, T. and Wijffels, R.H. (2014) Biofilm growth of *Chlorella sorokiniana* in a rotating biological contactor based photobioreactor. *Biotechnology and Bioengineering*, John Wiley & Sons, Ltd. 111, (12). https://doi.org/10.1002/bit.25301

[13.37] Nowack, E.C.M., Podola, B. and Melkonian, M. (2005) The 96-Well Twin-Layer System: A Novel Approach in the Cultivation of Microalgae. *Protist*, Urban & Fischer. 156, (2). https://doi.org/10.1016/J.PROTIS.2005.04.003

[13.38] Murphy, T.E. and Berberoglu, H. (2014) Flux balancing of light and nutrients in a biofilm photobioreactor for maximizing photosynthetic productivity. *Biotechnology Progress*, American Chemical Society (ACS). 30, (2). https://doi.org/10.1002/btpr.1881

[13.39] Naumann, T., Çebi, Z., Podola, B. and Melkonian, M. (2013) Growing microalgae as aquaculture feeds on twin-layers: A novel solid-state photobioreactor. *Journal of Applied Phycology*, Springer Netherlands. 25, (5). https://doi.org/10.1007/s10811-012-9962-6

[13.40] Ji, C., Wang, J., Zhang, W., Liu, J., Wang, H., Gao, L. *et al.* (2014) An applicable nitrogen supply strategy for attached cultivation of *Aucutodesmus obliquus*. *Journal of Applied Phycology*, 26, (1). https://doi.org/10.1007/s10811-013-0115-3

[13.41] Liu, T., Wang, J., Hu, Q., Cheng, P., Ji, B., Liu, J. *et al.* (2013) Attached cultivation technology of microalgae for efficient biomass feedstock production. *Bioresource Technology*, Elsevier. 127. https://doi.org/10.1016/J.BIORTECH.2012.09.100

[13.42] Cheng, P., Ji, B., Gao, L., Zhang, W., Wang, J. and Liu, T. (2013) The growth, lipid and hydrocarbon production of *Botryococcus braunii* with attached cultivation. *Bioresource Technology*, Elsevier. 138. https://doi.org/10.1016/J.BIORTECH.2013.03.150

[13.43] Boelee, N.C., Janssen, M., Temmink, H., Shrestha, R., Buisman, C.J.N. and Wijffels, R.H. (2014) Nutrient Removal and Biomass Production in an Outdoor Pilot-Scale Phototrophic Biofilm Reactor for Effluent Polishing. *Applied Biochemistry and Biotechnology*, Springer US. 172, (1). https://doi.org/10.1007/s12010-013-0478-6

[13.44] Abe, K., Matsumura, I., Imamaki, A. and Hirano, M. (2003) Removal of inorganic nitrogen sources from water by the algal biofilm of the aerial microalga *Trentepohlia aurea*. *World Journal of Microbiology and*

Biotechnology, Kluwer Academic Publishers. 19, (3). https://doi.org/10.1023/A:1023657310004

[13.45] Christenson, L.B. and Sims, R.C. (2012) Rotating algal biofilm reactor and spool harvester for wastewater treatment with biofuels by-products. *Biotechnology and Bioengineering*, John Wiley & Sons, Ltd. 109, (7). https://doi.org/10.1002/bit.24451

[13.46] Johnson, M. Ben. (2009) Microalgal Biodiesel Production through a Novel Attached Culture System and Conversion Parameters. Virginia Tech.

[13.47] Sekar, R., Venugopalan, V.P., Satpathy, K.K., Nair, K.V.K. and Rao, V.N.R. (2004) Laboratory studies on adhesion of microalgae to hard substrates. *Asian Pacific Phycology in the 21st Century: Prospects and Challenges*, Springer Netherlands, Dordrecht. p. 109–16. https://doi.org/10.1007/978-94-007-0944-7_14

[13.48] Gross, M.A. (2015) Development and optimization of biofilm based algal cultivation [Internet].

[13.49] Zhuang, L.-L., Yu, D., Zhang, J., Liu, F., Wu, Y.-H., Zhang, T.-Y. *et al.* (2018) The characteristics and influencing factors of the attached microalgae cultivation: A review. *Renewable and Sustainable Energy Reviews*, Pergamon. 94. https://doi.org/10.1016/J.RSER.2018.06.006

[13.50] Gao, F., Yang, Z.-H., Li, C., Zeng, G.-M., Ma, D.-H. and Zhou, L. (2015) A novel algal biofilm membrane photobioreactor for attached microalgae growth and nutrients removal from secondary effluent. *Bioresource Technology*, Elsevier. 179. https://doi.org/10.1016/J.BIORTECH.2014.11.108

[13.51] Wang, J.-H., Zhuang, L.-L., Xu, X.-Q., Deantes-Espinosa, V.M., Wang, X.-X. and Hu, H.-Y. (2018) Microalgal attachment and attached systems for biomass production and wastewater treatment. *Renewable and Sustainable Energy Reviews*, Pergamon. 92. https://doi.org/10.1016/J.RSER.2018.04.081

[13.52] Schnurr, P.J. and Allen, D.G. (2015) Factors affecting algae biofilm growth and lipid production: A review. *Renewable and Sustainable Energy Reviews*, Pergamon. 52. https://doi.org/10.1016/J.RSER.2015.07.090

[13.53] Cao, J., Yuan, W., Pei, Z.J., Davis, T., Cui, Y. and Beltran, M. (2009) A Preliminary Study of the Effect of Surface Texture on Algae Cell Attachment for a Mechanical-Biological Energy Manufacturing System. *Journal of Manufacturing Science and Engineering*, American Society of Mechanical Engineers Digital Collection. 131, (6). https://doi.org/10.1115/1.4000562

[13.54] Johnson, M.B. and Wen, Z. (2010) Development of an attached microalgal growth system for biofuel production. *Applied Microbiology and Biotechnology*, 85, (3). https://doi.org/10.1007/s00253-009-2133-2

[13.55] Guillard and Ryther. (2005) f/2 Medium for growing diatoms : 2L Recipe. *Techniques*, https://evolution.unibas.ch/people/dita/lab/methods_pdfs/f2_adjusted.pdf.

[13.56] Madhu, N. V., Jyothibabu, R., Balachandran, K.K., Honey, U.K., Martin, G.D., Vijay, J.G. *et al.* (2007) Monsoonal impact on planktonic standing stock and abundance in a tropical estuary (Cochin backwaters - India). *Estuarine, Coastal and Shelf Science*, Academic Press. 73, (1–2). https://doi.org/10.1016/j.ecss.2006.12.009

[13.57] Martin, G.D., Vijay, J.G., Laluraj, C.M., Madhu, N.V., Joseph, T., Nair, M. *et al.* (2008) Fresh water influence on nutrient stoichiometry in a tropical estuary, Southwest coast of India. Corvinus University of Budapest.

[13.58] Strickland, J.D.H. and Parsons, T.R. (1972) A practical handbook of seawater analysis. Bulletin 167. *Fisheries Research Board of Canada Bulletin*, Fisheries research board of Canada. 167.

[13.59] Besemer, K., Singer, G., Limberger, R., Chlup, A.K., Hochedlinger, G., Hödl, I. *et al.* (2007) Biophysical controls on community succession in stream biofilms. *Applied and Environmental Microbiology*, American Society for Microbiology (ASM). 73, (15). https://doi.org/10.1128/AEM.00588-07

[13.60] Unnithan, V. V., Unc, A. and Smith, G.B. (2014) Mini-review: A priori considerations for bacteria-algae interactions in algal biofuel systems receiving municipal wastewaters. *Algal Research*, Elsevier B.V. 4, (1). https://doi.org/10.1016/j.algal.2013.11.009

[13.61] Fisher, Wilcox and Graham. (1998) Molecular characterization of epiphytic bacterial communities on charophycean green algae. *Applied and Environmental Microbiology*, American Society for Microbiology (ASM). 64, (11).

[13.62] Comte, J., Jacquet, S., Viboud, S., Fontvieille, D., Millery, A., Paolini, G. *et al.* (2006) Microbial community structure and dynamics in the largest natural French lake (Lake Bourget). *Microbial Ecology*, Springer. 52, (1). https://doi.org/10.1007/s00248-004-0230-4

[13.63] Kim, I.G., Lee, M.H., Jung, S.Y., Song, J.J., Oh, T.K. and Yoon, J.H. (2005) *Exiguobacterium aestuarii* sp. nov. and *Exiguobacterium marinum* sp. nov., isolated from a tidal flat of the Yellow Sea in Korea. *International Journal of Systematic and Evolutionary Microbiology*, 55, (2). https://doi.org/10.1099/ijs.0.63308-0

[13.64] Stackebrandt, E., Frühling, A., Schumann, P., Hippe, H. and Sträubler, B. (2002) *Exiguobacterium undae* sp. nov. and *Exiguobacterium antarcticum* sp. nov. *International Journal of Systematic and Evolutionary Microbiology*, Microbiology Society. 52, (4). https://doi.org/10.1099/00207713-52-4-1171

[13.65] Dias, C., Borges, A., Saavedra, M.J. and Simões, M. (2018) Biofilm formation and multidrug-resistant *Aeromonas* spp. from wild animals. *Journal of Global Antimicrobial Resistance*, Elsevier Ltd. 12. https://doi.org/10.1016/j.jgar.2017.09.010

[13.66] Chung, K. and Okabe, S. (2009) Characterization of electrochemical activity of a strain ISO2-3 phylogenetically related to *Aeromonas* sp. isolated from a glucose-fed microbial fuel cell. *Biotechnology and Bioengineering*, John Wiley & Sons, Ltd. 104, (5). https://doi.org/10.1002/bit.22453

[13.67] Gilbert, P., Maira-Litran, T., McBain, A.J., Rickard, A.H. and Whyte, F.W. (2002) The physiology and collective recalcitrance of microbial biofilm communities. Advances in Microbial Physiology. Academic Press. p. 203–56. https://doi.org/10.1016/S0065-2911(02)46005-5

[13.68] Fiorentini, C., Barbieri, E., Falzano, L., Matarrese, P., Baffone, W., Pianetti, A. *et al.* (1998) Occurrence, diversity and pathogenicity of mesophilic *Aeromonas* in estuarine waters of the Italian coast of the Adriatic Sea. *Journal of Applied Microbiology*, John Wiley & Sons, Ltd. 85, (3). https://doi.org/10.1046/j.1365-2672.1998.853517.x

[13.69] Skwor, T., Shinko, J., Augustyniak, A., Gee, C. and Andraso, G. (2014) *Aeromonas hydrophila* and *Aeromonas veronii* predominate among potentially pathogenic ciprofloxacin- and tetracycline-resistant *Aeromonas* isolates from Lake Erie. *Applied and Environmental Microbiology*, American Society for Microbiology. 80, (3). https://doi.org/10.1128/AEM.03645-13

[13.70] Roostaei, J., Zhang, Y., Gopalakrishnan, K. and Ochocki, A.J. (2018) Mixotrophic Microalgae Biofilm: A Novel Algae Cultivation Strategy for Improved Productivity and Cost-efficiency of Biofuel Feedstock Production. *Scientific Reports*, Nature Publishing Group. 8, (1). https://doi.org/10.1038/s41598-018-31016-1

[13.71] Gross, M. and Wen, Z. (2014) Yearlong evaluation of performance and durability of a pilot-scale Revolving Algal Biofilm (RAB) cultivation system. *Bioresource Technology*, Elsevier. 171. https://doi.org/10.1016/J.BIORTECH.2014.08.052

[13.72] Orr, D., Lewis, A. and Adey, W.H. (2010) Algal Turf Scrubber (ATS) Algae to Energy Project.

[13.73] Park, J.-Y., Park, M.S. and Lee, Y.-C. (2015) Advances in direct transesterification of algal oils from wet biomass. *Bioresource Technology*, Elsevier. 184. https://doi.org/10.1016/J.BIORTECH.2014.10.089

[13.74] Dianursanti, Religia, P. and Wijanarko, A. (2015) Utilization of n-Hexane as Co-solvent to Increase Biodiesel Yield on Direct Transesterification Reaction from Marine Microalgae. *Procedia Environmental Sciences*, Elsevier B.V. 23, (Ictcred 2014). https://doi.org/10.1016/j.proenv.2015.01.059

[13.75] Ayetor, G.K., Sunnu, A. and Parbey, J. (2015) Effect of biodiesel production parameters on viscosity and yield of methyl esters: *Jatropha curcas, Elaeis guineensis* and *Cocos nucifera*. *Alexandria Engineering Journal*, Elsevier B.V. 54, (4). https://doi.org/10.1016/j.aej.2015.09.011

[13.76] Zhang, Y., Li, Y., Zhang, X. and Tan, T. (2015) Biodiesel production by direct transesterification of microalgal biomass with co-solvent. *Bioresource Technology*, Elsevier Ltd. 196. https://doi.org/10.1016/j.biortech.2015.07.052

[13.77] Johnson, M.B. and Wen, Z. (2009) Production of Biodiesel Fuel from the Microalga *Schizochytrium limacinum* by Direct Transesterification of Algal Biomass. *Energy & Fuels*, 23, (10). https://doi.org/10.1021/ef900704h

[13.78] Keera, S.T., El Sabagh, S.M. and Taman, A.R. (2011) Transesterification of vegetable oil to biodiesel fuel using alkaline catalyst. *Fuel*, Elsevier. 90, (1). https://doi.org/10.1016/J.FUEL.2010.07.046

[13.79] Atadashi, I.M., Aroua, M.K. and Aziz, A.A. (2010) High quality biodiesel and its diesel engine application: A review. *Renewable and Sustainable Energy Reviews*,. https://doi.org/10.1016/j.rser.2010.03.020

[13.80] Knothe, G. (2009) Improving biodiesel fuel properties by modifying fatty ester composition. Energy and Environmental Science. p. 759–66. https://doi.org/10.1039/b903941d

[13.81] Sharma, Y.C., Singh, B. and Upadhyay, S.N. (2008) Advancements in development and characterization of biodiesel: A review. *Fuel*, Elsevier. 87, (12). https://doi.org/10.1016/J.FUEL.2008.01.014

[13.82] Demirbas, A. (2009) Progress and recent trends in biodiesel fuels. *Energy Conversion and Management*, Pergamon. 50, (1). https://doi.org/10.1016/j.enconman.2008.09.001

[13.83] Knothe, G. (2011) A technical evaluation of biodiesel from vegetable oils vs. algae. Will algae-derived biodiesel perform?. https://doi.org/10.1039/c0gc00946f

[13.84] Xu, H., Miao, X. and Wu, Q. (2006) High quality biodiesel production from a microalga *Chlorella protothecoides* by heterotrophic growth in fermenters. *Journal of Biotechnology*, Elsevier. 126, (4). https://doi.org/10.1016/J.JBIOTEC.2006.05.002

[13.85] Francisco, É.C., Neves, D.B., Jacob-Lopes, E. and Franco, T.T. (2010) Microalgae as feedstock for biodiesel production: Carbon dioxide sequestration, lipid production and biofuel quality. *Journal of Chemical Technology & Biotechnology*, 85, (3). https://doi.org/10.1002/jctb.2338

[13.86] Islam, M.A., Magnusson, M., Brown, R.J., Ayoko, G.A., Nabi, M.N. and Heimann, K. (2013) Microalgal species selection for biodiesel production based on fuel properties derived from fatty acid profiles. *Energies*, 6, (11). https://doi.org/10.3390/en6115676

[13.87] Ramírez-Verduzco, L.F., Rodríguez-Rodríguez, J.E. and Jaramillo-Jacob, A.D.R. (2012) Predicting cetane number, kinematic viscosity, density and higher heating value of biodiesel from its fatty acid methyl ester composition. *Fuel*, Elsevier. 91, (1). https://doi.org/10.1016/j.fuel.2011.06.070

[13.88] Rodrigues, A., Bordado, J.C. and Dos Santos, R.G. (2017) Upgrading the glycerol from biodiesel production as a source of energy carriers and chemicals - A technological review for three chemical pathways. *Energies*, 10, (11). https://doi.org/10.3390/en10111817

[13.89] Harun, R., Davidson, M., Doyle, M., Gopiraj, R., Danquah, M. and Forde, G. (2011) Technoeconomic analysis of an integrated microalgae photobioreactor, biodiesel and biogas production facility. *Biomass and Bioenergy*, Pergamon. 35, (1). https://doi.org/10.1016/j.biombioe.2010.10.007

[13.90] Ramachandra, T. V. and Shruthi, B. V. (2007) Spatial mapping of renewable energy potential. Renewable and Sustainable Energy Reviews. Pergamon. p. 1460–80. https://doi.org/10.1016/j.rser.2005.12.002

14

Diatoms Characteristics and Mass Processing of Lipids for Biofuel Production

Tawaf Ali Shah*, **Zhihe Li**[†], **Zhiyu Li and Andong Zhang**

College of Agriculture Engineering and Food Science, Shandong University of Technology, Zibo, China

Abstract

Diatoms are ubiquitous microalgae that may be used as an excellent candidate for the biofuel industry because of their unique ability to produce and store lipid in their bodies. It is necessary to efficiently manage the primary cultivation, extraction, and subsequent downstream processing of diatom culture. It is also crucial to understand and research the mechanisms underlying diatoms' capacity to produce lipids, as these mechanisms will eventually determine the route for the synthesis of lipid content. Diatoms have many difficulties in their drive to establish themselves as a viable alternative to fossil fuels because of the enormous amount of energy they provide, but recently, they have offered a glimpse of the possibility of sustainable development in the energy industry. Diatom-based biofuel is included in the group of third-generation biofuel. Diatoms have all the necessary characteristics to be used as a substitute raw material for nanotechnology, biotechnology, and energy sources that will promote a sustainable economy. This chapter focuses on the uses of diatoms, the method and process by which their bodies generate lipids, and finally, their involvement in clean energy, especially for methane, hydrogen, ethanol, and other value-added hydrocarbons. In the future, diatom-derived biofuel will eventually advance the green economy and the clean energy industry.

Keywords: Biomass, energy, lipid, biotechnology, methane, hydrogen, anaerobic digestion

**Corresponding author*: tawafbiotech@yahoo.com

[†]*Corresponding author*: lizhihe@sdut.edu.cn

Vandana Vinayak and Richard Gordon (eds.) Diatom Cultivation for Biofuel, Food and High-Value Products, (377–398) © 2025 Scrivener Publishing LLC

14.1 Diatoms

Diatoms are eukaryotic, single-celled microorganisms that are distinguished by having a special, frustule-shaped cell wall that is silica-impregnated. Diatoms and algal cells are composed of an epivalve and a hypovalve, two equal portions that make up this frustule [14.24]. There are two valves in each frustule [14.11]. The older and newer valves each have a different set of identifying elements. The older valve is known as the epitheca, along with the girdle components (cingulum) that are connected to it, whereas the more recent valve is known as the hypotheca. The valve structure plays a significant role in the classification of diatoms. The occurrence of one or two longitudinal slits across the valve, which together make up the raphe system, is a defining characteristic of many diatom genera. The raphe's length, location, structure, and termination are significant taxonomic traits. The chromatophores, which are the most visible parts of the protoplast, are often brown because other pigments, including carotenoids, diatoxanthin, diadinoxanthin, and fucoxanthin, cover the color of the chlorophylls in diatoms [14.12]. High-energy lipids are produced by diatoms during the photosynthesis process. The aquatic food chain greatly benefits from the lipid stores formed by diatoms. Diatoms and other algae are very dependent on the availability of nutrients, particularly phosphorus and nitrogen. They are found all over the world and in a wide range of ecosystems with various abiotic and biotic environments. Due to its high pH tolerance and tendency to occur in tiny colonies, *Tabellaria flocculosa* (Kützig) appears to be one of the most prevalent diatom taxa in the polar regions. It is a good example of a cosmopolitan diatom species because it is only influenced by environmental factors [14.9]. The same species would therefore flourish anywhere provided the environmental factors remained the same. Rarer species have been discovered in the diatom groups. Diatoms respond to the surrounding habitat even in a matter of days because of their rapid life cycles and high species turnover in both space and time. Despite the fact that the majority of diatoms are pigmented and photosynthetic, several species of the genera *Nitzchia* and *Hantzschia* have the potential to be (facultatively) heterotrophic. The most diverse category of algae on Earth is composed of diatoms. Extrapolating from a diverse sample of genera and species complexes, it is likely that there are at least 30,000 diatom species, and most possibly as many as 100,000. Genomic phylogenetic analysis indicates that raphid diatoms actually evolved far earlier than pennate diatoms, which first emerged in the late Cretaceous and the Palaeocene, respectively. Diatoms are classified with other heterokont protists like Chrysophyceae and Xanthophyceae in the Stramenopiles in the most recent eukaryotic categorization scheme. Diatoms are further separated into two primary groups: *Coscinodiscophytina* and *Bacillariophytina*, based on the valve pattern bilateral organization vs. radial organization [14.8]. The oogamous (Mediophyceae) or isogamous (Bacillariophyceae) sexual reproduction of *Bacillariophytina* is distinguished by a bipolar or multipolar valve shape with a valve pattern radiating from a central circular or elongate annulus or from a sternum [14.10].

14.2 Reproduction

Asexual mitotic division is the main mode of reproduction for diatoms. Following division, each daughter cell's epitheca and hypotheca will develop into hypothece and epithecae in the following generation, respectively. One daughter cell has the same size as the mother cell and the other is smaller. The mean cell size of the population gradually shrinks as a result of repeated division. Auxospores are used to restore size. Typically, auxospore development is linked to sexual reproduction. A single, largest-sized cell eventually emerges from the auxospore. In terms of valve construction and form, this first cell could not be identical to the vegetative cell [14.11].

14.3 Ecology and Distribution

Diatoms are regulated by a number of abiotic and biotic factors, therefore there isn't just one factor that influences their ecology. On a larger scale, the diatom communities are shaped by climate, including the length of the ice cover, geology, bedrock, and dispersal history. On the other hand, at smaller sizes, the makeup of the algal species appears to be mostly determined by the water chemistry, which is influenced by the local environment. Moisture, nutrients, light, and space are the fundamental conditions that determine how diatom populations develop. Diatoms constantly compete with one another for these nutrients as well as with other species, particularly green algae [14.11]. Though limited, space seems to be a secondary resource to the availability of nutrients and light, which are determined by the size of the ecosystem. The euphotic zone, or the zone where photosynthesis takes place, is often large, including the profundal, making shallow, subarctic ponds with clear water the most likely habitat for diatoms [14.9]. In particular, through a relationship with light, the temperature has a direct and indirect impact on diatom communities. Since metabolism accelerates at higher temperatures, it has a direct impact on cell growth. Moreover, sunlight can quickly heat up the water in subarctic ponds, especially shallow ones, and have a significant impact on the diatom community. Subarctic ponds, on the other hand, might have radically diverse species compositions because they are frigid and nourished by groundwater. Water chemistry and geochemical qualities might change depending on the source of the water. This could have an effect on water acidity, which influences the physiology, longevity, and morphology of diatom communities. Diatoms may also be affected similarly by exceptionally high water pH levels. Diatoms can be divided into three major groups based on the pH of the water: acidophiles (pH 7), neutrophiles (pH 7), and alkaliphiles (pH >7), and they can be used to study freshwater acidification. While the majority of diatom species have a certain pH range in which they thrive, a small number of uncommon species are categorized as indifferent because they lack a distinct pH optimum [14.10] [14.17]. One of the most important factors in determining the composition of aquatic diatom communities is the availability of nutrients, particularly phosphorus and nitrogen. The amount of nutrients available at surrounding locations might vary significantly, which can have a substantial impact on the species richness and community structure of the diatoms. The increase in nutrient availability in harsh, subarctic conditions seems more likely to boost species richness than to decrease it, as may be the case in existing eutrophic waters. Based on four broad substrate types, total phosphorus is often the limiting nutrient circumpolarly. It is clear that the substrate offers diatoms an unstable environment because epipelic and epipsammic substrates regularly collect or dissipate very quickly. Subarctic ponds' substrate types are less important than they are in streams, where turbidity and continual disturbance have an impact on the microhabitats. The most prevalent diatoms in the subarctic are epilithic and epipsammic, which are typically nutrient- and light-limited due to the absence of macrophytes there. In general, the subarctic's abundance of light is what drives the quick seasonal change in diatom communities. Rapid continuous climate change will have a direct impact on aquatic environments and consequently a secondary impact on diatom communities. Diatoms are mostly transported by the water column in lakes and streams and aquatic ecosystems, but they can also be dispersed over wider areas by fish, birds, mammals, air, and insects. Yet, because diatoms' cells dry up while in transportation, several of these transport strategies are unusual in this group of organisms. The various methods of dissemination can help to explain how unique species emerge. Competition among benthic invertebrates and grazing may result in an increase in the number of rare or unique species in the habitat.

For instance, Lowe demonstrated that pennate diatoms are more resistant to disturbance when they are attached to a substrate. Their unique shape enables a stronger adhesion while still maintaining enough flexibility to shield the cell from damage. Because of this, epilithic diatom sampling offers the optimum substrate in terms of technique [14.10] [14.17].

14.4 Morphology and Identification

Diatoms are distinguished by an exterior silica frustule that enables long-term storage in watery environments or deeper in sediments. The diatom is well preserved and protected since the cell is divided into multiple pieces. The parent cell predetermines the size and shape of the hypotheca. Normally, the size of the cell declines following cell splitting, but occasionally, because of the flexible girdle bands, the cell can maintain its size and shape. Diatoms vary in size, shape, and exterior cell texture, which is referred to as their morphology [14.11]. On the basis of a mix of valve traits and phylogenological data, taxonomists have split the diatoms into many groups. Yet, as knowledge on the subject expands, new species are being found, and researchers are beginning to rearrange and rename diatom groups, which supports the notion that form and symmetry are ineffective identification indicators. In contrast to academic journals, many diatom species are now available online, and as more research is done, more species should appear. Diatom taxa need to be better arranged in order to facilitate identification that is quicker, more universally accepted, and easier. The crucial elements for identification are the location, density, and number of striae, the quantity, type, and length of the raphe, as well as the general shape and size of the frustule. All species in a genus have the same fundamental morphologies because they typically produce the frustule, spines, girdle bands, striae, and raphe in the same way [14.10] [14.17].

14.5 Diatom Age, Diversity and Ecological Functions

Diatoms are single-celled microalgae organisms that are members of the Stramenopiles phylum. A silica cell wall, or frustule, is their distinctive feature. They can be found everywhere in a variety of aquatic or moist terrestrial environments. According to valve symmetry, they have historically been split into two groups: pennates with bilateral symmetry and centrics with radial symmetry. Most species-rich category of algae are made up of diatoms, which are a reasonably young class of species that likely go back to the early Cretaceous era. There have been 14,575 species described as of today, but this number still appears to undervalue the exact number of the entire diversity. The globe still has many unexplored habitats and places, including epipelon, subtidal marine epipsammic diatoms, and freshwater and aerophytic diatoms in tropical locales [14.10] [14.17]. Diatoms continue to evolve quickly despite having such a large number of species because their rRNA coding regions undergo rapid mutation. Diatoms, one of the primary classes of marine planktonic organisms, produce a large amount of biomass, which provides them with significant ecological value. They are factories that produce oxygen, contribute to around 20% of the world's total primary production, and have a significant impact on the atmosphere's CO_2 concentration. Diatoms plan an important role in the transition and recycling of nitrogen, carbon, and silica on a global scale [14.11].

Diatoms are divided into three classes according to their modes of adaptation: attached, suspended, and motile. Benthic organisms are those that are suspended and motile and are found near the surface. Planktonic diatoms are those that live suspended in the water column. They swim passively, but because the silica frustule is so heavy and they can't move actively, they might sink. Several species evolve modifications that allow them to survive in more well-lit surface waters. They may make chitin fibers that slow the sinking, form spatially expansive colonies, like *Asterionella formosa* Hassall, or may include low-density substances into their vacuoles, as several marine species do (Figure 14.1). Diatoms that live close to the substrate are known as benthic diatoms. Both mobile and immobile species are included. Since each sort of surface is quite distinct, diatoms are typically categorized in accordance with that. Epipelon, endopelon, epipsammon, endopsammon, epilithon, endolithon, epixylon,

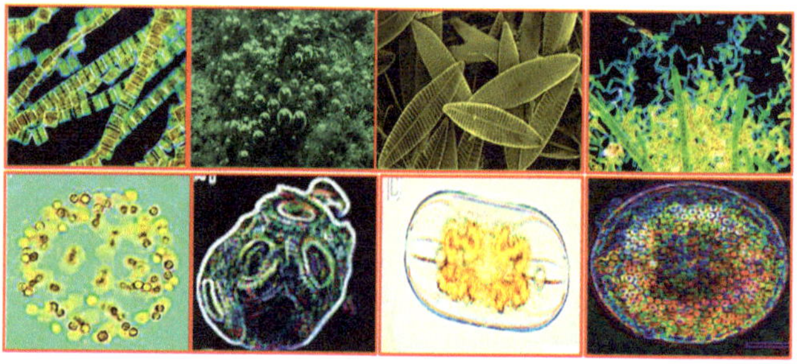

Figure 14.1 Diatom demonstrates a number of structural properties, including a raphe that can move, spin, and change direction.

epizoon, endozoon, endophyton, and epiphyton are the different types of benthic algae. Fine sediments coexist with epipelic and endopelic diatom populations. They cover the mud's surface, where they can produce films that are brown in color. While diatoms are typically biraphid and hence capable of physiologically driven circadian or daily migrations up and down, there are no boundaries between species on or in the sediments. According to popular consensus, diatoms travel down later in the day to come in contact with higher concentrations of nutrients since nutrients are easier to obtain under hypoxic or anoxic conditions, while moving up during the day ensures better illumination. Diatoms are frequently buried by disturbances and can only remain active in the top few millimeters of the sediment, which suggests that movement is a key attribute in sediment environments. In the sandy environment, diatoms can be found living within the substrate or affixed to a sand grain. Diatoms can also exist on or within living things. Epizoons are known for many different species; they can be found on freshwater and marine turtles, sperm whales, tiny crustaceans, and a variety of other animals. Animals and algae serve as a vector for the spread of diatoms. And last, diatoms can also coexist with plant matter surface [14.10] [14.17].

14.6 Biofuel Production and Types of Biofuels

Biofuels are widely categorized into many types based on the material used to produce them (Figure 14.2). According to the biomass feedstock, they are separated into the following categories [14.22]:

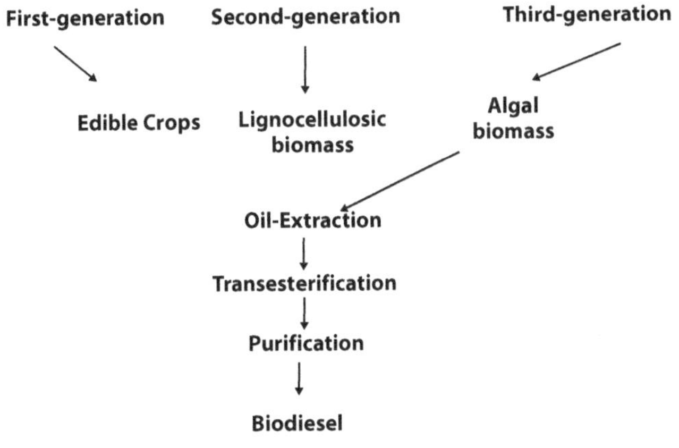

Figure 14.2 Third-generation biodiesel generation from diatoms and algae.

1) First-generation biofuels
2) Second-generation biofuels
3) Third-generation biofuels

14.6.1 First-Generation Biofuels

Human dietary components or grains/crops used as animal feed are the main sources of the first-generation biofuels. These fuels are referred to as "conventional biofuels" because they are made utilizing a variety of well-established methods and processes, including fermentation and transesterification [14.6].

14.6.2 Second-Generation Biofuels

The production of second-generation biofuels uses all non-food feedstocks, including agricultural waste, waste lignocellulosic feedstock, and forest waste. The second generation of biofuel production is an upgrade over the first generation because it focuses on producing biofuel that is even more affordable and creating secondary waste carbon that can be converted into value-added material. It is an economically viable technology because, in contrast to the first generation type, it focuses on producing value-added fuels while lowering the cost of energy as a whole and the amount of waste generated [14.6].

14.6.3 Third-Generation Biofuels

Third-generation biofuels are produced from microalgae by transesterfying or hydrotreating the algal oil [14.5]. These methods can significantly raise the yield of ethanol each year compared to first-generation biofuels that use conventional crops. The second and third generations of biofuels are referred to together as advanced biofuels because of ongoing research and development. The best sources contain useful substances that don't interfere with the food chain, are practical, simple to obtain, and flexible to shifting environmental conditions. The majority of these sources include microalgae, used cooking oil, animal oils, fish oils, and other substances. The possibility of reducing the waste management facility load and water pollution is another significant development [14.6]. Diatom biomass that lies in this category could be used in bio-oil extraction reactions. In the future, new and advanced biofuels and industrial biochemicals will be synthesized due to novel microalgae and diatom species and genetic engineering processes [14.15]. Biofuels are available in three different forms:

solid, liquid, and gaseous. Solid biofuels can be used to produce power, heat buildings, or cook meals since they are made from non-fossil, organic materials like plant biomass, animal waste, fuelwood, charcoal, wood pellets, wood scraps, and municipal trash. These solid organic carbon biomass biofuels are used to generate energy such as heat and electricity [14.3]. Liquid biofuels are better in many respects than solid and gaseous biofuels because of their high energy density, making them the ideal option for transportation, storage, and retrofitting. One or two of the most significant major examples of liquid biofuels is bioethanol, followed by biodiesel and bio-oil. Triglycerides and lipids of diatoms, biodiesel, and biogasoline are other varieties of liquid biofuels. Whereas, the lignocellulosic-based biofuels are biogas and ethanol [14.3]. The three gaseous biofuel types are biogas, biohydrogen, and biosyngas. Various processes like pyrolysis, gasification, or anaerobic fermentation are reported for the production of solid, liquid and gaseous biofuels [14.20]. A brief conceptual representation of each of the three types of biofuels and the production of biodiesel from diatom lipids is presented in Figure 14.2.

14.6.4 Diatoms Mass and Lipids for Biofuel

In the context of growing environmental pollution and global warming, biofuel is the planet's best hope for a sustainable energy source. Diatoms are a third-generation biofuel source. Diatoms' lipids can be removed to synthesize bio-oil, or their entire biomass can be used to make biogas, biohydrogen and biocrude [14.25]. Diatoms are one of the important microalgae that is currently most abundant in the oceans. They make up a large percentage (about 50%) and a sizable portion (roughly 40%) of marine primary production. Due to their high productivity, diatoms are a significant group of oceanic algae that produce about half of the marine food web and have been recognized as a possible source of biofuel. Opportunities for commercial production of diatoms depend on obtaining robust marine strains that can produce triacylglycerols and high value-added metabolites. Diatoms, or any microalga, could be used to produce biofuels in one of two ways: directly extracting the lipid, or digesting the microalga to produce oil. Oleic acid (56.6%) and palmitic acid (25.9%), which make up the majority of the fatty acids in HA-1, are excellent raw materials for biofuels. Diatoms contain large amounts of long-chain C14-C22 fatty acids. Thus, the species of diatoms can accumulate high lipids in the thylakoid layers because of their high growth rate and speed of photosynthetic activity. It is feasible to efficiently manage the primary cultivation, harvesting, and subsequent downstream processing of diatom culture for oil production.

A potential technique to improve the production of biofuel from diatoms and other algae species is to include a molecular antenna into the diatom Cy5, which increases cell density and photosynthetic oxygen production by 49% and 40%, respectively. Diatoms can grow in wastewater as substrate solution by obtaining nutrients, and after harvesting, the same biomass of diatoms can be used for biofuel. The synthesis of additional bioproducts and biofertilizers can be achieved with diatom sources that can possibly reduce the cost of producing fuel in the future. Diatoms are quick-growing, can make lipids of long chains with only a few cells in their body structure, and can be grown in reactors supplied with carbon dioxide and nutrients from waste biomass or waste water to produce energy and value-added materials while also cleaning the environment. Therefore, proper planning and management are required to better utilize them [14.1].

According to geochemists, algal lipids have been the primary feedstock for petroleum for the past 70 years. Analysis of lipids that serve as biomarkers and are persistent for millions of years in the petroleum reservoir can frequently be used to determine the origin of the algae. The primary indicators for diatoms in petroleum are 24-norcholestanes, highly branched isoprenoid (HBI) alkenes, and the ratio of steranes with 28 and 29 carbon atoms. Several of the best oil fields in the world contain these biomarkers. Lipids, particularly triacylglycerides (TAGs), are their main source of carbon storage, and under appropriate growth conditions, approximately 15 and 25% of their biomass is made up of fatty acids, which is a useful substrate for biodiesel (Figure 14.3) [14.11]. The diatoms species primarily synthesize 13–21–carbon fatty acids (FAs), primarily saturated and

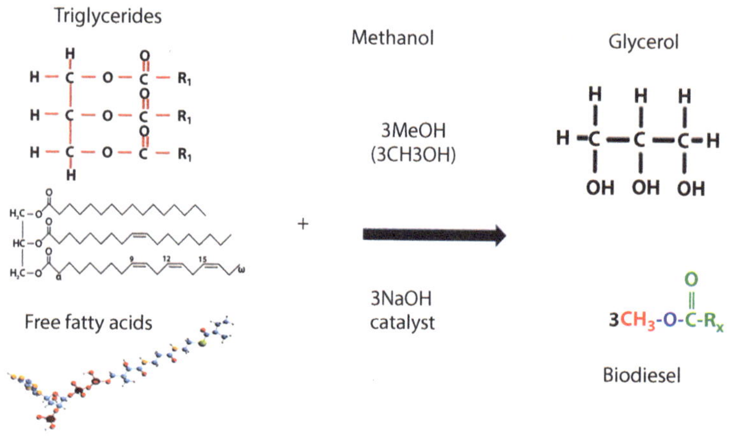

Figure 14.3 Transesterification reaction and biodiesel production from TAG.

monounsaturated fatty acids (MUFAs) (Table 14.1), which provide more oxidative energy when oxidized than polyunsaturated fatty acids (PUFAs) with the same number of carbons. In marine and freshwater diatoms, 16:1 and 16:0 are the two most common fatty acids (FAs); collectively, they make up up to 70% of the lipid profile of the cell. Depending on the circumstances, 14:0 fatty acids (FAs) can make up as much as 32% of the cell's total lipids (Table 14.1). Eicosapentaenoic acid, the fourth most prevalent fatty acid (FA), is well-known for its commercial significance in the food and health industries. The characteristics of the FAs used to make biodiesel have a direct impact on its qualities. Several biodiesel properties, including

Table 14.1 Fatty acids profiles and components of fatty acids in different diatom species.

Diatoms species	Major fatty acids contents
Phaeodactylum tricornutum	16:0, 16:1, 20:5
Chaetoceros sp.	14:0, 16:1
Navicula muralis	16:0, 16:1
Nitzschia closterium	16:0, 16:1, 20:5
Thallasiosira psuedonana	14:0, 16:12, 20:5
Navicula inserta	16:0, 16:1, 20:5
Nitzschia palea	16:0, 16:1, 20:5
Nitzschia longissima	16:0, 16:1
Nitzschia frustulum	16:0, 16:1
Navicula pelliculosa	16:1, 20:5
Nitzschia closterium	16:1, 20:5
Nitzschia ovalis	16:0, 16:1
Amphora exigua	16:0, 16:1
Cyclotella cryptica	16:0, 16:1, 16:3, 20:5
Biddulphia aurita	14:0, 16:1, 20:5
Fragilaria sp.	14:0, 16:1, 20:5
Amphora sp.	16:0, 16:1, 20:5

cetane number, amount of emissions, cold flow, oxidative stability, stickiness, and lubricity, are impacted by the length and saturation level of the fatty acids (FAs) [14.11]. The requirements for different features of biodiesel are frequently in conflict. For instance, the inclusion of polyunsaturated fatty acids (PUFAs) enhances the fuel's cold temperature characteristics but decreases its oxidative stability and raises its nitrogen oxide (NOx) emissions. Although it seems that no natural fatty acid profile can produce the perfect biodiesel, using feedstock with high levels of monounsaturated fatty acids (MUFAs), such as palmitoleate (16:1) or oleate (18:1), and low levels of saturated fatty acids and polyunsaturated fatty acids (PUFAs), can nonetheless yield biodiesel with close to ideal properties. The profile of fatty acids (FAs) could become more valuable economically by being optimized [14.11]. As previously accomplished with soybeans that had higher levels of oleic acid, the use of metabolic engineering techniques could enable this alteration. Algal biodiesel can be used in conventional engines with little to no modification because its physical and chemical characteristics are comparable to those of petroleum-based diesel fuels. Yet, there are major challenges to be overcome in order to be economically competitive with fossil fuels, starting with choosing the most effective strains to optimize growing, collecting, separating, and refining. To compete with petroleum on the global market, algae biofuel production costs per barrel must be reduced. Technical solutions are needed to address these issues.

14.6.5 Growth, Biomass and Lipid Extraction

To produce fatty acids and lipids, diatoms are grown in both autotrophic and heterotrophic environments. Autotrophic growth is ensured by illumination at the (light/dark) photoperiod, whereas heterotrophic growth is accomplished by culturing cells entirely in the dark and by adding glucose as an additional organic carbon source to the medium. Gently bubbling sterile air is used to agitate cultures. In autotrophic environments, daylight fluorescent tubes with a 14:10 h light: dark photoperiod are used to artificially light the diatoms' growth, whereas two pulses of 1g/L glucose can be added to each culture depending on the requirement of the experiments and then the cultures are incubated in the dark under heterotrophic conditions [14.16]. Nitrates ($NaNO_3$), phosphates (NaH_2PO_4), and silicate (Na_2SiO_3) are added as nutrients to support the growth of cells. The growth of diatoms can be monitored by using a microscope. Centrifugation at a rate of 10,000 g for 10 min is used to harvest the cells. On a lyophilized biomass culture, pellets are lyophilized, frozen, and their dry weight is estimated. The 4,40-dihydroxybenzophenone (DHBP) (1 mg mL^{-1}) internal standard is used to conduct lipid extraction using the methyl-tert-butyl ether (MTBE) procedure. The DHBP and MeOH are specifically used to suspend the dry cell pellet. To enable extraction at room temperature with continual shaking after vortexing, MTBE is added. After 10 min of centrifugation at 10,000 g, the organic extracts are collected (upper phase). To determine the lipid content gravimetrically, the extract is dried under nitrogen flow and weighed (mg L1 culture). The chemicals used, volumes, cell mass and other conditions depend on the diatom species and laboratory experimental plan (Figure 14.4). Using GC-MS/1H-NMR and comparing standard fatty acid methyl esters (FAMEs), it is possible to quantify the total fatty acid contents and lipid compositions from the diatom cells' organic extracts. Another strategy for diatom mass and lipids extraction is outdoor cultivation process. In this method, the diatoms are fed with natural sea or wastewater as a culture media in open-pond bioreactor cultivation conditions. In order to maintain a pH of 9.0, CO_2 gas can be provided. A general illuminance meter is set up near the open-pond bioreactors to measure the solar irradiance. In order to achieve better growth, the culture media may also be agitated.

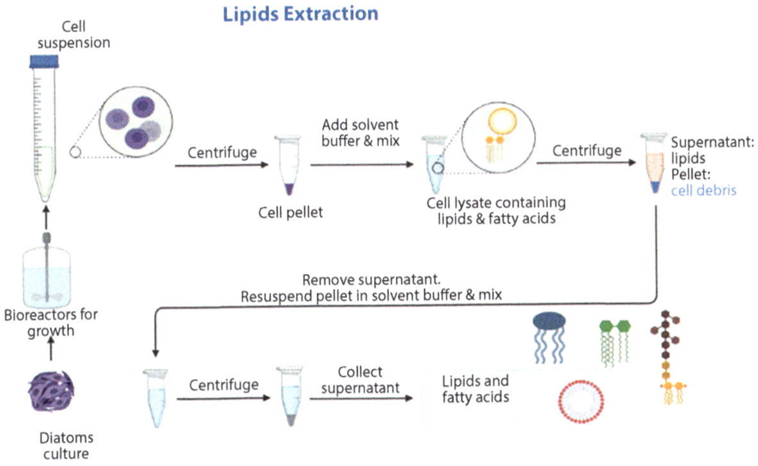

Figure 14.4 A generalized method and procedure of lipids extraction from diatom biomass.

14.7 Different Methods of Lipid Extraction for Biofuel

Diatom biofuels could be produced in two ways. The entire biomass fraction can be thermochemically converted into bio-crude oil, or a direct removal of the lipid content for use as biofuel. A diatom cell's oil content percentage, the types of unsaturated fatty acids and saturated fatty acids, the optimum method, a suitable strain that can produce a high yield, and additives in the growing medium are the main issues for lipid content and utilization. If the main objective is to produce bio-crude, the organic carbon content of diatoms, not their lipid content, can be used [14.16]. The issue is that whereas bio-oil (an upgraded lipid) just requires minimal processing before it can be used as fuel, bio-crude would need H2 input for upgrading to a fuel item, which would make it expensive. Various methods are reported for diatoms cultivation and lipids extraction, a few of which are listed below.

14.7.1 Plastic Bubble Wrap for Diatom Cultivation

A recent study that used a newly developed plastic bubble wrap technique to produce biofuel from diatoms grown in used plastic trash reached the conclusion that diatoms are the best source for biofuel. The primary objective of the aforementioned study was to cultivate diatom cells in a closed system. To do this, the reactor rim was tightly sealed using bubble wrap made of plastic and inexpensive plastic material that is used in the shipping and storage of goods. Various plastic wraps discarded from a plastic business were first sent to be tested for their impermeability to water loss and permeability to gases. As a result, it was discovered that out of all the plastic bubble wrap varieties tested, the use of low-density polyethylene (LDPE) used to seal glass containers as photobioreactors allowed for the highest cell count, maximum lipid, and maximum CO_2 absorbance, as well as 40 days of nutrient uptake with almost no water loss being noticed. *Haematococcus pluvialis,* a red-green microalga, showed potential to be expanded for astaxanthin synthesis using leftover bubble wrap plastic in order to test the suitability of LDPE on other microalgae. The findings of this study might help develop a new strategy for reducing plastic waste and for using diatoms to produce biofuel [14.7].

14.7.2 Spontaneous Oozing

Researchers found a strain of diatom called *Diadesmis confervacea* that not only accumulates oil (14.6%), but also naturally extracts oil practically on day 31 of *in-vivo* culture, when the cells reach maturity, which is an important aspect for cheap biofuel production. It was also suggested that diatoms be kept in solar panels throughout their stationary development phase, where they would continue to produce oil droplets, also known as lipid droplets or oleosomes, that contained organic components suitable for biofuel. While separating the oil from the diatom cells used to be an expensive and fuel-intensive procedure, spontaneous oozing dramatically lowers the cost of algal fuels. Also, it is now simpler to construct diatom biofuel solar panels as a result of spontaneous oozing. The standard method for performing spontaneous oozing involves collecting samples, followed by culturing and then inoculation onto cultivated solid agar medium. The exocytosis of oil by the diatom cells is observed on each day after their inoculation [14.13].

14.7.3 Mechanical Pressure

Lipid extraction procedures have been upgraded with the use of ultrasound. Diatom cells are famous for their distinctive characteristics, including being encased in a hydrated silicon dioxide cell wall known as a frustule. Mechanical approaches may offer an incredibly strong strategy to facilitate the discharge of high-value molecules outside the cell surface because of the frustule's hardness. To perform this procedure, selected diatom cells must first be harvested. They must then be placed on a stiff surface and mechanical pressure applied to release oil from the cell surface [14.4].

14.7.4 High-Pressure Homogenization

In this process, an oily phase and an aqueous phase of two liquids are dispersed using a high-pressure homogenizer. Finely split solids in a liquid must be forced through a small input orifice at a high pressure of approximately 500–5000 psi. Because of the considerable turbulence and hydraulic shear this causes, very small suspended particles are produced. The efficiency of high-pressure homogenization in diatom cells varies between different species and may decline in proportion to the rigidity of the cell wall [14.4].

14.7.5 Ball Milling

A ball mill is a hollow, rotating metal cylinder loaded with magnetic beads that acts as a crushing frame, providing collisions by the rapidly rotating metallic beads, causing cell lysis. Without any prior biomass preparation, damage from metallic beads can trigger cell lysis in a matter of minutes. The benefits of this method are that it is easy and fast. The success of the ball mill as a pretreatment step for lipid content extraction can be applied to a few microalgae species. The data showed that lipid extraction from *Botryococcus* sp. using a ball mill was 20% more efficient than solvent extraction and ultrasound [14.13].

14.7.6 Microwave Oven

Microwaves with a range in the frequency of 300 MHz to 300 GHz are used to extract lipids from microbial cells by breaking the cell wall, which releases lipid molecules that are then drained from the cell as a result of the waves' interaction with molecules and the generation of heat. Because heat is transmitted by radiation rather than conduction or convection, a microwave-aided heating system heats a room more quickly than a standard heating system. This procedure is a preferable choice for polar solvents, such as water, which generates steam that breaks down cell walls and releases intracellular contents, leading to an efficient method of lipid withdrawal. Some products, including fatty acids and lipids, occasionally experience deterioration as a result of high temperature generation. In that case, it's critical to speed up the process and use cooling systems to prevent the degradation of a number of bioproducts. The microwave method requires fewer solvents and takes less time to extract, although it is expensive for its scale-up process because it requires a lot of energy [14.14].

14.7.7 Transesterification

Transesterification is the process through which triacylglycerol (TAG) in oils and an alcohol (often methanol, ethanol, propanol, or butanol) react chemically to create glycerol and alkyl esters of fatty acids, also known as biodiesel. Triacylglycerol (TAG) is a glycerol compound generated from three fatty acids. a large part of many algae, particularly diatoms, that serve as storage. By combining algal biomass, a catalyst, which is often an acid, and an alcohol, are mixed and heated at high temperatures, and the cell's lipids are converted into biodiesel [14.2].

14.8 Benefits of Diatoms

Research that primarily aims at assessing the viability of diatoms for large-scale biofuel production is limited, and very few results are available under favorable cultural conditions. To learn more about the flaws and potential outcomes caused by ecological environments, it is crucial to conduct studies comparing the productivity of diatom cells under favorable culture conditions. Thorough research and experimental studies are required on certain diatom strains that can be utilized for biofuel production. Transportation contributes to an increase in greenhouse gas emissions because of the pollutants in fossil fuels [14.23], but diatom-based biofuels have the potential to greatly minimize air quality problems. Diatom-based biofuel is expected to be less hazardous to handle than conventional fossil fuel due to its low volatility and has the possibility of decreasing cancer-causing chemical emissions by up to 85%. Diatom silica can be produced in a monoculture to obtain silica from diatoms with added attractive features that may be used in nanotechnology. Diatom cells can be used in a production process to produce a useful, marketable silica product. So, in addition to producing biofuel, diatoms may also offer advantages such as particulate filtration. Diatom frustules are practical sieve materials that can be used to separate very small particles due to their porous nature. Silica is recognized for its hygroscopic properties and can be used as in natural pesticides and neutral wormicides in addition to being utilized in the cosmetic industry to synthesis amino acids. The abrasive properties of diatoms, on the other hand, make them suitable for manufacturing toothpaste, metal polish, and amino acid synthesis. So, producing biodiesel from diatoms on a wide scale could offer a number of advantages in the current competitive market [14.18].

14.9 Genetic Engineering and Metabolic Pathway Engineering

Diatoms are diploid and only occasionally have a phase of sexual recombination; therefore, breeding techniques or traditional genetics are not practical. However, genetic engineering techniques have opened up possibilities for getting raw sequence data from the genomes and annotated genomes using transformation approaches to improve the efficacy of diatom lipid production. For instance, it's been noted that overexpressing two plant thioesterases (C14-TE and C12-TE) from *Cinnamomum camphora* increased the accumulation of short saturated chain length fatty acids (FAs) in

P. tricornutum, and that insertion of the acetyl-CoA carboxylase gene (ACC1) into *Cyclotella cryptica* led to increased activity of the enzyme. It is obvious that fourth-generation diatom-based biodiesel is still in its infancy, however, targeting nitrogen absorption, the glutamine synthetase/glutamine oxoglutarate aminotransferase pathway, and TAG production seem to be potential areas for genetic modification [14.11]. It is proposed that changing those branch points while identifying the ideal growing environments for the resulting strains will get us closer to mass producing biofuel. The system's sustainability and energy return on investment (EROI) may also be improved by increasing resistance to biotic stressors (competitors, parasites, and pathogens) through the introduction of new genes (such as herbicides) [14.19].

14.10 Future Prospects

Since it is widely known that the first and second generations of biofuels, which are mostly produced by terrestrial crops, are insufficient to meet the world's energy needs, researchers are looking for substitute sources of biofuel [14.21]. Also, the demand for agricultural land and fertilizers will rise as a result of the increased demand for cultivating terrestrial crops. Because they can grow on non-arable land and produce a high oil yield per area, third-generation biofuels derived from microalgae are receiving a lot of attention right now. Because diatom species differ in terms of morphology and physiology, these cells can respond quickly to any stressful situations [14.11]. The choice of the strain or species to be used for cultivation, the culture parameters (such as nitrogen stress conditions offered, light intensity required for growth, framework selected for recovering biomass concentration, and especially protocol selected for oil extraction) all affect the oil content of diatom cells. The strain chosen, the type of their cell wall, whether nonpolar or polar lipid is present, as well as operating cost and energy expenditures, all have an impact on the optimal method for cell lysis and the efficiency of the extraction strategy. However, the extraction of biofuel requires a large energy input and is a challenging scale-up process. Microwave oven and solvent-aided ultrasound are considered as effective processes for cell disruption. The important technique for building up a diatom lipid biofuel production system is to obtain it at an affordable price by selecting the best strain for growing and developing a better growth plan that allows for excellent lipid yield [14.19]. The main feedstock for the manufacturing of biodiesel is terrestrial harvests such corn, rapeseed, and soybeans. Regrettably, using food crops to make biodiesel creates pressure between using agricultural land for production of food and using it to make biological fuel, leading to an increase in food costs and a possible loss of biodiversity and habitat. When improving the quality of the biofuel and decreasing the quantity of energy and time used in the extraction processes, it is essential to devote effort to extraction techniques that are environmentally sustainable as well, with a lesser proportion of solvent utilization. This will enable the production of valuable, viable, and greener diatom fuels [14.18]. To improve the practicality of biorefinery methods and build a bioeconomy based on renewable resources, solvents must be carefully analyzed while considering the most efficient method for extracting lipid content. Consequently, it is possible to synchronize the production of both biodiesel and bioethanol. By doing this, it is possible to make bioethanol from defatted biomass and biodiesel from the fatty acids of microalgae without using petroleum-based solvents (the methyl solvent is switched out for the ethyl solvent during the transesterification process). For a successful economic, ecological, and social production of various products over individual processes, under the concept of a biorefinery, the use of green solvents for the extraction of microalgae and the synthesis of biofuel is therefore important in future consideration.

14.11 Conclusion

The main sources of modern global petroleum are fossilized diatoms. Diatoms' increase in ecological significance over the last 34 million years is due to their outstanding ability to convert sunlight into usable energy and their speedy uptake and digestion of nutrients. Profiting from the natural selection of these organisms that have grown to have such high photosynthetic energy levels with lipids as their main storage product would be advantageous. Algal feedstocks provide promising alternatives for biofuel production. Here, we focused on diatoms, a particular taxon of algae that may someday take the place of fossil fuels. We discussed previous efforts to find suitable diatom strains, made predictions about their genetic flexibility and presented genetic modification options to boost production. However, in order for diatoms to successfully replace fossil fuels, the cost of producing algae biofuels needs to be equal to that of petroleum-based fuels, and the lack of stable public policies and funding that value these renewable fuels and promote their adequate development needs to be addressed.

Acknowledgment and Funding

No funding was received to assist with the preparation of this chapter; however, the author is grateful to Professor Li Zhihe for providing the post-doctoral position and facility in the Shandong University of Technology.

Data Availability

The authors confirm that all data are included in this published article.

References

[14.1] Dhanker, R., Kumar, R., Tiwari, A. and Kumar, V. (2022) Diatoms as a biotechnological resource for the sustainable biofuel production: a state-of-the-art review. *Biotechnology and Genetic Engineering Reviews* 38(1), 111-131.

[14.2] Ehimen, E., Sun, Z. and Carrington, C. (2010) Variables affecting the in situ transesterification of microalgae lipids. *Fuel* 89(3), 677-684.

[14.3] Ghosh, S.K. (2016) Biomass & bio-waste supply chain sustainability for bio-energy and bio-fuel production. *Procedia Environmental Sciences* 31, 31-39.

[14.4] Halim, R., Harun, R., Danquah, M.K. and Webley, P.A. (2012) Microalgal cell disruption for biofuel development. *Applied Energy* 91(1), 116-121.

[14.5] Hildebrand, M., Davis, A.K., Smith, S.R., Traller, J.C. and Abbriano, R. (2012) The place of diatoms in the biofuels industry. *Biofuels* 3(2), 221-240.

[14.6] Kassaye, S., Pant, K.K. and Jain, S. (2017) Hydrolysis of cellulosic bamboo biomass into reducing sugars via a combined alkaline solution and ionic liquid pretreament steps. *Renewable Energy* 104, 177-184.

[14.7] Khan, M.J., Gordon, R., Varjani, S. and Vinayak, V. (2022) Employing newly developed plastic bubble wrap technique for biofuel production from diatoms cultivated in discarded plastic waste. *Science of The Total Environment* 823, 153667.

[14.8] Kollár, J., Fránková, M., Hašler, P., Letáková, M. and Poulíčková, A. (2015) Epiphytic diatoms in lotic and lentic waters-diversity and representation of species complexes. *Fottea* 15(2), 259-271.

[14.9] Letáková, M., Cantonati, M., Hašler, P., Nicola, A. and Poulíčková, A. (2016) Substrate specificity and fine-scale distribution of epiphytic diatoms in a shallow tarn in the Brenta Dolomites (south-eastern Alps). *Plant Ecology and Evolution* 149(2), 144-156.

[14.10] Letáková, M., Fránková, M. and Poulíčková, A. (2018) Ecology and applications of freshwater epiphytic diatoms. *Cryptogamie, Algologie* 39(1), 3-22.

[14.11] Levitan, O., Dinamarca, J., Hochman, G. and Falkowski, P.G. (2014) Diatoms: a fossil fuel of the future. *Trends in Biotechnology* 32(3), 117-124.

[14.12] Maeda, Y., Nojima, D., Yoshino, T. and Tanaka, T. (2017) Structure and properties of oil bodies in diatoms. *Philosophical Transactions of the Royal Society B: Biological Sciences* 372(1728), 20160408.

[14.13] Mubarak, M., Shaija, A. and Suchithra, T. (2015) A review on the extraction of lipid from microalgae for biodiesel production. *Algal Research* 7, 117-123.

[14.14] Pasquet, V., Chérouvrier, J.-R., Farhat, F., Thiéry, V., Piot, J.-M., Bérard, J.-B., Kaas, R., Serive, B., Patrice, T. and Cadoret, J.-P. (2011) Study on the microalgal pigments extraction process: Performance of microwave assisted extraction. *Process Biochemistry* 46(1), 59-67.

[14.15] Paul, D., Arora, A. and Verma, M.L. (2021) Advances in Microbial Biofuel Production. *Frontiers in Microbiology*, 2768.

[14.16] Peng, L., Fu, D., Chu, H., Wang, Z. and Qi, H. (2020) Biofuel production from microalgae: a review. *Environmental Chemistry Letters* 18, 285-297.

[14.17] Poulíčková, A., Letáková, M., Hašler, P., Cox, E. and Duchoslav, M. (2017) Species complexes within epiphytic diatoms and their relevance for the bioindication of trophic status. *Science of The Total Environment* 599, 820-833.

[14.18] Pragya, N., Pandey, K.K. and Sahoo, P. (2013) A review on harvesting, oil extraction and biofuels production technologies from microalgae. *Renewable and Sustainable Energy Reviews* 24, 159-171.

[14.19] Saade, A. and Bowler, C. (2009) Molecular tools for discovering the secrets of diatoms. *Bioscience* 59(9), 757-765.

[14.20] Saleem Aimen, Hafiz Umar, T.A., Shah, R. and Tabassum (2020) Fermentation of simple and complex substrates to biohydrogen using pure Bacillus cereus strains. *Environmental Technology & Innovation*, 100704.

[14.21] Shah, A., Favaro, L., Alibardi, L., Cagnin, L., Sandon, A., Cossu, R., Casella, S. and Basaglia, M. (2016) *Bacillus* sp. strains to produce biohydrogen from the organic fraction of municipal solid waste. *Applied Energy* 176, 116-124.

[14.22] Shah, T. and Ullah, R. (2019) Pretreatment of wheat straw with ligninolytic fungi for increased biogas productivity. *International Journal of Environmental Science and Technology*, 1-12.

[14.23] SHAH, T.A., Zhihe, L., Zhiyu, L., Zhang, A., Lu, D., Fang, W. and Xuan, H. (2022) Whole Cell of Pure *Clostridium butyricum* CBT-1 from Anaerobic Bioreactor Effectively Hydrolyze Agro-Food Waste into Biohydrogen.

[14.24] Wang, J.-K. and Seibert, M. (2017) Prospects for commercial production of diatoms. *Biotechnology for Biofuels* 10, 1-13.
[14.25] Yi, Z., Xu, M., Di, X., Brynjolfsson, S. and Fu, W. (2017) Exploring valuable lipids in diatoms. *Frontiers in Marine Science* 4, 17.

Index

Also of Interest

Check out these published and forthcoming titles in "Diatoms: Biology and Applications" series from Scrivener Publishing

Diatom Cultivation for Biofuel, Food, and High-Value Products
Edited by Vandana Vinayak and Richard Gordon
Published 2025. ISBN 978-1-394-17448-5

Diatom Ecology
Edited by Nora I. Maidana, Magdalena Licursi and Eduardo Morales
Published 2024. ISBN 978-1-394-17445-4

Diatom Photosynthesis: From High-Value Molecules to Primary Production
Edited by Johannes Wilhelm Goessling, João Serôdio and Johann Lavaud
Published 2024. ISBN 978-1-119-84208-8

Mathematical Macroevolution in Diatom Research
By Janice L. Pappas
Published 2023. ISBN 978-1-119-75043-7

The Mathematical Biology of Diatoms
Edited by Janice L. Pappas
Published 2023. ISBN 978-1-119-74985-1

Diatom Microscopy
Edited by Nirmal Mazumder and Richard Gordon
Published 2022. ISBN 9781119711537

Diatoms Morphogenesis
Edited by Vadim V. Annenkov, Joseph Seckbach and Richard Gordon
Published 2021. ISBN 978-1-119-48795-1

Diatoms Motility
Edited by Stanley A. Cohn, Kalina M. Manoylov and Richard Gordon
Published 2021. ISBN 978-1-119-52635-3

Diatoms
Fundamentals and Applications
Edited by Joseph Seckbach and Richard Gordon
Published 2019. ISBN 978-1-119-37021-5

Printed in the USA/Agawam, MA
September 16, 2026

910661.181